Dams and Public Safety

A Water Resources
Technical Publication

By
Robert B. Jansen

U. S. DEPARTMENT OF THE INTERIOR
Bureau of Reclamation
1983

As the Nation's principal conservation agency, the Department of the Interior has responsibility for most of our nationally owned public lands and natural resources. This includes fostering the wisest use of our land and water resources, protecting our fish and wildlife, preserving the environmental and cultural values of our national parks and historical places, and providing for the enjoyment of life through outdoor recreation. The Department assesses our energy and mineral resources and works to assure that their development is in the best interests of all our people. The Department also has a major responsibility for American Indian reservation communities and for people who live in Island Territories under U.S. Administration.

Mission of the Bureau of Reclamation

The Bureau of Reclamation of the U.S. Department of the Interior is responsible for the development and conservation of the Nation's water resources in the Western United States.

The Bureau's original purpose "to provide for the reclamation of arid and semiarid lands in the West" today covers a wide range of interrelated functions. These include providing municipal and industrial water supplies; hydroelectric power generation; irrigation water for agriculture; water quality improvement; flood control; river navigation; river regulation and control; fish and wildlife enhancement; outdoor recreation; and research on water-related design, construction, materials, atmospheric management, and wind and solar power.

Bureau programs most frequently are the result of close cooperation with the U.S. Congress, other Federal agencies, States, local governments, academic institutions, water-user organizations, and other concerned groups.

M
SI METRIC

REVISED REPRINT 1983
REPRINTED 1992
REPRINTED 2000
REPRINTED 2004
REPRINTED 2008

UNITED STATES GOVERNMENT PRINTING OFFICE
DENVER: 1983

Preface

In the United States, and in many other countries, general interest in the safety of dams and reservoirs has grown appreciably in recent years. Protection of the public from the consequences of dam failures has taken on increasing importance as populations have concentrated in limited and vulnerable areas.

The Bureau of Reclamation alone has created more than 300 dams, and continues to be responsible for their safety. In safeguarding these structures and their reservoirs, the Bureau must take prudent measures and must conduct its engineering in a way that merits public confidence. This demands professional practices that incorporate the lessons of the past and conform to the most advanced state of the art.

The history of dams and their misfortunes is essential reading for those who carry such responsibilities. Dams have influenced the course of civilization since early times. Their benefits are everywhere to be seen. Yet the vital services that they provide may be accompanied by serious hazards.

In the continuing endeavor to change the earth in the interest of human progress, the imperfections of man's work and of nature itself have not always been recognized. While the potential effects of failure have been feared by many, they have been unknown or even ignored by few. Natural defects and human mistakes have been combined to cause great tragedies. Some of the most frightening have involved the collapse of dams. History reveals that for each 1,000 dams existing today, approximately ten others have failed.

Each disaster is viewed with public dismay and receives a measure of official scrutiny. A more sustained effort to collect, analyze, and remember the lessons from failures of dams has been made by the professionals charged with their care. Various authors and societies have contributed to preservation of the attendant records. But these are stored in many places. This book attempts to bring accounts of significant happenings together for examination in a common perspective.

A history of dams is included to emphasize knowledge which had very early origins. Many of the experiences in ancient lands are still relevant today. The most intelligent of our ancestors valued their waterworks and gave them careful attention. They recognized that dams could be vulnerable to natural forces, and they learned how to protect them. Over the many centuries, some of the lessons were forgotten, and then learned again. The experiences of the early peoples showed the way to safer dams.

This book is not a theoretical reference on the engineering of dams. Mathematical or experimental treatises on the related subjects are available in many other publications. The present effort is intended to collect the lessons and to explain practical methods that are of value in the care and treatment of dams. While special emphasis is placed on structures and experiences of the Bureau of Reclamation, the material presented is drawn from sources worldwide. Some of the practices described or recommended are derived from theory and others are empirical, but all are directed toward the effective management of the problems that may develop at dams and reservoirs. This is part of the Bureau's continuing effort to bring to its personnel, and to share with others, the latest technology and experience in this important field of engineering.

This book was written in its entirety by Robert B. Jansen, who served the Bureau of Reclamation as Director of Design and Construction; Assistant Commissioner for Engineering and Research; and Assistant Commissioner for Dam and Structural Safety. It is based upon experience in the engineering of dams in both the public and private sectors. Mr. Jansen was Chairman of the U.S. Committee on Large Dams for the term 1979–81, and is presently a consulting engineer specializing in dams.

Preparation of this work for printing was supervised by the editor, James M. Tilsley, under the general direction of Warren E. Foote, Chief of the Technical Publications Branch. The author also wishes to acknowledge and thank Marilyn J. Cochran and Edna J. Hunsinger for their editorial assistance, and to express his gratitude to H. J. Ribbeck for research assistance, and to the personnel of the Structural-Architectural-Mechanical Section of the Drafting Branch who prepared the drawings for the text.

Contents

PART IV — SIGNIFICANT ACCIDENTS AND FAILURES

Page

PART V—PREVENTIVE AND REMEDIAL ENGINEERING

PART V—PREVENTIVE AND REMEDIAL ENGINEERING—Continued

Page

PART V—PREVENTIVE AND REMEDIAL ENGINEERING—Continued

Page

PART V—PREVENTIVE AND REMEDIAL ENGINEERING—Continued

TABLE

FIGURES

FIGURES - Continued

6 7 8 8 8 8 8 8 8 8 8

CONTENTS

FIGURES—Continued

Figure   *Page*

1-52 Grande Dixence Dam, cross section 81
1-53 Contra Dam, cross section 82
1-54 Contra Dam 83
1-55 Nurek Dam, cross section 84
1-56 Rogun Dam, cross section 85
1-57 Inguri Dam, cross section 86

3-1 A hydraulic fill under construction (San Pablo Dam in
     Calif.) 107

4-1 Baldwin Hills Reservoir after failure 122
4-2 Bouzey Dam, cross section 127
4-3 Bradfield Dam, cross section 129
4-4 Fontenelle Dam 141
4-5 Frías Dam after failure, downstream view 144
4-6 Frías Dam after failure, upstream view 145
4-7 Machhu II Dam after failure 155
4-8 Malpasset Dam, right abutment, showing remnant 159
4-9 Malpasset Dam, right abutment, showing gap caused by
     pivoting 161
4-10 St. Francis Dam before failure 172
4-11 St. Francis Dam after failure 173
4-12 St. Francis Dam, distribution of fragments 175
4-13 Surviving monolith of St. Francis Dam 176
4-14 St. Francis Dam, plan and profile 178
4-15 South Fork Dam, cross sections (1 of 2) 186
4-15 South Fork Dam, cross sections (2 of 2) 187
4-16 Teton Dam failure sequence 192
4-17 Teton Dam, downstream view after failure 195
4-18 Teton Dam, upstream view after failure 197
4-19 Teton Dam, plan and cross sections (1 of 2) 198
4-19 Teton Dam, plan and cross sections (2 of 2) 199
4-20 Teton Dam, prefailure leakage—location of visual events .. 206
4-21 Teton Dam, section along approximate path of failure 207
4-22 Teton Dam, key trench erosion diagrams 209
4-23 Vaiont Dam after failure, showing massive slide upstream
     of dam 215
4-24 Vaiont Dam after failure, looking upstream at slide mass ... 217
4-25 Vaiont Dam showing damage to top of dam and spillway 218
4-26 Lower Van Norman (San Fernando) Dam after earthquake
     of February 9, 1971 221
4-27 Vega de Tera Dam before failure 224
4-28 Vega de Tera Dam after failure 225
4-29 Walter Bouldin Dam after failure 228

5-1 Sheffield Dam following failure 239

xi

FIGURES - Continued

Part I

DAMS FROM
THE BEGINNING

Introduction

The engineering of dams is a vital part of the story of civilization. Reservoirs for water supply were undoubtedly among the earliest structures devised by mankind. The role that dams have played over the ages is documented in many records of ancient lands. Dams have been linked closely to the rise and decline of civilizations, especially to those cultures highly dependent upon irrigation.

Dams have served people for at least 5,000 years, as evidenced in the cradles of civilization, in Babylonia, Egypt, India, Persia, and the Far East. The remains of these ancient structures exist in both the old and the new worlds, marking the attainments of societies which have long since died. Many of the outstanding waterworks of antiquity eventually declined into disuse because the knowledge of their designers and builders was not preserved by the generations who inherited them. And without water the civilizations which it had supported faded away.

History does not record exactly when irrigation systems and dams were first constructed. Study of ancient China, India, Iran, and Egypt does reveal that such work in these lands was begun thousands of years ago, and provided lifelines on which their civilizations depended. Menes, the first Pharaoh of Egypt, ordered irrigation works to draw from the River Nile. In China, construction of impressive dams was accomplished on the Min River for flood control and diversion of water to nearby farm lands. The sacred books of India cite the very early operation of dams, channels, and wells; evidence that this land may have been the birthplace of the art. The Persians of ancient times recognized the importance of irrigation to the sustenance of civilization. By excavating underground water tunnel and gallery systems (quanats) and by constructing many dams, they accomplished projects which rank among the greatest in history. In the ruins at Sialak, near Kashan, are to be seen traces of irrigation channels which are considered to be as much as 6,000 years old, suggesting that irrigation was practiced there from very early times, even before the arrival of the Aryans in the land now known as Iran.

The Period B.C.

The remote history of dams is not well known. Most dates of events earlier than 1000 B.C. can be only estimated. This is particularly true of early Egypt, whose peculiar chronology sometimes sheds only dim light on the many dynasties and their engineering achievements.

Ruins of ancient works in India and Sri Lanka (Ceylon) offer some evidence of how water reservoirs were created by early peoples. A common method of construction involved the placement of earth barriers across streams. Some of the lakes formed were of vast area. The materials for the embankment were transported in baskets or other containers. Compaction was accomplished incidentally by the trampling feet of the carriers. Even today in some countries where labor costs are relatively low, this procedure is still used.

Turning to the most available materials, the ancient dam builders made liberal use of soils and gravels. Since they had only the slightest understanding of the mechanics of materials or of floodflows, their methods were haphazard, and their works often failed. Embankment dams were low on the scale of public confidence for many centuries.

One of the earliest accounts of any major engineering work relates to the founding of Memphis on the River Nile (fig. 1-1), which can only be estimated at sometime between 5700 B.C. and 2700 B.C. The historian Herodotus attributed this construction to Menes, the first king of the initial Egyptian dynasty. According to some interpretations of the accounts of Herodotus, King Menes had a masonry dam constructed on the Nile at Kosheish, about 20 kilometers (12 miles) upstream from the site of his planned capital at Memphis.

This version, considered by some historians to be no more than legend, says that before founding the capital, Menes altered the course of the Nile to the east side of the valley rather than the west. One of his purposes reportedly was to assure enough space for the city west of the river. The location provided a better defense perimeter on the east, whence his enemies usually approached. To accomplish this, he is reported to have constructed an immense dam across the river near the Libyan Hills, diverting the stream to a new channel.

Some translations of the writings of Herodotus suggest that the dam was composed of cut-stone masonry. It is reported to have reached a height of about 15 meters (50 feet) and a crest length of 450 meters (1475 feet). The skepticism of modern historical analysts toward this account stems from the magnitude of the project, which they judge to have been beyond the capability of builders of that time.

Elsewhere in Egypt, well-preserved remains of other masonry barriers can still be seen. The abutments of what some archeologists regard as one of the oldest dams in the world still survive in the normally dry channel of the Wadi el-Garawi near Helwan, about 32 kilometers (20 miles) south of Cairo. At some time — perhaps as early as the reign of Khufu (King of Egypt about 2900-2877 B.C.) — the Sadd el-Kafara Dam was built in the wadi to impound water for workmen in the nearby quarries.

The dam had a crest length of about 107 meters (350 feet). Its height was 11 meters (37 feet). The faces were formed by rubble-masonry walls, each 24 meters (78 feet) thick at the base and extending to the top of the dam. The total volume of these two rock walls was about 22 900 cubic meters (30 000 cubic yards). At the base, the walls are separated by a distance of 36 meters (118 feet). Evidently the dam did not have the benefit of a cutoff

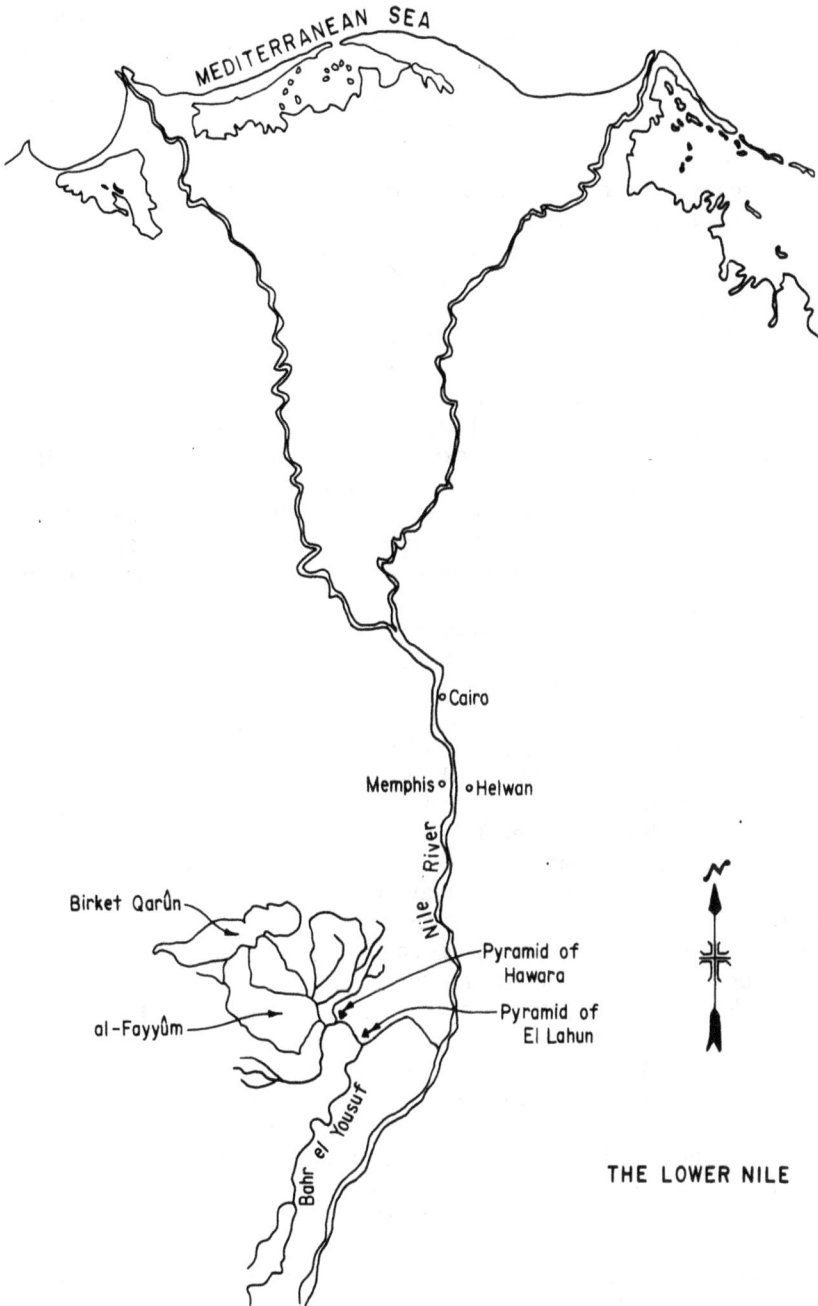

Figure 1-1.—The Lower Nile. P-801-D-79287.

trench excavated in the foundation. The core was filled with approximately 54 400 metric tons (60 000 tons) of gravel and other stones and probably some earth. The exposed face of the upstream wall was lined with stepped rows of roughly cut limestone blocks, evidently set with unmortared joints. The stones, reportedly having an average weight of approximately 23 kilograms (50 pounds), were placed in steps about 0.3 meter (1 foot) high on a slope of 3 vertical on 4 horizontal. The massive section had a thickness from face to face of 84 meters (274 feet) at the base and nearly 61 meters (200 feet) at the crest. There is evidence that the top of the dam sloped longitudinally toward the center, causing overflow to be concentrated at that point.

The major deficiency in this dam apparently was its lack of a spillway. This was a serious omission, since the Nile watershed is subject to cloudbursts that cause damaging floods in the tributary wadis. The reservoir capacity of only about 570 000 cubic meters (460 acre-feet) was insufficient to provide significant flood detention. Evidently, the dam was overtopped and its central section was broken away soon after completion of construction, since there is no sign of siltation in the reservoir.

Although its builders may have expected the lower crest elevation at the middle to serve as a spillway, the core at that point was inadequately protected from erosion by overtopping waters. This primary mistake was made at many dams in other areas in later history. The Sadd el-Kafara failure probably discouraged the early Egyptians from constructing other dams of the same composite section.

One of the outstanding reclamation projects in history was created by the Theban Dynasty of 2000 to 1788 B.C., which converted the great desert basin of al-Fayyûm into fertile farmland west of the lower Nile. In the valley of the Nile, less than 90 kilometers (56 miles) upstream from Memphis, there is a gap in the Libyan Hills leading to this immense depression, whose bottom is much lower than the Nile. The basin, roughly 80 kilometers (50 miles) in length and 48 kilometers (30 miles) in width, now contains a lake called the Birket Qarûn. A narrow, rocky gorge connects the depression with the west branch of the Nile known as Bahr el Yousuf (Canal of Joseph). In ancient times, there may have been a natural overflow into al-Fayyûm from the Nile when the river was passing extreme floods. Some scholars believe that the Theban kings enlarged and controlled this channel to divert the waters for land reclamation.

Dams also were constructed across ravines leading into the basin, apparently to capture the runoff during the wet season. One of these dams was a barrier across the Wadi Gezzaweh, a ravine about 73 meters (240 feet) wide at the site. The dam was 44 meters (143 feet) wide at the base and had a height of approximately 11 meters (36 feet). It had a composite embankment consisting of a lower zone of irregular stones embedded in clay, an intermediate rockfill zone of undressed limestone blocks, and an upper section composed of cut stones laid in steps.

Many of the ancient dams such as this did not have separate spillways. During overflow the stepped courses of stone on the slope tended to dissipate the energy of the falling water and to protect the structure from scour. Eventually, the middle of the barrier across the Wadi Gezzaweh was broken loose by a flood. Remnants of the dam can still be seen at the abutments.

The monarch credited with the plan for diverting part of the Nile flow into the enormous al-Fayyûm was Amenemhat of the 12th dynasty. The Greeks call him Moeris. According to some reports, Amenemhat could see the potential of this depression as a reservoir for the surplus floodwaters, and he had a channel dug which provided conveyance from the Bahr el Yousuf. Although a connection is believed to have existed between the river and the basin as far back as the reign of King Menes of the first dynasty, Amenemhat reportedly widened and deepened the canal, thus facilitating diversion of the excess floodwaters of the river. This may have been not long after 2000 B.C.

Herodotus gave a first account of the lake in about 430 B.C., saying "Now the Labyrinth being such as I have described, the lake named that of Moeris causes still greater astonishment, on the bank of which the Labyrinth was built.

"The water in the lake is not derived from local sources, for the earth in that part is exceedingly dry and waterless, but it is brought in from the Nile by a canal. It takes six months filling and six months flowing back ***."

Strabo, writing in 20 B.C., added: "It has also a remarkable lake, called the lake of Moeris, large enough to be called a sea, and resembling the open sea in colour.

"Thus the lake of Moeris is from its size and depth capable of receiving the overflow of the Nile at its rising, and preventing the flooding of houses and gardens; when the river falls, the lake again discharges the water by a canal at both mouths, and it is available for irrigation. There are regulators at both ends for controlling the inflow and outflow."

Recent scholars have questioned these accounts of the system's operation, suggesting that the diverted waters more likely were applied directly to irrigation of the slopes of the depression, before reaching storage. Investigators have not all agreed in their estimates of the size of the lake. Some have believed that only the lower levels of al-Fayyûm impounded water. Others contend that the whole depression was inundated except for a few high points, and that the depth of the lake may have been as great as 91 meters (300 feet).

According to Sir William Willcocks, who studied the area painstakingly, "Lake Moeris *** had a surface of 1700 million of square metres, a capacity of some 50,000 million cubic metres and, being drained back into the Nile and kept at a low level, it was able to take from a flood 13,000 million cubic metres of water, and 3000 million of cubic metres extra for every year it was not used. It was capable of reducing a very high flood to one of moderate dimensions; and, if injudiciously or maliciously opened in an ordinary flood, it was capable of depriving a great part of Lower Egypt of any basin irrigation at all, for such irrigation utilised only the surface waters of the Nile flood."

The Bahr el Yousuf carried the floodwaters which were diverted. It was part of a complex system of natural channels, canals, and dams. The connecting canal has been reported to have been 13 kilometers (8 miles) long, 49 meters (160 feet) wide, and 9 meters (30 feet) deep. The dams in the project were constructed using both earth and masonry, and the flow of the water was controlled by gates.

The Ha-Uar of the Hyksos (1788-1580 B.C.) now called Hawara, is where the pyramid of the Labyrinth stands and where the Labyrinth and regulating dams diverted the Nile's waters. The main regulators were two earth dams 10 kilometers (6 miles) apart, closing the gap between the river and the lake. In that era, the Nile evidently flowed in two channels opposite the intake of the canal. The Bahr el Yousuf of the 20th century at Lahoun was in those times either the main channel of the Nile as it was in King Menes' day or of such large capacity that the cutting of the two dams at Hawara Eglan and Hawara el Makta diverted a large part of the Nile's flow.

Much of Egypt in those days was under basin irrigation and depended for its life on the river being held high enough to be diverted into the distribution systems. When the Nile was dangerously high, floodflows were diverted through the canal into Lake Moeris. If the barriers were breached during lower riverflows, the Nile in Lower Egypt could be lowered so drastically that a famine could ensue. Egyptian history tells of four famines of long duration. One was the famine of Joseph's time, about 1730 B.C. There were two kingdoms then in Egypt, the Hyksos of Lower Egypt and the Egyptians of Upper Egypt. The frontier was at the canal into Lake Moeris. By capturing Lower Egypt's frontier fort and breaching the dams controlling discharge into the depression, the King of Upper Egypt reportedly produced Joseph's famine. Retaking of the fort and restoration of the barriers brought the drought to an end.

The decline of Lake Moeris 1500 years after Joseph's time has been attributed to the gradual diminution of the Lahoun branch of the Nile due to the less frequent use of the wasteway as the irrigation systems in lower Egypt were established. Eventually, the branch became so small that the diversion had little effect on the Nile.

In Babylonia and Assyria, irrigation was extensively developed in the Tigris and Euphrates Valleys (fig. 1-2) as early as 2100 B.C. This reached its peak much later in Sassanian times. On the Tigris River, two great canals diverted from the final rapids near Beled. These were the Nahrwan Canal, extending for 250 kilometers (155 miles) on the left bank and an equally wide but shorter canal on the right bank, known as the Dijail Canal.

Traces of the Nahrwan Canal, which was estimated to be as much as 122 meters (400 feet) wide and 5 meters (15 feet) deep, can still be identified. To facilitate desilting, the canal had two intakes, each with sufficient capacity to serve the system while the other was shut down for maintenance. The upper intake diverted water from the Tigris at Dura, and the lower one joined the canal about 60 kilometers (37 miles) downstream at Kudesieh, where there were large regulators.

At some time during the operation of the Nahrwan Canal, its diversion from the Tigris River may have been accomplished by an earth dam. However, the ruins found at the river near the ancient headworks are of massive rubble masonry. Stoneworks were also used to divert tributary streams such as the Atheim (also called Adheim or Adhaim) into the Nahrwan Canal. Parts of the Atheim Dam were still in evidence at the beginning of the 20th century. The Atheim River goes through the Hamrin hills about 80 kilometers (50 miles) from the Tigris. Evidently a masonry dam 17 meters (56 feet) high was erected on the river in this vicinity for diversion into two canals, the Nahr Rathan and the Nahr Batt. These served water on

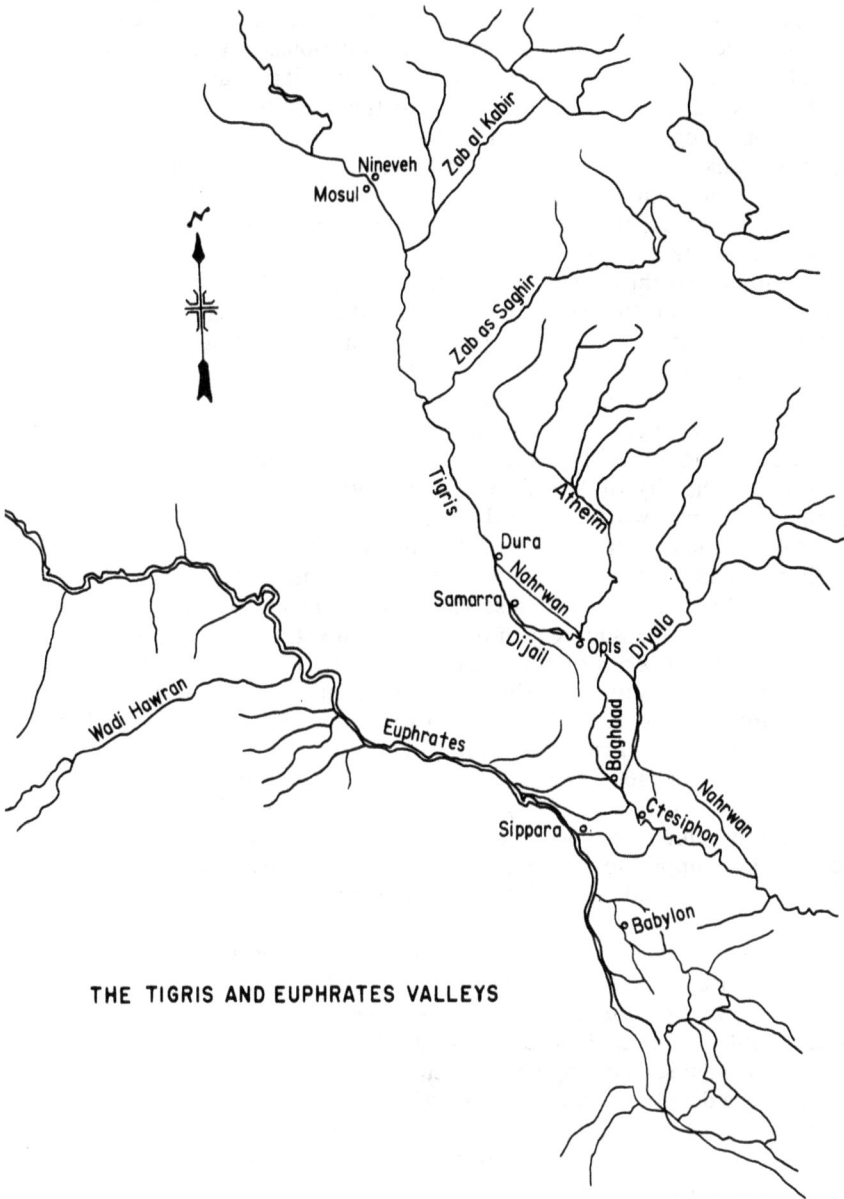

THE TIGRIS AND EUPHRATES VALLEYS

Figure 1-2.—The Tigris and Euphrates River Valleys. P-801-D-79288.

both sides of the Atheim. The Nahr Batt joined the Nahrwan Canal at a regulator made of masonry.

Another notable old dam was the Marduk Dam on the Tigris River north of Baghdad and south of Samarra. It survived the Assyrian, Chaldean, Persian, Greek, Roman, and Sassanian dominations; but it breached and was left in ruin in the 13th century A.D. This dam was constructed of materials including "reeds," according to the inscription on a clay tablet dating from roughly 2000 B.C. The barrier got its name from the Babylonian Marduk, whose history is closely tied to the biblical Nimrod. Tradition holds that Nimrod, the reputed builder of the city of Nineveh, put a large earthfill across the Tigris and thus elevated its level about 12 meters (40 feet) to create a major diversion.

The dam across the ancient bed of the Tigris above Opis is still known as Marduk Dam. The "reeds" referred to on the clay tablet were probably timbers placed as a kind of cofferdam to protect the embankment during construction. These wood members may have been left in the fill in the hope that they would provide some resistance to erosion.

Some observers believe that Nimrod also may have built diversion works farther upstream on the Tigris to supply a large canal along the edge of the valley above the city of Mosul, across the river from the site of Nineveh. Remnants of these works can still be seen.

One of the most impressive ancient water systems was developed in Judah by King Solomon (1018-978 B.C.). Some of these facilities have been rehabilitated and have provided service to Jerusalem as they did many centuries ago. The water source is in the hills southwest of the city. Solomon's system included a series of three reservoirs constructed in a valley among those hills. The basins were shaped with essentially straight boundaries, each having four sides, with the lowest reservoir additionally divided into two chambers by a transverse wall. Dimensions of the ponds vary from about 110 meters (360 feet) to 146 meters (480 feet) in length and from approximately 8 meters (27 feet) to 19 meters (63 feet) in depth. Part of the water supply came from springs at the reservoirs. One of these fed a tank at the side of the uppermost impoundment, where flow was controlled so that delivery could be made either into storage or into the aqueduct extending to Jerusalem.

The southern corner of Arabia, where the Red Sea meets the Gulf of Aden, embraces a land famed for its fertility. This region, known today as Yemen, was occupied in ancient times by several kingdoms, including Saba (Sheba) and Qataban. Records on stone slabs, plaques, and monuments attest to the high level of engineering attained in the area. Irrigation systems of impressive detail and extent were developed by the South Arabian people. Their lands supported a thriving agriculture for at least 2,000 years. The Sabaeans, or people of Sheba, were concentrated mostly around Marib, the hub of an important network of water supply.

Marib may have been the capital governed by the Queen of Sheba in about 950 B.C. This city was probably considerably older than the Qatabanian cities in Beihan. There appears to be little doubt that it was founded in the second millennium B.C. and was occupied continuously until the seventh century A.D. The flourishing economy in the city and its environs was made possible by large dams which impounded the runoff from the hills and provided soil conservation and irrigation. Such structures existed at Adraa (Edraa or Aedraa), Adschma (Adshma), and Marib.

Until recently, only a few trained observers had examined the site of the largest of these, the Marib Dam. This barrier, called Sudd-al-Arim by the Moslems, is ranked as the largest of the ancient dams in southern Arabia. According to one report, it was located on the Wadi Sadd (Saba) near Marib and roughly 320 kilometers (200 miles) north of Aden. Approximations of dates related to the dam, as well as its size, vary among the works of historians. It is mentioned in an inscription dating from the year 750 B.C. Some investigators believe that Lokman, King of the Sabaeans in about 1700 B.C., built this dam on a wadi at a site approximately 10 kilometers (6 miles) from the city of Saba. Others say it was constructed on the Wadi Dhana (Danna or Denne) in the period around 1000 to 700 B.C., as the Kingdom of Sheba approached its peak.

One account described the dam as 3.2 kilometers (2 miles) long, 37 meters (120 feet) high, and 152 meters (500 feet) wide at the base, with a volume of several million cubic meters of rock. But another, much more plausible, version tells of an embankment only a fraction as large and composed of earth. This presumably is the dam whose remains can still be seen today, at a site on the Wadi Dhana about 5 kilometers (3 miles) upstream from Marib. Arabs who examined the ruins in 1936 and 1947 described this structure as 650 meters (2130 feet) long, with five "spillways." Also 14 irrigation channels were said to be associated with the reservoir. The explorers were impressed with the exceptional quality of the masonry of the diversion works.

Some of the apparent discrepancies in the early historical records are possibly explained by the existence of several dams. Evidently, there was a series of barriers that regulated flows in the wadis which drain the eastern slope of a mountain range in Yemen. The principal structure in the series was said to be the Marib Dam, which functioned as the central control for distribution of the mountain waters. The hillsides at the site were excavated to form intakes for diversion of water to the fields in the vicinity. The role that the Marib Dam played in the prosperity of the region is clear from the reference in the Koran to "great gardens of the Sabaeans."

Around 500 B.C., the dam on the Wadi Dhana was enlarged to make it about 7 meters (23 feet) high and 610 meters (2000 feet) long. Each face was built on a 1 to 1 slope. The upstream face was protected with a lining of mortared masonry. A later major enlargement raised the Marib Dam's height to 14 meters (46 feet). This was accomplished evidently at some time after the Sabaean rule ended in 115 B.C., giving way to the power of the Himyarites.

Major canal intakes at the abutments of the dam are well preserved. The outlet facility at the right abutment served a hillside canal, while the works at the left end of the dam diverted water into a canal built on an embankment. This transitioned to a more conventional aqueduct farther downstream. Remains of these conveyance facilities are still visible at the site. Between the left outlet structure and the left abutment was a wall which evidently functioned as a spillway. The joints in its masonry appear to have been filled with a bituminous material. The stone diversion works were formed by large blocks so neatly trimmed that they fitted closely at the joints. These hewn stones were placed crosswise at intervals to assure interlocking of their courses. The stones were further fastened by pins (lead rods) about 100 millimeters (4 inches) long and 1000 square millimeters in section. These pins were inserted into holes drilled in the stones to a depth of approximately 50 millimeters (2 inches). The adjoining block in the next higher course was drilled and positioned so that the upper end of the pin extended into it as the stones were brought together.

Some accounts say that no mortar was used in the masonry joints, even though the engineers and artisans in Yemen were acquainted with the use of mortar. This does not agree with another report which described masonry so tightly bonded by mortar that not a single stone could be pulled out. In fact, a mortar coating reportedly was placed on the crest of Marib Dam for weatherproofing. These careful construction measures were evidently successful to some degree. Remnants of the diversion works more than 15 meters (50 feet) in height have survived the attack of the elements. Other sections of the structure are gone, perhaps destroyed by a violent storm during Abyssinian rule in the sixth century A.D.

King Sharahbil Yafur reportedly had the dam rehabilitated in 449 A.D., but in 450 A.D. floods again ruptured the structure. The dam was restored. Then in 542 A.D., during the rule of the Abyssinian Viceroy Abraha, another major breach occurred. The last known inscription relating to the structure was made in that same year. It reported that the Viceroy had ordered repairs of the dam and had requested large quantities of provisions for the many workers, including 200,000 sheep and goats, 50,000 sacks of flour, and 26,000 crates of dates. Evidently the reconstruction was completed expeditiously. Historians tend to agree that the final disaster struck the dam soon thereafter, and the plain of Saba reverted to desert. There are some scholars who say that the last failure of the dam was between 542 and 570 A.D. Others believe that this happened in the seventh century A.D. The loss of the dam has been ascribed to various causes, ranging from volcanic activity to earthquake to neglect. The last may be most likely.

Siltation of the reservoir was undoubtedly a problem. Some evidence of a rock structure has been uncovered at the foundation of the dam near the middle of the valley that indicates a facility for sluicing or low-level diversion. Presumably, sediments were removed by this means or by crews of laborers. Some observers, on examining the layers of deposits in the reservoir, have suggested that water storage capacity was eventually so reduced by silt encroachment that the dam was overtopped and breached.

About the dam's death, the *Encyclopedia of Islam* says "There is hardly any historical event in pre-Islamic history that has become embellished with so much that is fanciful, and related in so many versions, as the bursting of the Marib dam (Sudd-al-Arim)." The Koran recalls that "the people of Saba had beautiful gardens with good fruit. Then the people turned away from God, and to punish them, He burst the dam, turning the good gardens into gardens bearing bitter fruit."

Some analysts of ancient times in the Middle East have ascribed the fall of the South Arabian kingdoms to the breaking of the Marib Dam. More likely, the decline of these governments had begun many years earlier. The dam was neglected and suffered from leakage. The loss of the vital water facility was a critical adversity for an already weakened people.

While the Marib Dam is recognized as the outstanding structure among the waterworks of ancient Yemen, the ruins of other impressive barriers can be seen in the region. Two which deserve mention, at Adraa and Adschma, were built in narrow gorges; and each may have been as high as 15 to 20 meters (50 to 65 feet). These were made of stones and soil, confined by nearly vertical masonry walls at the upstream and downstream faces. The cut stones in the outer walls of the dam near Adraa were stepped and plastered. It also had two parallel core walls made of uncut stones. These walls were evidently erected to the full height of the barrier. The gap of 2 to 3 meters (6 to 10 feet) between them was filled with clayey soil, apparently without any stones. A composite section such as this, with outer embankment zones built against a relatively impervious core, would indicate that the constructors had some understanding of the basic requisites of design. This view is also supported by the still existing outlet works with the intake on the upstream side connected to a conduit placed through the masonry walls. Here, as well as at the dam near Adschma, diversion was made into hillside canals. Remains of these canals can be seen today. The histories of these dams are incomplete. Their origins are unknown, but both structures may have failed in the seventh century A.D. when war and religious strife swept the land.

In Iraq (Mesopotamia), some of the earliest dams are attributed to the Assyrian King Sennacherib (705-681 B.C.), who ordered their construction to serve his capital city of Nineveh. Among these were two masonry structures at Ajilah on the Khosr River. The more important one had a length of about 240 meters (787 feet) and a height of at least 3 meters (10 feet). The Khosr River was also dammed farther upstream near Qayin, and another barrier was built at Bavian on the Atrush River for diversion of its waters to the Khosr.

Soon after the beginning of the sixth century B.C., the Babylonian King Nebuchadrezzar II, the Nebuchadnezzar of the Bible, constructed a dam at Abbu Habba, south of Baghdad. He is also credited with construction of the "Royal Canal" running between the Tigris near Ctesiphon and the Euphrates at Sippara. This channel was reported to be of such large dimensions that it accommodated any of the ships that sailed in those days. Herodotus recorded that a basin at Sippara was almost 8 kilometers (5 miles) around and walled with stone. Other important canal projects were accomplished under the direction of Nebuchadrezzar, and important advances were made in design. Babylon's eastern canal was "walled up from the bottom," indicating that the lining of canals was accepted practice in that era. Stone dams were erected to turn the river water into the canals, and gates were operated to control the flow.

In 539 B.C., Cyrus the Great, King of Persia, defeated the Babylonian Army commanded by Crown Prince Belshazzar, son of Nebuchadrezzar. According to generally accepted accounts, Cyrus then built an earth dam on the Diyala, a tributary of the Tigris, to create diversion works for irrigation. He reportedly had 30 canals excavated to establish an extensive water distribution network.

In Persia, the Achaemenians built dams on the River Kur south of Persepolis. Most of this work was done in the sixth or fifth centuries B.C., when their power was at its peak. The Persian King Darius the Great

(521-485 B.C.) had three gravity dams built on the River Kur near his palace at Persepolis.

The Kingdom of Qataban in South Arabia covered the region now called Bayhan. Its capital was at Timna. The Qatabanians thrived during the first five centuries B.C. Their prosperity was owed largely to waterworks of remarkable scope. A canal about 24 kilometers (15 miles) long served between the village of Beihan al-Qasab and a point north of Hajar bin Humeid. This was one of several such facilities for conservation of local runoff. These conveyance works were operated in conjunction with masonry reservoirs and stone gate structures which regulated flows to the delivery systems.

When Qataban was at its zenith, the countryside bordering the Wadi Beihan was nourished by the waters of a skillfully engineered irrigation system. The chronology of this vital network has been determined closely from inscriptions on some of the regulating structures, with dates of the facilities varying from about the fifth century B.C. to the first century A.D. This evidently means that even after the fall of the Kingdom of Qataban, after 25 B.C., the water project on the Wadi Beihan was maintained by its successors, the states of Hadhramaut, Saba, and Dhu-Raidan. When these kingdoms in turn faded away, the irrigation systems also crumbled.

Other dams created in antiquity were of impressive dimensions. On the island of Ceylon, for example, Sinhalese engineers established daring precedents in earthfill construction. Many of the ancient waterworks in Ceylon were built in its northern and eastern regions. Ruins of some of these facilities can still be seen near the old capitals of Anuradhapura, Polonnaruwa, and Tissamaharama. The Sinhalese kings aggressively advanced water development in lands under their control. After their immigration in the fifth century B.C., the Sinhalese implemented irrigation plans which supported a flourishing economy until these people were overcome by new invaders in about 1200 A.D.

Embankments of great length were constructed by the Sinhalese to form reservoirs or "tanks" of large capacity. The Kalabalala Tank was formed by an earthfill 24 meters (79 feet) high and about 6 kilometers (3.5 miles) long. Its perimeter measured 60 kilometers (37 miles). The storage was used to supply irrigation systems around the city of Anuradhapura.

Among the oldest reservoirs in Ceylon are those of Basawakkulam (430 B.C.), Tissa (307 B.C.), and Nuwara (first century B.C.) near Anuradhapura. The earthfills forming these impoundments were relatively low but very long. They were restored during the latter part of the 19th century A.D.

In 331 B.C., Alexander the Great led his forces into the Valley of the Tigris. The records of his campaign indicate that dams on the river had to be partially removed to permit passage of his fleet. These have been described as massive rubble-masonry weirs which served as diversion works for canal intakes. Other accounts refer to Alexander's removal of embankments which blocked his way on the Tigris. Possibly, he encountered both kinds of barriers during his advance. As he consolidated his gains, he presumably had the dams repaired. His chroniclers have given enthusiastic narratives about the irrigation of the conquered land.

In Baluchistan (Pakistan), ruins of pre-Aryan dams have been discovered near Lakorian Pass and in the Mashkai Valley in the southern region of that country. In the centuries following the Aryan invasions in the middle of the second millennium B.C., irrigation on the subcontinent was expanded. One of the outstanding structures was the Sudarsana Dam built near Girnar in Kathiawar during the reign of Chandragupta, the first emperor of India (322-298 B.C.). Rock inscriptions describe this "pleasant looking" barrier and record two disasters which struck it. In one of these "by a breach, four hundred and twenty cubits long, just as many broad (and) seventy-five cubits deep, all the water flowed out, so that (the lake), almost like a sandy desert, (became) extremely ugly (to look at)." The Sudarsana Dam is known to have survived until at least 457 A.D., but its fate after that is obscure.

In about 240 B.C., a stone-crib dam about 30 meters (98 feet) high and almost 300 meters (1000 feet) long was built on the Gukow River in the Shansi Province of China. However, not many other dams were constructed in that country in the early centuries.

The Nabataeans built dams in the Negev desert in the vicinity of the present Israeli-Jordanian border. Outstanding examples were a rockfill 14 meters (46 feet) high near the old capital city of Petra (Jordan) and gravity dams in the Wadi Kurnub, 38 kilometers (24 miles) southeast of Beersheba (Israel). Two of the gravity structures remain intact today. Most of the others constructed in the region in that era were allowed to deteriorate and fall into disuse. The dependent farms then faded into barren wasteland.

Today the central Negev has many ruins that testify to the effective water projects that sustained its farms in Nabataean and Roman times. Agriculture thrived there between the second century B.C. and the seventh century A.D. Low dams and irrigation ditches intercepted and distributed the limited runoff.

The ancient city of Ovdat (Aboda) is in a farming district in the central Negev. It was built by the Nabataeans in about the second century B.C., but most remnants of the first construction have been erased by Roman and Byzantine and later activity. Many thousands of dams reportedly can be found in an area of about 130 square kilometers (50 square miles) surrounding Ovdat. Practically every controllable wadi in the vicinity was dammed. Most of these barriers were small, up to about 2 meters (6 feet) high. But in the larger ravines, there are traces of dams that must have measured as much as three times this height.

The dams in the Negev served various functions. First among these was the diversion and storage of water. Such structures were erected on the wider and deeper wadis. For example, in the Wadi Ovdat there are ruins of a 4-meter (14-foot) wide barrier composed of large stones and having a curved axis.

Thousands of low dams in the area served as sediment interceptors. These generally were in the smaller wadis. A typical barrier would have been about 2 meters (6 feet) high, 2.5 meters (8 feet) wide, and 46 meters (150 feet) long. These dams were usually stone and earth embankments, with slopes protected by stepped courses of masonry. A series of such structures would extend along a wadi, at a spacing of about 40 meters (130 feet). Eventually, the accumulations of silt between the barriers would form a continuous stairway of alluvium that could be placed in cultivation. Then, by flood irrigation and periodic deposition of more silt, these fields were turned into productive farms. In this way the people of the Negev effected significant accretions to their minimal acreage of arable land.

At about the same time, important waterworks were being developed in Spain. After the Romans gained control of Toledo in 193 B.C., they undertook the transport of water to that city in an aqueduct. A reservoir was created by construction of the Alcantarilla Dam, which was 20 meters (66 feet) high and at least 550 meters (1800 feet) long. The dam is believed to have been built in the second century B.C., on the Arroyo del Guajaraz at a site 20 kilometers (12 miles) south of Toledo. It was essentially a masonry and concrete mass buttressed on its downstream side by an earthfill. Today the dam is in ruin.

The dam of Alcantarilla is considered to be the oldest in Spain and is possibly the earliest Roman dam. It is of cruder construction than those of the same type built later at Mérida, such as the Proserpina Dam. Unlike the latter, the Alcantarilla Dam had no buttresses supporting its upstream face. On that side there were two parallel rubble-masonry walls each approximately 1 meter (3 feet) thick, separated by a space of about 0.6 meter (2 feet) which was filled with concrete. Cut-stone blocks protected the upstream face. To bolster the composite wall against the waterload, its downstream face was buttressed with an earthfill 14 meters (46 feet) thick at the crest of the wall and with a slope of 3 to 1. Deep openings at each abutment are evidence that substantial spillways were incorporated in the structure.

Today both the masonry wall and the embankment of the Alcantarilla Dam are breached over a central distance of about 200 meters (650 feet). This suggests that the masonry mass may have been toppled upstream by the pressure of the earthfill as the reservoir level was being lowered.

The Romans were also active in southern France, where Glanum was one of their very early settlements. An aqueduct conveyed water to Glanum from a reservoir impounded by a low curved dam. This structure is believed to date from the first century B.C., or possibly a little later. In 1891 A.D., a new dam was constructed on the ruins of the Roman barrier, concealing evidence of the earlier works.

The Roman dam near Glanum was reported to be approximately 6 meters (20 feet) high, with a crest length of possibly 9 meters (30 feet). It was a composite of two masonry walls, each slightly thicker than 0.9 meter (3 feet), and separated by a space of about 1.5 meters (5 feet), which was probably filled with soil and stones. The thin section, only about 3.6 meters (12 feet) thick and rising to a height of 6 meters (20 feet), must have been dependent upon its curvature in plan to assure its stability against water forces.

The First Millennium A.D.

In Spain, a little more than 100 years later, work was begun on some of the finest Roman dams. These were near the town of Mérida, which is noted today for its impressive Roman ruins. Six kilometers (4 miles) north of Mérida, early in the second century A.D., the Romans built the Proserpina Dam (fig. 1-3), 19 meters (62 feet) high above its foundation and 427 meters (1400 feet) long.

The Proserpina Dam has been ranked as a classic among structures of its type. An upstream section is comprised of a concrete core sandwiched

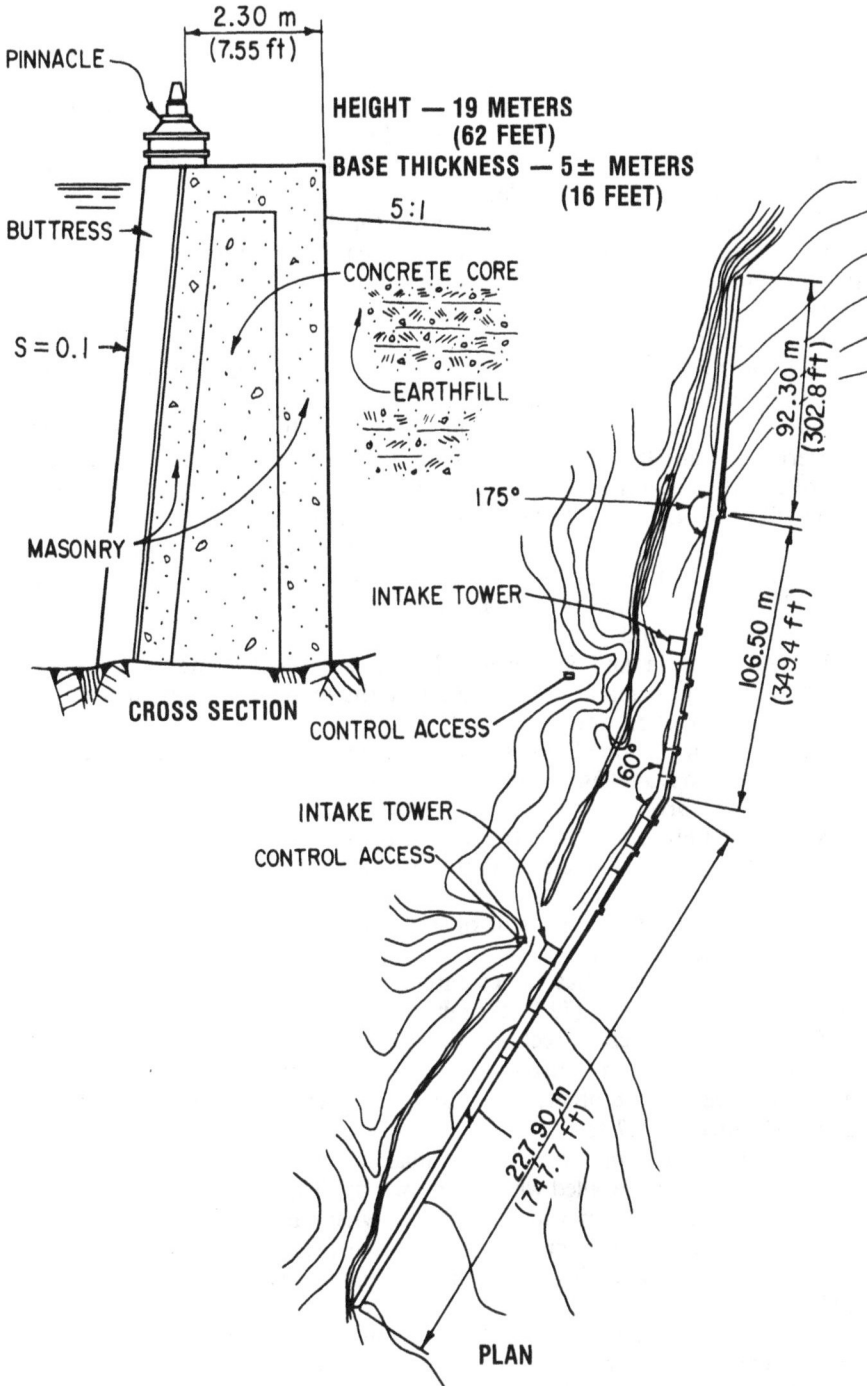

Figure 1-3.—Proserpina Dam, plan and cross section. P-801-D-79289.

between two masonry walls. The original thickness of the composite wall has been estimated to be about 3.75 meters (12.5 feet) at the top. Some reports indicate that this may have been as little as 2.1 or 2.4 meters (7 or 8 feet) in a few places. At the foundation, the thickness may have been as much as 5 meters (16 feet). The upstream face is battered steeply at 1 to 10 while the downstream masonry face is vertical. The wall extends about 6 meters (20 feet) into the foundation. Following the precedent set at Alcantarilla Dam, an earthfill was placed against the downstream side of the wall. This slopes from the crest to intersect the natural ground at a maximum distance of about 60 meters (200 feet).

A basic difference between the Proserpina and Alcantarilla Dams is that masonry buttresses were erected at the upstream face of Proserpina to provide resistance against overturning. Lack of this feature was the weakness that led to the collapse at Alcantarilla.

The records do not reveal much about the nearly 2,000-year lifespan of Proserpina Dam. It is assumed to have suffered long periods of neglect. There is information on repairs and modifications accomplished in the years 1617, 1689, and 1791. This would indicate that the reservoir has been in possibly continuous service for nearly 400 years. It has not suffered serious impairment by siltation. Major repairs were made in 1942, including rehabilitation of the masonry. Water is still supplied via a Roman aqueduct between the dam and Mérida.

The Alcantarilla and Proserpina Dams had large capacity spillways, testifying to the Roman understanding of the vulnerability of such structures to uncontrolled floods.

The water supply for Mérida was enhanced at a later date by construction of the Cornalbo Dam (fig. 1-4) 16 kilometers (10 miles) northeast of the city on the Río Albarregas. It is a more sophisticated structure than its predecessors in that area. Although the history of this dam is fragmentary, there appears to be general agreement that it was inoperable for extended periods in medieval times. Then in the 18th century it was placed back into service. It is one of the oldest dams still operational and among the largest constructed by the Romans.

The Cornalbo Dam has an essentially straight longitudinal axis. It is approximately 24 meters (79 feet) high above its foundation and 200 meters (650 feet) long. In cross section, the structure is trapezoidal. The crest thickness is flared from 7 meters (22 feet) at one abutment to more than 12 meters (40 feet) at the other. The maximum thickness at the foundation is about 118 meters (387 feet).

The core of the dam is made up of masonry walls which form interconnected boxes that were filled with stones or clay. This core was then enclosed in an earth embankment with a 3 to 1 downstream slope and a 1½ to 1 upstream slope with masonry revetment for protection against wave wash. Much of this original facing was replaced when the dam was rehabilitated in 1936.

The Romans built many stone dams throughout their empire. These were usually composed of mortared cut-stone masonry of great durability and impermeability. The one built at Subiaco, about 50 kilometers (30 miles) east of Rome, by Emperor Nero in the first century A.D., lasted nearly 1,300 years, as testified by an official account of its failure.

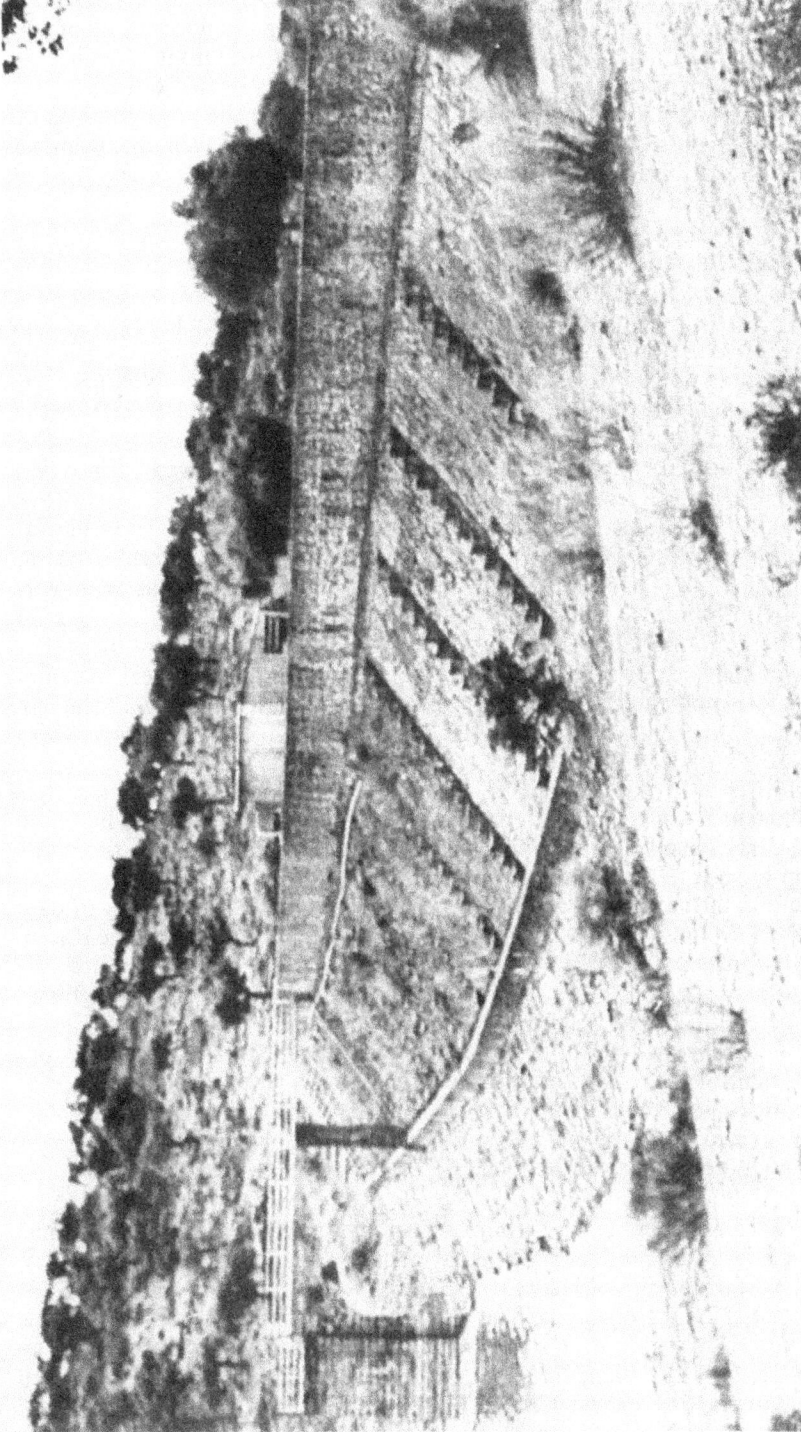

Figure 1-4.—Cornalbo Dam (Courtesy, Comité Nacional Español, ICOLD). P-801-D-79290.

The Romans also recognized the need for soil conservation. Near the coast of Tripolitania (Libya), several wadis draining the north slope of the Jebel Nefuza Mountains discharge large volumes of silt into the Mediterranean during the flood season. Three important cities of the Roman era — Leptis Magna, Oea, and Sabratha — were built near the lower reaches of these watercourses. The complex of masonry dams which the Romans erected in the wadis was designed for two purposes — water supply for the cities and protection of the land from erosion.

A Roman dam deserving mention was built in about the second century A.D. near Kasserine, 217 kilometers (135 miles) southwest of Tunis. It is a masonry-faced structure with a core evidently composed of earth and rubble. Cut-stone blocks with mortared joints were used in the facing. The upstream face is vertical for its full height of 10 meters (33 feet), while the downstream side is stepped down from the crest through six courses of masonry and then vertical in the remaining 3.8 meters (12.5 feet) to the base. The thickness varies from roughly 4.9 meters (16 feet) at the crest to 7.3 meters (24 feet) at the base. In plan, the structure is curved but not in a true circular arc. It is about 150 meters (500 feet) long.

Two noteworthy Roman dams were built in Turkey, the one at Orükaya, 190 kilometers (118 miles) northeast of Ankara, and the Cavdarhisar Dam, 210 kilometers (130 miles) south of Istanbul. The former was 16 meters (52 feet) high and 40 meters (131 feet) long; and the latter was 7 meters (23 feet) high and 80 meters (262 feet) long. These structures were of similar design, each comprising an earth core enclosed by two vertical masonry walls. Attempt was made to seal the joints between the stones with lead. In each case, the total thickness of the composite wall from face to face was about 5.5 meters (18 feet).

Another early gravity barrier was the Al-Harbaqa Dam 70 kilometers (43 miles) southwest of Palmyra (Syria). It is 18 meters (59 feet) high and 198 meters (650 feet) long. The structure has endured through the many centuries, but the reservoir is filled with silt.

Among the nations of the Orient, Japan has a continuous history of dam building dating far back into antiquity. Its oldest notable dam is the Kaerumataike earth embankment constructed in 162 on the Yodo River near the one-time capital city of Nara. Its dimensions: 17 meters (56 feet) high and 260 meters (853 feet) long.

About 100 years later, the Persian King Shapur I (241-272) undertook improvement of the irrigation projects in Khuzestan (Iran) using the labor of Roman soldiers whom he had captured. One of the most impressive structures erected by these prisoners was the dam-bridge extending approximately 550 meters (1800 feet) across the Karun River near Shushtar, in about 270.

Other dams built by the Romans have been found in Syria. The most impressive early construction in that country was 13 kilometers (8 miles) southwest of the city of Homs where the River Orontes was dammed in 284 to create the Lake of Homs. The core of the barrier was of basaltic rubble masonry, cemented with a strong mortar. Cut basalt stones were placed on both faces of the dam, and the joints were mortar sealed. The finished structure was 6.1 meters (20 feet) high, with a thickness varying from 7 meters (23 feet) at the top to approximately 20 meters (66 feet) at the bottom. This Roman dam survived and provided service for 17 centuries. In 1934 a new and larger dam was superimposed on the ancient one.

The engineering of dams also continued to advance in India and Ceylon. Many earthfills were built to create irrigation reservoirs, or "tanks". One of the largest was the Kalaweva Tank in Ceylon, built in 459. Its embankment was about 19 kilometers (12 miles) long.

Byzantine Emperor Justinian (527-565) gave impetus to development of the water supply for Constantinople (Istanbul). Eventually, eight dams were built in the vicinity. Four are still in operation. The largest of these impounds the 617 000-cubic-meter (500-acre-foot) Büyuk Bent on a tributary of the Kiathene Deresi River about 14 kilometers (9 miles) north of the city. This is a rubble-masonry gravity dam faced with cut stone. It has a height of 12.5 meters (41 feet), a length of 76 meters (250 feet), and a base thickness of 10 meters (33 feet).

On the Turkish-Syrian border near the city of Daras, the engineer Chryses from Alexandria, under the direction of Justinian, built a dam which was noteworthy for its curvature in plan. It is regarded by some historians as the first known arch dam. Evidence of its existence was documented in about 560 when Daras was a strategic place at the frontier with the Persian empire. The records indicate that a flood control dam was constructed on a tributary of the Khabur River just outside the walls of the town. The steep abutments were notched to assure firm anchorage of the arch, and gates were provided for flood regulation. However, nothing in the documents or at the river site gives any useful clue as to the structural dimensions. The remnants eroded away long ago.

One of the earliest dams in China is believed to have been built in the year 833. Its name is Tashanyan and it was built on the Zhang Xi River near Ningbo. The dam is a gravity type structure 27 meters (88 feet) high and is still being used for irrigation.

Many of the other dams built in the Middle Ages have lasted a long time. One of the more remarkable is a rubble-masonry weir on the Río Guadalquivir in Córdova, Spain. Dating back to about 900, it is regarded as probably the oldest remaining Moslem dam in that country. The weir has deteriorated significantly. Pieces of its masonry are scattered in the river channel. But enough of the structure survives to mark its zigzag alinement, totaling about 427 meters (1400 feet), across the stream immediately downstream from the Puente Romano (Roman Bridge). In addition to its original functions of water supply and mill operation, the pool at this dam has protected the bridge piers from erosion. The parts of the weir still standing suggest that its height and its thickness were each about 2.5 meters (8 feet).

In the year 960, the Band-i-Amir Dam was erected on the River Kur in Persia. The masonry structure still stands, but its function has been impaired by siltation. It has a height of 9 meters (30 feet) and a length of 76 meters (250 feet). The downstream slope is about 1 to 1. The dam evidently was built entirely of cut stones with mortared joints reinforced by iron bars anchored in lead. This followed the same practice used in the dam on the Atheim River near Baghdad.

At about the same time, important dams were being developed in southern India. The Moti-Talab Dam, an earthfill structure near Mandya (Mysore), was built in the 10th century and is still functioning today. It is 24 meters

(79 feet) high and 157 meters (515 feet) long. The unique cross section has a broad crest about 27 meters (90 feet) wide and has steep slopes of 2 to 3 upstream and 1 to 1 downstream.

The Period 1000 to 1600 A.D.

The engineers of India developed a design for earthfill dams with relatively steep slopes protected by cut-stone facing. An outstanding example is the 16-kilometer (10-mile) long Veeranam Dam, erected in the period 1011 to 1037.

Also in the 11th century, Indian engineers created a reservoir with an area of 650 square kilometers (250 square miles) in a valley about 32 kilometers (20 miles) southeast of the city of Bhopal (Madhya Pradesh), in central India. Evidence still remains of this great Bhojpur Lake, impounded by two earth dams covered on both slopes with immense blocks of cut stone. The project reportedly was developed by Raja Bhoj of Dhara. Flow of the holy River Betwa was augmented by diverting the River Kaliasot into it.

Two natural gaps existed in the circle of hills that enclosed the basin. The width of opening was about 90 meters (300 feet) at one site and 460 meters (1500 feet) at the other. To close these gaps, the Raja had barriers constructed that were said to be impressively watertight. These earthfill dams were faced with unmortared stones fitted together skillfully by the Indian masons. The higher embankment, which filled the smaller gap in the hills, had a height of about 27 meters (90 feet) and was 92 meters (300 feet) long, with a base width of 92 meters (300 feet). In the wider opening, the Raja's forces built a dam approximately 12 meters (40 feet) high and 30 meters (100 feet) wide at the crest.

A spillway was excavated in the rock of a saddle in the hills. Water stains on the spillway sides mark the maximum reservoir level at about 1.8 meters (6 feet) below the top elevation of the dams. The design of the spillway and its successful functioning for several centuries attest to the talents of the engineers assigned to the project.

Five hundred years after the reservoir was placed in operation, Shah Hussain breached the higher dam to drain the lake so that its fertile bed could be opened to cultivation. According to traditional accounts by natives of the area, a large work force took 3 months to remove the barrier. The longer dam survived to continue diversion of the Kaliasot into the Betwa River.

In the first 15 centuries A.D., the Japanese built about 30 dams higher than 15 meters (49 feet). All were earthfills. One of the most outstanding was the Daimonike Dam, erected in 1128 near Nara. It was about 32 meters (105 feet) high and 79 meters (259 feet) long.

In the same century, the Sinhalese were setting new records in Ceylon. An earthfill structure length of 18 kilometers (11 miles) was attained at Padawiya Dam 60 kilometers (37 miles) northeast of Anuradhapura. The embankment was built to a height of approximately 21 meters (70 feet). The crest was 9 meters (30 feet) wide, and the maximum base width was about 61 meters (200 feet). Slope facing consisted of cut stone.

While the Asians were making important advances in embankment construction, many European engineers continued to build masonry barriers. One of these was the Almonacid de la Cuba Dam (fig. 1-5) on the Río Aguavivas about 40 kilometers (25 miles) south of Zaragoza, Spain. It probably dates from the 13th century A.D. and is judged by some historians to be the oldest surviving Christian-built dam in Spain. The original structure was made of rubble masonry bonded with crude lime mortar and enclosed by a facing of cut stones. Its estimated height was 29 meters (95 feet), and its length was about 85 meters (280 feet). The downstream face was composed of large stone blocks placed in tiers and set in mortar.

At an unrecorded date, the Almonacid de la Cuba Dam was heightened. Evidently, this was necessitated by the continuing effects of silting, which eventually encroached on most of the storage capacity. The added section is a mass of crude rubble masonry set in a matrix of stones, earth, and lime mortar. This vertical-faced wall increased the height to about 30 meters (98 feet) and the length to over 100 meters (328 feet). The maximum crest thickness is about 25 meters (82 feet).

The spillway for the original dam was in rock at the left abutment. As part of the enlargement, the spillway crest was elevated by a curved weir.

In 1258, Hulagu Khan led his Mongols into Baghdad and eliminated Arab rule. In the confusion of the years which ensued, most of the ancient public works in that region were reduced to ruin. Nimrod's earth dam on the Tigris was breached, lowering the river level about 8 meters (25 feet). The Nahrwan and Dijail Canals were left inoperable. The rich farmlands bordering the upper Tigris reverted to desert. The ancient barrier on the Sakhlawia branch also was breached, and the irrigation works of western Baghdad fell into disuse.

During the same century, at a site southwest of Tehran in Persia, the Saveh Dam was erected for irrigation of the Mongol-dominated land. The date of its construction has been estimated as between 1281 and 1284. Apparently its main claim to fame is that its reservoir has never held water, although the dam has survived more or less intact. It is a crude rubble-masonry gravity structure without any cut-stone facing. With a height of 18 meters (60 feet), and a length of 46 meters (150 feet), it might have been an effective barrier if built on a sound foundation. However, it was placed on river alluvium. The first water entering the reservoir evidently found its way under the dam and left it standing high and dry.

Another dam of the same Mongol period in Persia spans a narrow gorge on the Kebar River, about 24 kilometers (15 miles) south of the town of Qum and 170 kilometers (105 miles) southwest of Tehran. Dating from about 1300, it is regarded as the oldest known surviving arch dam. The constant radius of curvature of its downstream face is 38 meters (125 feet). Its height is about 26 meters (85 feet), and its length is 55 meters (180 feet). The crest has a practically constant thickness of 5 meters (16 feet). Materials used in the construction were cemented rubble masonry with mortared stone block facing. The arch was keyed into the rock abutments.

In the year 1384, an arch-gravity dam was built about 5 kilometers (3 miles) west of the town of Almansa in Albacete Province in Spain. This is regarded as the first known arched dam in that country. The original Almansa Dam (figs. 1-6 and 1-7) was about 14.6 meters (48 feet) high and curved to a radius of about 26 meters (85 feet) at the downstream side of the crest. Its thickness has been estimated to be approximately 10 meters (34 feet) at the base and about 4 meters (13 feet) at the top. The structure is composed of rubble masonry with a facing of stone blocks. It is anchored securely into the rock foundation.

The Almansa Dam was enlarged in 1586 and again in the years 1736 and 1921. The most prominent addition is an angled wall superimposed upon the top of the original barrier. The wall is 6 meters (20 feet) high, 3 meters (10 feet) thick at the top, and nearly 3.6 meters (12 feet) thick at the base. The lengths of the straight walls on each side of the angle are about 36 meters (118 feet) and 53 meters (174 feet). Almansa Dam is now officially listed as 25 meters (82 feet) high and 90 meters (295 feet) long. The structure is still in sound condition.

A structure generally regarded as the oldest significant dam existing in Italy is the Cento Dam on the Savio River about 30 kilometers (19 miles) south of Ravenna. The year of its construction is estimated to be 1450. It is 71 meters (234 feet) long and about 14 meters (45 feet) thick at the base. The original gravity dam has a vertical upstream face approximately 6 meters (19 feet) high and a crest thickness of just a few feet. A parapet wall added at a later date increased the height by about 1.4 meters (4.5 feet). The Cento Dam is noteworthy for its construction of bricks set in lime mortar within a framework of wood poles.

Until about 500 years ago, very few earthfill dams approached 24 meters (80 feet) in height. Among these were the early embankments in India and Ceylon. Then in 1500, the Mudduk Masur Dam, with a height of 33 meters (108 feet), was constructed in Madras Province in southern India. This embankment height was unsurpassed for about 300 years.

Among masonry structures, an equally impressive record was held by the monumental Alicante Dam (figs. 1-8 and 1-9) on the Río Monegre 18 kilometers (11 miles) northwest of the town of Alicante in Spain. Construction of this barrier, also known by the name of the nearby village of Tibi, was begun in 1580 and then suspended due to lack of funds. Work was resumed several years later and was completed in 1594. Then, for nearly three centuries, this dam was the highest in the world, measuring 41 meters (135 feet) from base to top.

The damsite in the gorge of Tibi is only about 9 meters (30 feet) wide at the bottom. The structure is composed of rubble masonry set in mortar and faced with large cut stones. The plan is curvilinear, with a total crest length of about 80 meters (262 feet). Structural thickness varies from approximately 20.5 meters (67 feet) at the top to 33.7 meters (111 feet) at the base. Volume of the mass is 36 400 cubic meters (47 600 cubic yards).

The Alicante Dam was provided with a desilting system and a vertical shaft outlet works. These have been closed permanently. Originally there was no separate spillway. Water discharged over the crest and down the stepped masonry face. In 1697, a flood damaged the face; the dam was rehabilitated in 1738. A side channel spillway was constructed later. However, this has not operated often in recent years and has been closed with stoplogs. The dam was enlarged in 1943. Its present height is listed as 46 meters (151 feet).

Figure 1-5.—Almonacid de la Cuba Dam (Courtesy, Comité Nacional Español, ICOLD). P-801-D-79291.

PLAN

HEIGHT — 25 METERS
(82 FEET)
BASE THICKNESS — 10 METERS
(33.7 FEET)

CROSS SECTION

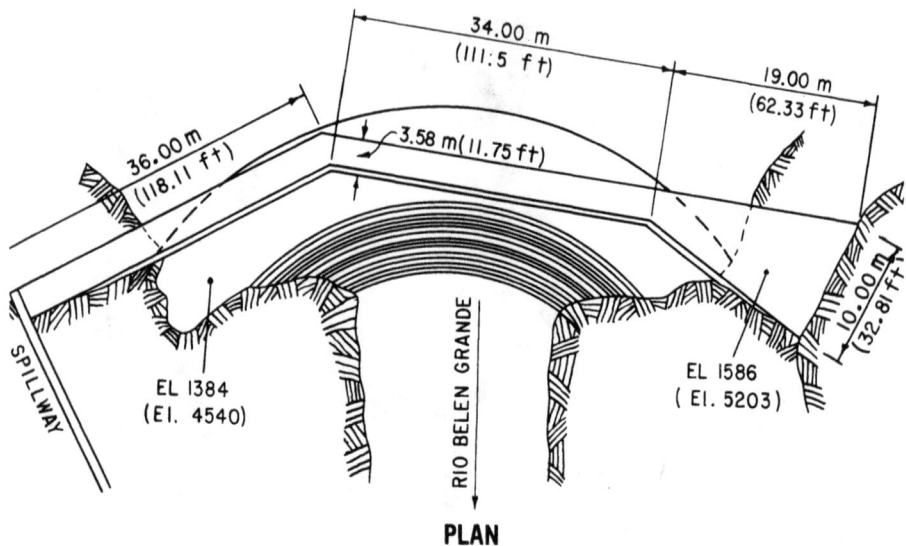

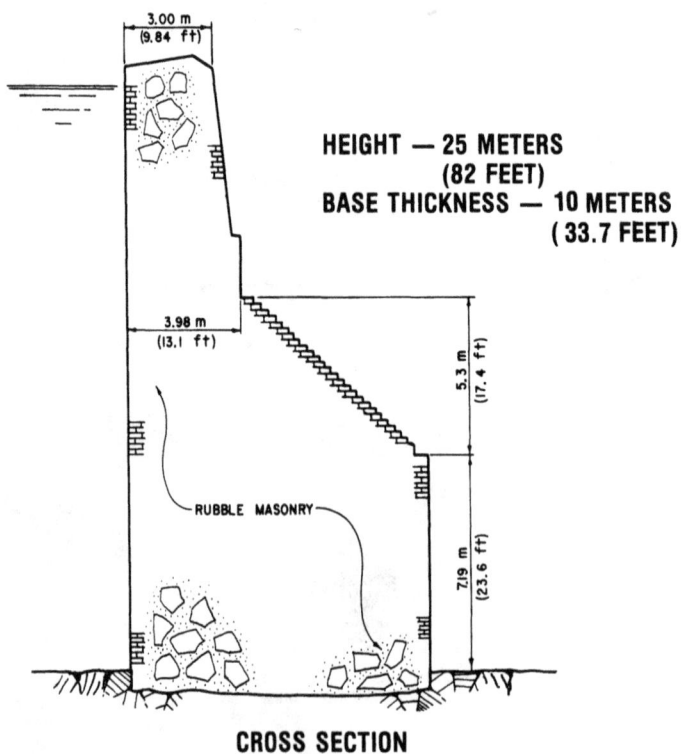

Figure 1-6.—Almansa Dam, plan and cross section. P-801-D-79292.

Figure 1-7.—Almansa Dam (Courtesy, Comité Nacional Español, ICOLD). P-801-D-79293.

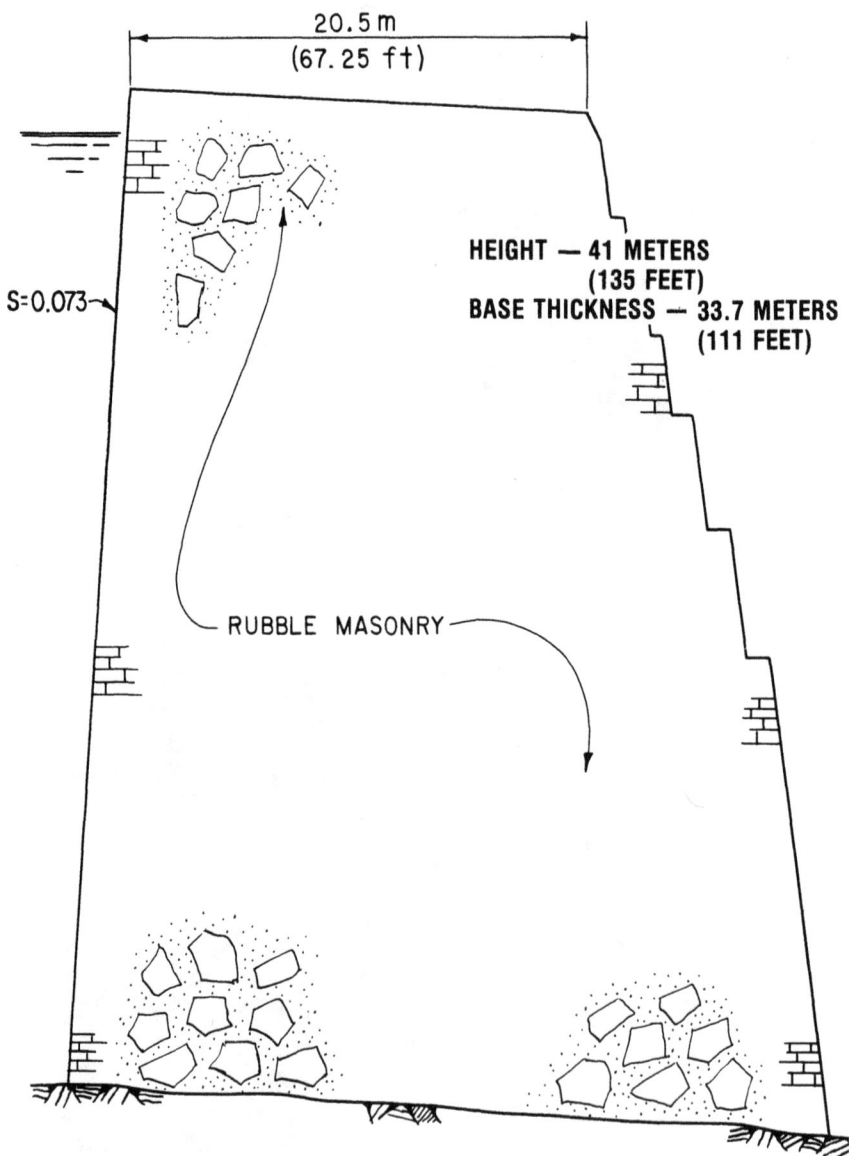

ORIGINAL DAM AS CONSTRUCTED—1594

Figure 1-8.—Alicante (Tibi) Dam, cross section. P-801-D-79294.

Figure 1-9.—Alicante (Tibi) Dam (Courtesy, Comité Nacional Español, ICOLD). P-801-D-79295.

27

The Period 1600 to 1800 A.D.

Not far from the Alicante site, the first true arch dam in Spain was built on the Río Vinalopó near the town of Elche in the mid-17th century. The Elche Dam (figs. 1-10 and 1-11) was constructed of the traditional rubble masonry with cut-stone facing. It was enlarged in 1842. The plan is curved, with a mean radius of 62.6 meters (205 feet) and a crest length of about 70 meters (230 feet). Its height is 24 meters (79 feet). The arch thickness varies from about 9 meters (30 feet) at the top to 12 meters (39 feet) at the base. As at the Alicante Dam, the crest of the Elche structure has served as a spillway and has been damaged by floods.

Another arched structure built in the 17th century was the Relleu Dam (fig. 1-12) on the Río Amadorio northeast of Alicante. Its first stage was 28 meters (92 feet) high, with a practically constant thickness of 10 meters (33 feet) and a crest length of about 24 meters (80 feet). The arch was laid out on a mean radius of 65 meters (213 feet). A second stage was added in 1879. This consisted of a crest wall 3.85 meters (12.6 feet) high and 5 meters (16.4 feet) thick, extending the length to 34 meters (112 feet). The many years of overflow have eroded the top of Relleu Dam and dislodged much of the stone surfacing which protected the rubble masonry.

Early in the 17th century, the Ternavasso Dam was built about 30 kilometers (19 miles) southeast of Turin in Italy. This is an earthfill about 7.6 meters (25 feet) high and 335 meters (1100 feet) long. The top width varies from 5.2 meters (17 feet) to 6.7 meters (22 feet). The vertical upstream face consists of a brick wall approximately 0.6 meter (2 feet) thick, coated with mortar and braced at intervals by buttresses about 1.8 meters (6 feet) thick.

Another Italian structure of the same period is the Ponte Alto Dam, a thin arch on the River Fersina just east of Trento. Regarded as the first arch dam in Italy, it is located on the site of an earlier wood barrier, reportedly erected in 1537 and washed out by a flood in 1542. The initial stage of the Ponte Alto Dam, made of masonry blocks with unmortared joints, was under construction from 1611 to 1613. It was about 5 meters (16 feet) high and 2 meters (6.5 feet) thick, with a radius of approximately 14 meters (46 feet). In 1752, the addition of a second stage increased the height to 17 meters (56 feet). Subsequent enlargements in 1825, 1847, 1850, and 1887 created a dam which is now 38 meters (125 feet) high. The integrity of the structure is assured by closely fitted stone blocks joined with iron bars.

About 19 kilometers (12 miles) north of Istanbul in the Belgrad Forest are several small masonry dams which are part of the water system for Istanbul. Most of them are hundreds of years old and serve even more ancient conduits delivering water to Istanbul. The first water conveyance from the Belgrad Forest to Istanbul (Constantinople) was accomplished by the Romans, but many of their works are said to have been replaced some time following the Turkish conquest in 1453.

Some of the first dams in the northern American colonies were built for impoundment of water to run gristmills and sawmills. A dam was erected in 1623 to operate the first sawmill in America, on the Piscataqua River at South Windham, Maine. Another provided water for the first gristmill at Portsmouth, New Hampshire.

Meanwhile, Persia was approaching its second peak in dam engineering. During the reign of Shah Abbas II (1642-67), reservoirs were built near Mashhad and Kashan. The famous bridge-dam of Pul-i-Khadju was erected in the same period. Its slotted weir is about 6 meters (20 feet) high, 30 meters (100 feet) thick, and 141 meters (462 feet) long. The dam and the arched bridge are constructed of cut-stone blocks.

During the period 1667 to 1675, the St. Ferreol Dam was built on the River Laudot about 50 kilometers (30 miles) southeast of Toulouse in France. It is an earthfill 36 meters (118 feet) high and 780 meters (2560 feet) long. Three parallel masonry walls extend the full length of the dam, one at each face and one in the center. The upstream wall has a height of nearly 14.6 meters (48 feet) and an average thickness of about 6 meters (20 feet). The one on the downstream side is approximately 18 meters (60 feet) high and has a thickness varying from about 5 meters (17 feet) at the crest to 9 meters (30 feet) at the base. The central wall is 5 meters (17 feet) thick and rises to the full height of the embankment to form the dam crest. The fill between the walls is composed of stones and earth. All materials in the structure evidently were hand-placed.

Between 1714 and 1721, the first large dam in Germany, the Oderteich Dam, was constructed in the Oberharz. It is composed of two stone-block face walls confining a central zone of sand. The wall joints were apparently calked with earth and moss. The structure is 22 meters (72 feet) high and about 151 meters (495 feet) long and has a width varying from roughly 16 meters (52 feet) at the top to 44 meters (144 feet) at the foundation.

In 1747, the Almendralejo Dam was built about 51 kilometers (32 miles) south of Badajoz, Spain. It is also known as the dam of Albuera de Feria. This rubble-masonry buttress dam has survived without any significant deterioration. The original structure was approximately 20 meters (65 feet) high, with a thickness varying from 10 meters (32 feet) to 12 meters (40 feet) from top to base. Later successive enlargements increased the height finally to 23.5 meters (77 feet). Buttresses provide support at the downstream face. The structure is 170 meters (558 feet) long.

The design concepts of the early Spanish dam engineers were conveyed to the colonies in America. However, in some of these lands, water projects had been developed before the conquest. Near Teotihuacan, Mexico, and in the Nepena and Canete Valleys in Peru there are still signs of ancient dam projects.

Hundreds of masonry dams were erected by the Spanish in Mexico. Some of the most remarkable were in the State of Aguascalientes, about 480 kilometers (300 miles) north of Mexico City, where a rubble-masonry buttressed type was popular. These are believed to date from the 18th century. Outstanding examples of these Mexican structures are the Pabellón, (fig. 1-13), Presa de los Arcos (fig. 1-14), and San José de Guadalupe Dams (fig. 1-15).

The Pabellón (San Blás) damsite is on the Río Pabellón about 40 kilometers (25 miles) north of the city of Aguascalientes. The dam (fig. 1-16) was constructed as a buttressed masonry wall 177 meters (580 feet) long and 23.5 meters (77 feet) high. Originally the dam height was about 17.7 meters (58 feet) and the crest thickness was nearly 4.5 meters (15 feet). That the structure was enlarged is suggested by a change in the masonry design at a level several feet below the crest. The structure was extended by

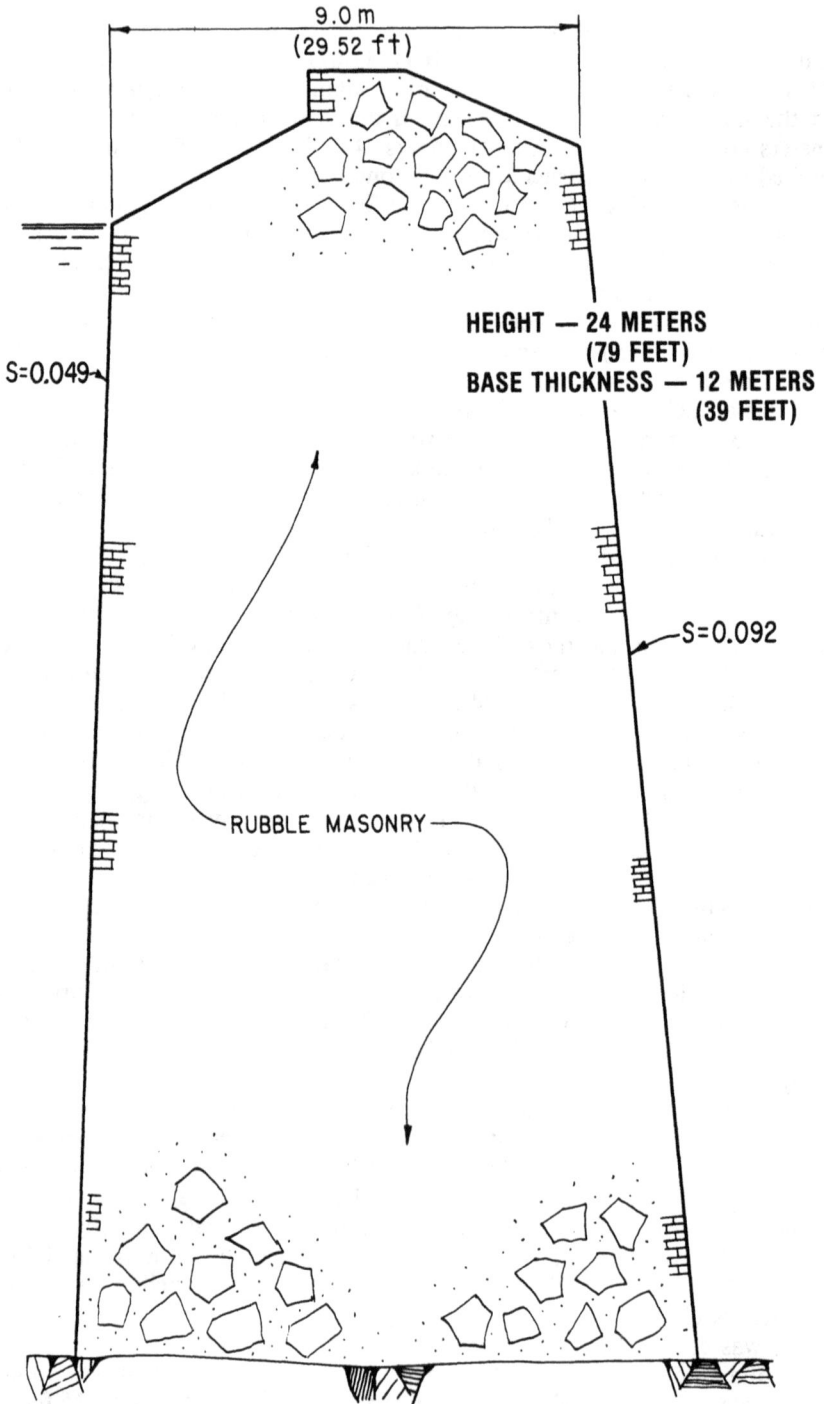

9.0 m
(29.52 ft)

HEIGHT — 24 METERS
(79 FEET)
BASE THICKNESS — 12 METERS
(39 FEET)

S=0.049

S=0.092

RUBBLE MASONRY

Figure 1-10.—Elche Dam, cross section. P-801-D-79296.

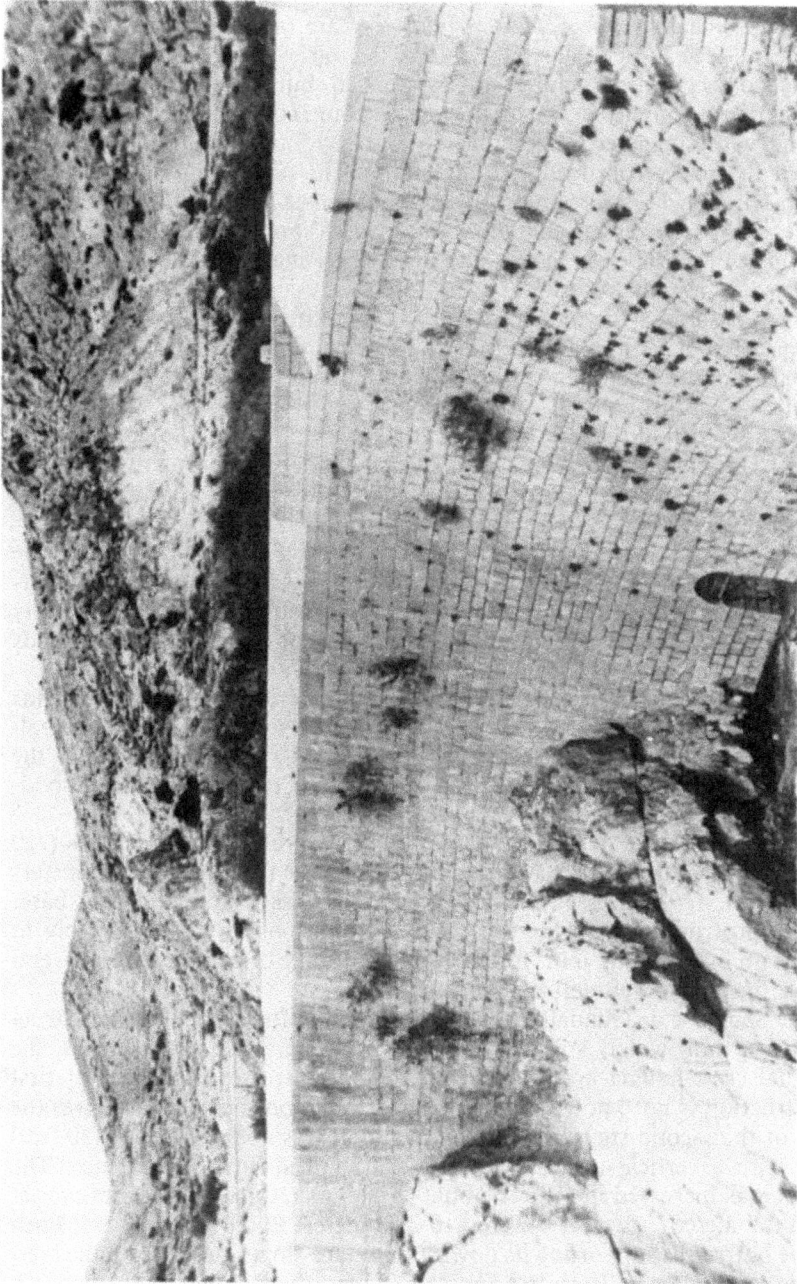

Figure 1-11.—Elche Dam (Courtesy, Comité Nacional Español, ICOLD). P-801-D-79297.

addition of a wall of triangular cross section topped by a parapet. The dam's upstream face is vertical. Maximum base thickness of the masonry mass is approximately 7.15 meters (23.5 feet). Its largest buttresses extend about 6.35 meters (21 feet) from the downstream face and are each roughly 2 meters (7 feet) thick. The buttress spacing varies.

The stones in this structure were not dressed, but some of them evidently were squared roughly in the quarrying. In some courses, they were placed on end with staggered joints. The rock is rhyolitic, and the mortar was reportedly made from local hydraulic lime. The noted American engineer Julian Hinds, who wrote about the Aguascalientes dams in the 1930's, was impressed by the remarkable durability of this mortar. He pointed to the example of the Pabellón Dam's spillway, where many years of spilling had caused only minimal erosion.

Hinds described the local practice of cementing vertical stones in holes dug in the rock to anchor a structure to its foundation. He surmised that this custom had been handed down from earlier generations and that the Pabellón Dam probably had this feature.

When he visited the site, he found the reservoir silted to a considerable depth. He judged the dam to be overstressed and speculated that the accumulations of sediment may have tended to seal any cracks caused by the overload.

Floods are allowed to pass over the top of the Pabellón Dam, since the stoplogs in the spillway on the right abutment are apparently left permanently in place. The crest was built several inches lower at the spillway end than in the middle so that small flows discharge there, but during floods the entire right half of the dam serves as a spillway.

Direct overpour is also the method of flood operation at the Presa de los Arcos on the Río Morcinique, 11 kilometers (7 miles) east of Aguascalientes. The spillway there comprises three small stoplog openings at the right end. However, the logs are reportedly never removed, causing floodwaters to go over the dam.

The Presa de los Arcos (fig. 1-17) is a structure about 220 meters (720 feet) long and 20 meters (65 feet) high, with a base thickness of 7.75 meters (25.4 feet). Its buttresses extend 4 meters (13 feet) from the dam base. Stones protruding from the top courses of masonry were presumably to enable an interlocking joint for an additional stage of construction. Several of the buttresses were left uncompleted.

The San José de Guadalupe Dam, also on the Río Morcinique, is a structure with thin, widely spaced buttresses. It was erected in two stages, the original crest having been at the top of the buttresses. The date of first construction is not known precisely, but a tablet on the dam face gives the date of the second stage as 1865. Originally it was about 9 meters (30 feet) high with a vertical upstream face and a stepped downstream face. The height was increased in 1865 to about 11 meters (36 feet). Thickness varies from 0.8 meter (3 feet) at the crest to 2.75 meters (9 feet) at the base. Each of the buttresses supporting its downstream face slopes down from the crest of the original dam. Each is 1.5 meters (5 feet) thick. The spacing between centers of buttresses is about 8.75 meters (29 feet). The entire dam is composed of rubble masonry bonded by hydraulic lime mortar.

Figure 1-12.—Relleu Dam (Courtesy, Comité Nacional Español, ICOLD). P-801-D-79298.

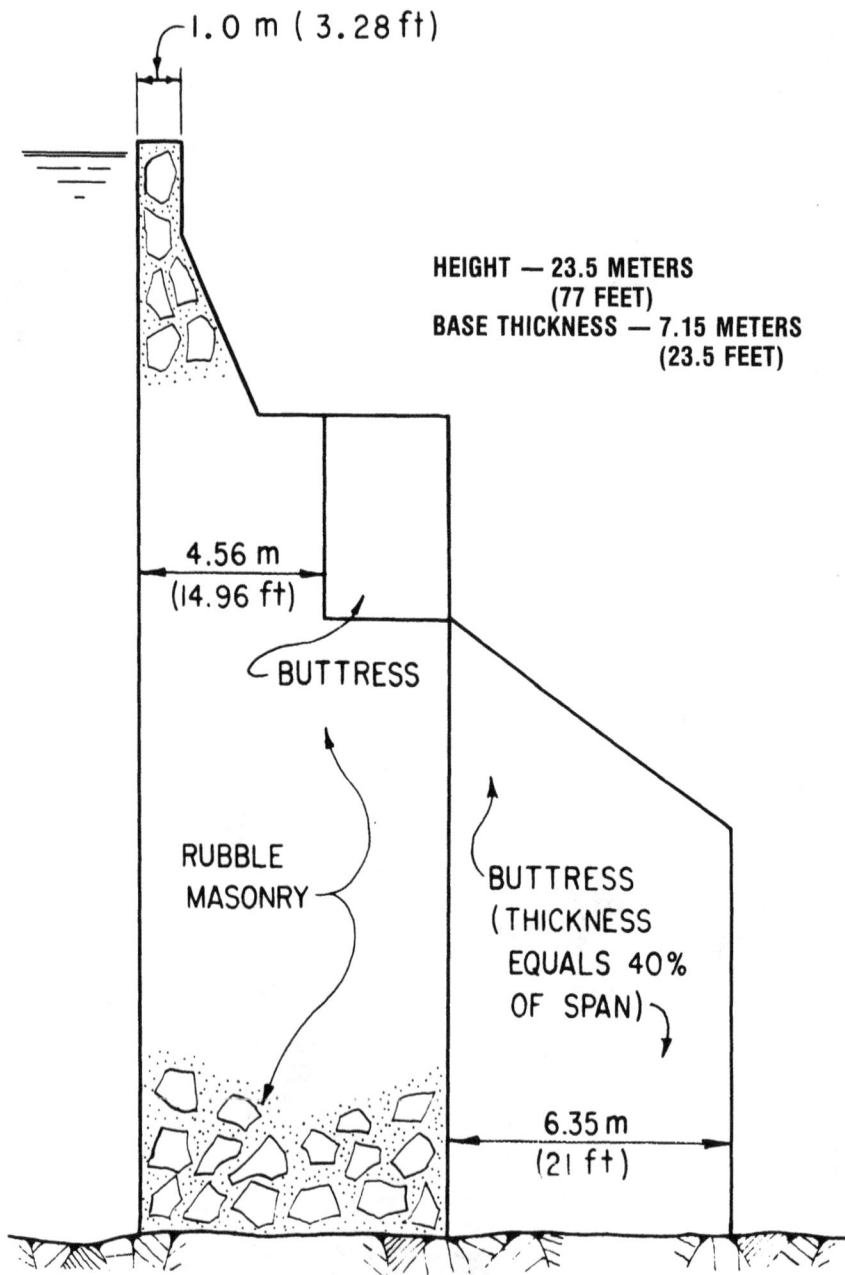

Figure 1-13.—Pabellón (San Blás) Dam, cross section. P-801-D-79299.

Figure 1-14.—Pabellón (San Blás) Dam (Courtesy of Julian Hinds). P-801-0-79300.

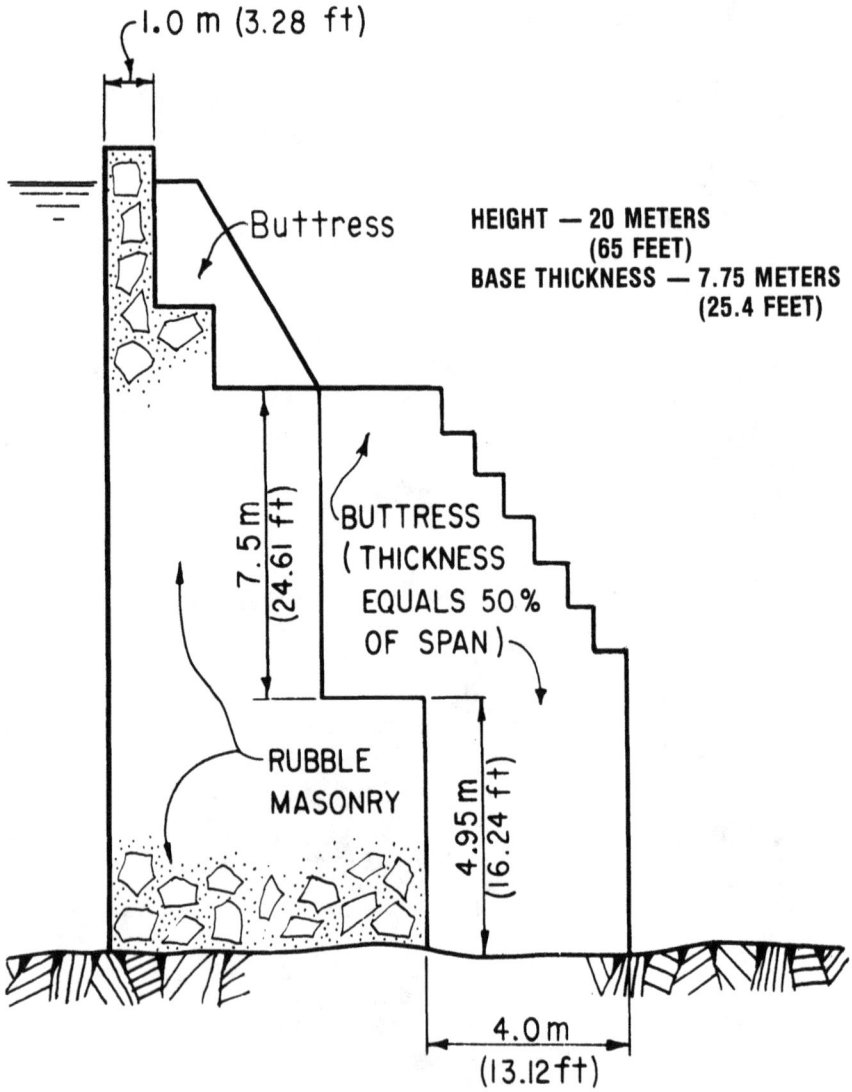

1.0 m (3.28 ft)

Buttress

HEIGHT — 20 METERS
(65 FEET)
BASE THICKNESS — 7.75 METERS
(25.4 FEET)

7.5 m
(24.61 ft)

BUTTRESS
(THICKNESS
EQUALS 50%
OF SPAN)

RUBBLE
MASONRY

4.95 m
(16.24 ft)

4.0 m
(13.12 ft)

Figure 1-15.—Presa de los Arcos, cross section. P-801-D-79301.

Figure 1-16.—Presa de los Arcos. P-801-D-79302.

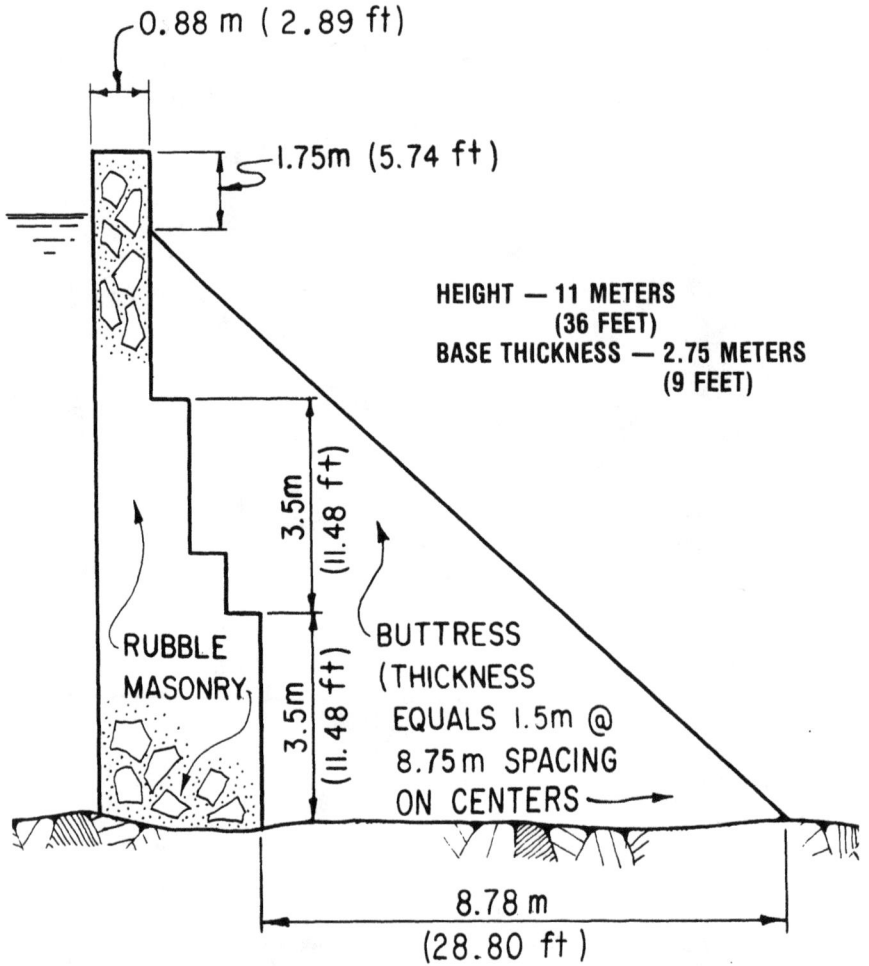

0.88 m (2.89 ft)

1.75m (5.74 ft)

HEIGHT — 11 METERS
(36 FEET)
BASE THICKNESS — 2.75 METERS
(9 FEET)

3.5m
(11.48 ft)

3.5m
(11.48 ft)

RUBBLE
MASONRY

BUTTRESS
(THICKNESS
EQUALS 1.5m @
8.75m SPACING
ON CENTERS

8.78 m
(28.80 ft)

Figure 1-17.—San José de Guadalupe Dam, cross section. P-801-D-79303.

The river section of the San José de Guadalupe Dam is of more massive proportions than the remainder of the structure and has a stoplogged sluiceway extending from foundation to crest. A spillway was built at the right end of the dam, but its stoplogs are left in place during floods. These are removed in the dry season to release water for diversion downstream.

Julian Hinds was amazed that dams of the type seen around Aguascalientes had weathered the centuries and were in such sound condition. Failures had occurred only where basic engineering principles had been conspicuously ignored. Among about fifty such structures with which he was familiar, three failures were known; two of these were attributed to inadequate foundations. He concluded that well-executed work on a good foundation can compensate for many deficiencies in design.

Contemporary with these pioneering achievements in Mexico, the Jesuits introduced dam building to California. One of their first missions received water from the Old Mission Dam (fig. 1-18) erected on the San Diego River in 1770. This long barrier, which eventually fell into ruin, was composed of mortared rubble masonry and was about 1.5 meters (5 feet) high.

Until the middle of the 19th century, few rational criteria for dam design had any acceptance. Problems encountered in construction were attacked by trial and sometimes by error. The failure of the Puentes Dam (fig. 1-19) on the Río Guadalentín in Spain in 1802 illustrated the weakness of some empirical methods. This 50-meter (164-foot) high rubble-masonry gravity dam was intended to be built on rock, but discovery of a deep crevice in the channel foundation led to use of piling in the alluvial fill under the central part of the structure. Inevitably, after 11 years of service, the inadequate underpinning blew out under reservoir pressure. A new dam was constructed just downstream of this location in 1884.

Another of Spain's 18th century projects was the Valdeinfierno Dam (fig. 1-20), which set a new record for masonry mass in that country. This dam impounded water for the environs of Lorca in the province of Murcia. The dam was built 14 kilometers (9 miles) upstream from the Puentes site, in the canyon of the Río Luchena, a tributary of the Río Guadalentín. It was about 35.5 meters (116 feet) high and 87 meters (285 feet) long in a series of seven chords arching upstream (virtually circular). Structural thickness was 12.5 meters (41 feet) at the crest, increasing abruptly at an elevation 4.5 meters (15 feet) under the crest level to about 30 meters (98 feet), and then varying to about 42 meters (137 feet) at the dam base. The reservoir eventually became filled with silt.

The Nineteenth Century

The first true multiple-arch water barrier recorded is the Meer Allum Dam, built in about 1800 near Hyderabad in India. This mortared masonry structure is about 12 meters (40 feet) high and measures approximately 762 meters (2500 feet) along its curved axis. Arch thickness is 2.6 meters (8.5 feet). The spans of the 21 vertical arches vary up to a maximum of 45 meters (147 feet). Each buttress is 7.3 meters (24 feet) thick and 12.8 meters (42 feet) long. Although a spillway was provided, some floodflow passes over the crest of the arches. The dam has endured this without detriment.

Figure 1-18.—Old Mission Dam (San Diego). P-801-D-79304.

10.89 m
(35.73 ft)

HEIGHT — 50 METERS
(164 FEET)
BASE THICKNESS — 44.2 METERS
(145 FEET)

RUBBLE MASONRY

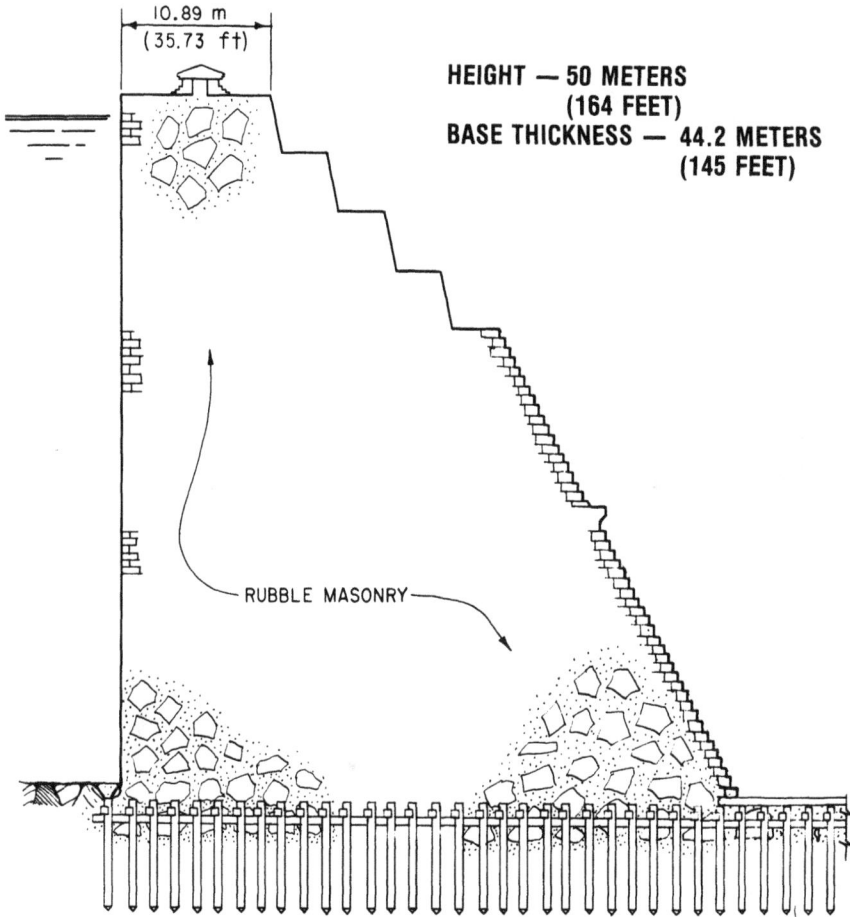

Figure 1-19.—Puentes Dam, cross section. P-801-D-79305.

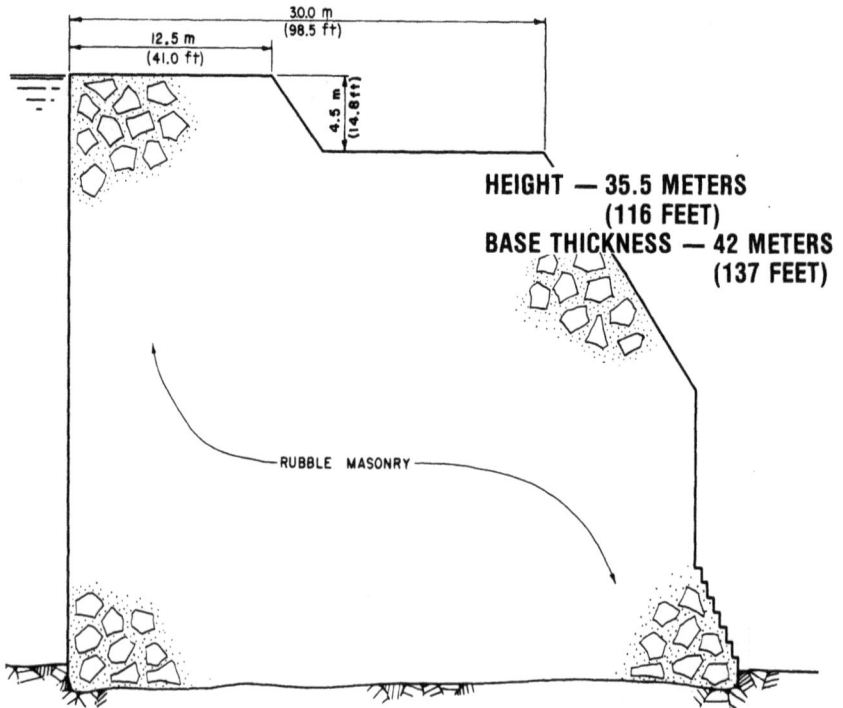

30.0 m
(98.5 ft)

12.5 m
(41.0 ft)

4.5 m
(14.81ft)

HEIGHT — 35.5 METERS
(116 FEET)
BASE THICKNESS — 42 METERS
(137 FEET)

RUBBLE MASONRY

ORIGINAL DAM AS CONSTRUCTED—1791

Figure 1-20.—Valdeinfierno Dam, cross section. P-801-D-79306.

In 1811, the Couzon Dam was completed near St. Étienne in France. It is a masonry-walled embankment patterned after the St. Ferréol Dam. The masonry core is 33 meters (108 feet) high and 218 meters (715 feet) long, with thickness varying from 4.9 meters (16 feet) at the top to a little more than 6.8 meters (22 feet) at the base. The upstream retaining wall is 10.7 meters (35 feet) high. The one at the downstream face is essentially a low toe block about 5.2 meters (17 feet) thick. Since its rehabilitation in 1896 to control seepage and sliding, the Couzon Dam has remained in service without trouble.

One of the largest of several old dams serving Istanbul, Turkey, is Yeni Dam (also known as the Sultan Mahmut Dam), built in 1839 by the decree of Sultan Mahmut II. This curved masonry gravity structure is 16 meters (52 feet) high and 93 meters (305 feet) long, with a crest thickness of 7 meters (23 feet) and a base thickness of 9.5 meters (31 feet). At the left abutment is a spillway about 1 meter (3 feet) long and 0.5 meter (1.5 feet) high.

During the 19th century, the art of masonry dam construction made important advances. European gravity dams developed architectural form and finish quite in contrast to their crude predecessors. French dam design began to incorporate rational approaches to analysis of forces. In 1853, M. de Sazilly, a French engineer, advocated that pressures within a dam be held to specific limits and that the structure be dimensioned to preclude sliding. However, he did not recognize the concept of keeping the resultant of forces within the middle third of each horizontal plane. This was emphasized about 25 years later by W. J. M. Rankine of England, who also sought a relationship between pressures on different planes in the structure. Lacking a mathematical solution, he suggested assuming different allowable unit pressures for the upstream and downstream faces. The ideas of M. de Sazilly and Rankine showed the way toward logical analysis of dams.

At about the same time, engineers in the eastern part of the United States of America were starting to build some notable dams. Among them were the Old Croton Dam, completed in 1842, on the Croton River for water supply to New York City; Mill River Dam in 1862 for New Haven, Connecticut; Lake Cochituate Dam in 1863 for Boston, Massachusetts; and the Druid Lake Dam in 1871 for Baltimore, Maryland.

During the development of irrigation in the arid western United States, starting in about 1850, many earth dams as high as 38 meters (125 feet) were built, but the number of failures was alarming. Mining also gave impetus to dam building. Discovery of gold in California in 1848 led to extensive placer workings which necessitated the use of dams and conduits. At first, the reservoirs for hydraulic mining were created with stone-filled log cribs. A later development was the dumped rockfill confined by dry rock walls at the faces and lined with two or more layers of wood planking.

Prior to 1850, many small American dams had been fabricated of wood by skilled millwrights. These dams led to the timber crib filled with rock and faced with planking. Such methods were admittedly primitive, but they made effective use of local materials in a land where transportation was difficult.

While these relatively crude works were being constructed by the western American pioneers, European engineers were engaged in more sophisticated

43

projects. In France, the Zola Dam (figs. 1-21 and 1-22) regarded as the first arch dam built by the French, was completed in 1854, with the unprecedented arch height of 42 meters (138 feet). This rubble-masonry structure has cut-stone faces which are still intact. Its crest is curved to a mean radius of 51 meters (168 feet) and is 6 meters (20 feet) thick. At the base, the thickness of the arch is approximately 13 meters (43 feet). The total length of the structure is 66 meters (216 feet).

Also in France, the Gouffre d'Enfer Dam (fig. 1-23) on Le Furan River was completed in 1866 to a height of 60 meters (197 feet). The crest of this rubble-masonry gravity dam is 100 meters (328 feet) long and is curved to a radius of about 252 meters (828 feet). Structural thickness varies from 5 meters (16.4 feet) at the top to 49 meters (161 feet) at the base. The dam continues to provide water to St. Étienne.

The French also constructed several large dams in Algeria. One of the earliest was El Habra Dam (fig. 1-24), completed in 1873. Failure of El Habra Dam in 1881 stimulated intensive study of tensile and shear stresses in gravity dams.

Meanwhile, the Mutha Canal project in India (1869-79) included some record-breaking masonry dams. Outstanding among these were the Poona Dam (fig. 1-25), 40 meters (131 feet) high, and the Bhatgarh Dam, 38.7 meters (127 feet) high and 1.6 kilometers (1 mile) long. In 1892, the Tansa Dam was completed to serve Bombay. The dimensions include a length of 2.8 kilometers (1.74 miles) and a height of 41 meters (134 feet). Another Indian structure, the Periyar Dam (1887-97) was built to a height of 54 meters (177 feet). It is a gravity dam (fig. 1-26) containing about 141 000 cubic meters (184 000 cubic yards) of concrete made of hydraulic lime, sand, and broken stone. Both faces were built of uncoursed rubble masonry.

In 1875, the first stage of the Lower San Leandro (Chabot) Dam was completed to serve the communities on the east side of San Francisco Bay in California. It was built by Anthony Chabot, who had engineered San Francisco's first public water supply. A special feature of this earthfill structure is a central foundation trench excavated 9 meters (30 feet) below the streambed. In the bottom of the trench, three parallel concrete cutoff walls were built, each 0.9 meter (3 feet) thick and 1.5 meters (5 feet) high, with about half this height anchored in the foundation and half protruding. The fill contains a core zone which is about 27 meters (90 feet) wide at its bottom in the foundation trench. The slopes of the primary original embankment are 3 to 1 upstream and 2½ to 1 downstream. Within these limits the earth material was dumped by wagons and then sprinkled. Compaction was accomplished by the wagon wheels and by a band of horses which were led back and forth on the fill. A sluiced zone of earth and rock was placed against the 2½ to 1 slope on the downstream side, which provided a final outer slope of 6.7 to 1.

Reportedly, more than 800 Chinese workmen were employed in the sluicing operation. When completed in 1875, the dam rose 35 meters (115 feet) above the streambed and contained a total of approximately 415 000 cubic meters (543 000 cubic yards), of which about 30 percent was sluiced. The dam was enlarged in the 1890's and its height above foundation is now recorded as 47 meters (154 feet) and its length as 137 meters (450 feet).

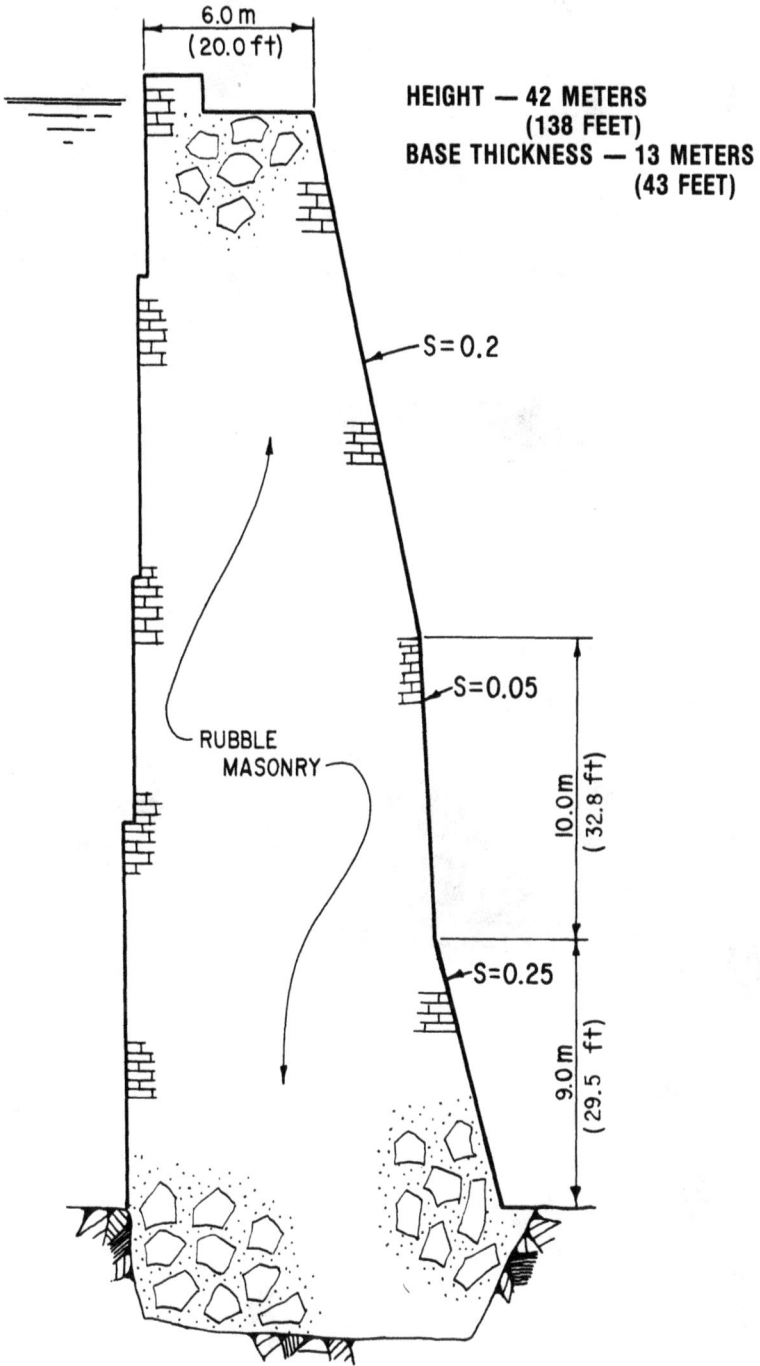

6.0 m
(20.0 ft)

HEIGHT — 42 METERS
(138 FEET)
BASE THICKNESS — 13 METERS
(43 FEET)

S=0.2

S=0.05

RUBBLE
MASONRY

10.0 m
(32.8 ft)

S=0.25

9.0 m
(29.5 ft)

Figure 1-21.—Zola Dam, cross section. P-801-D-79307.

45

Figure 1-22.—Zola Dam (Courtesy, Comité Français des Grands Barrages, ICOLD). P-801-D-79308.

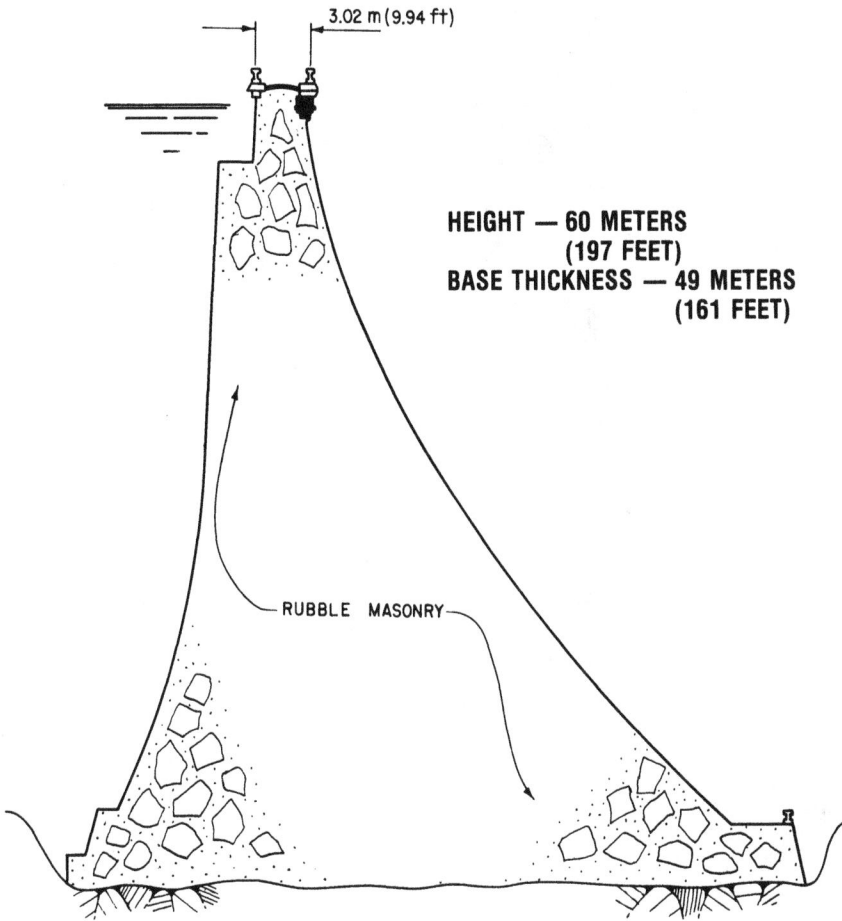

3.02 m (9.94 ft)

HEIGHT — 60 METERS
(197 FEET)
BASE THICKNESS — 49 METERS
(161 FEET)

RUBBLE MASONRY

Figure 1-23.—Gouffre d'Enfer Dam, cross section. P-801-D-79309.

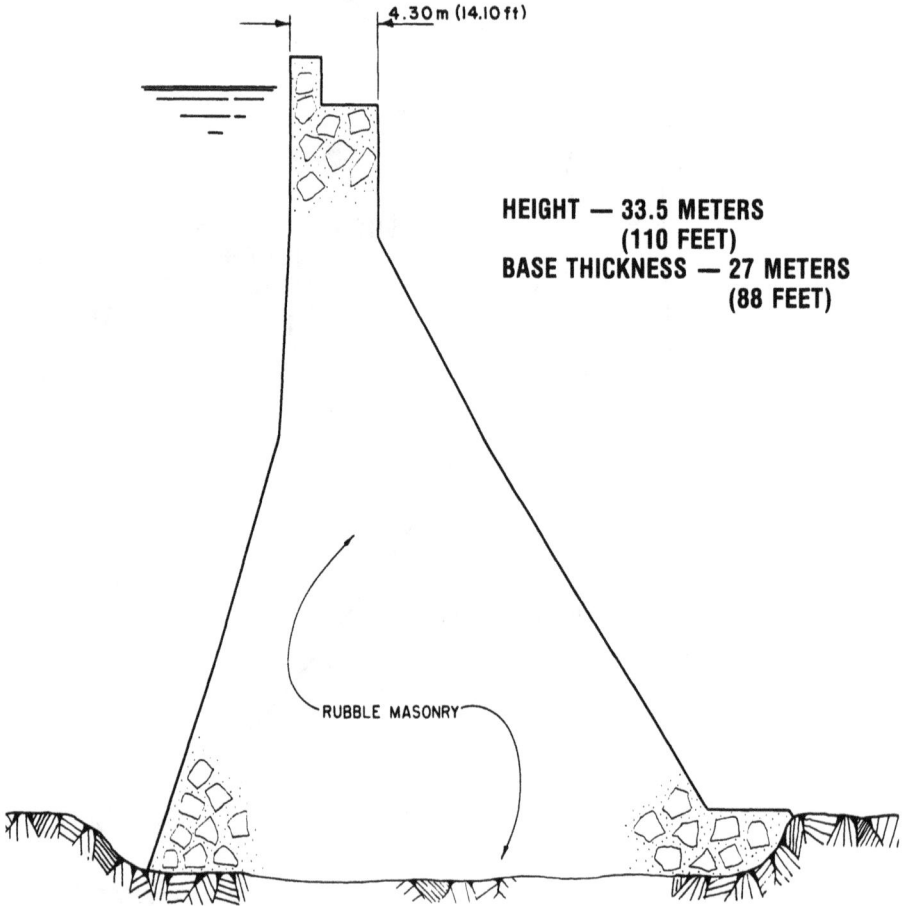

4.30 m (14.10 ft)

HEIGHT — 33.5 METERS
(110 FEET)
BASE THICKNESS — 27 METERS
(88 FEET)

RUBBLE MASONRY

Figure 1-24.—El Habra Dam, cross section. P-801-D-79310.

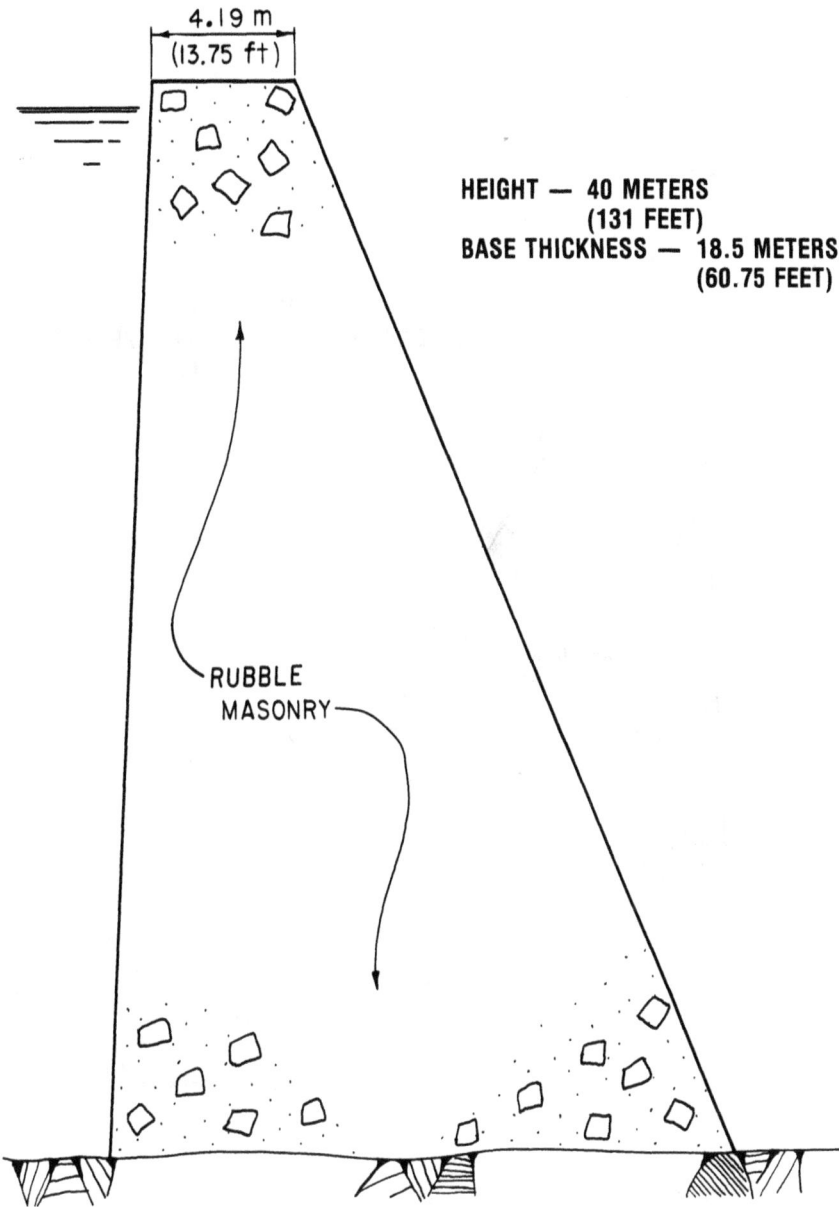

Figure 1-25.—Khadakwasla (Poona) Dam, cross section. P-801-D-79311.

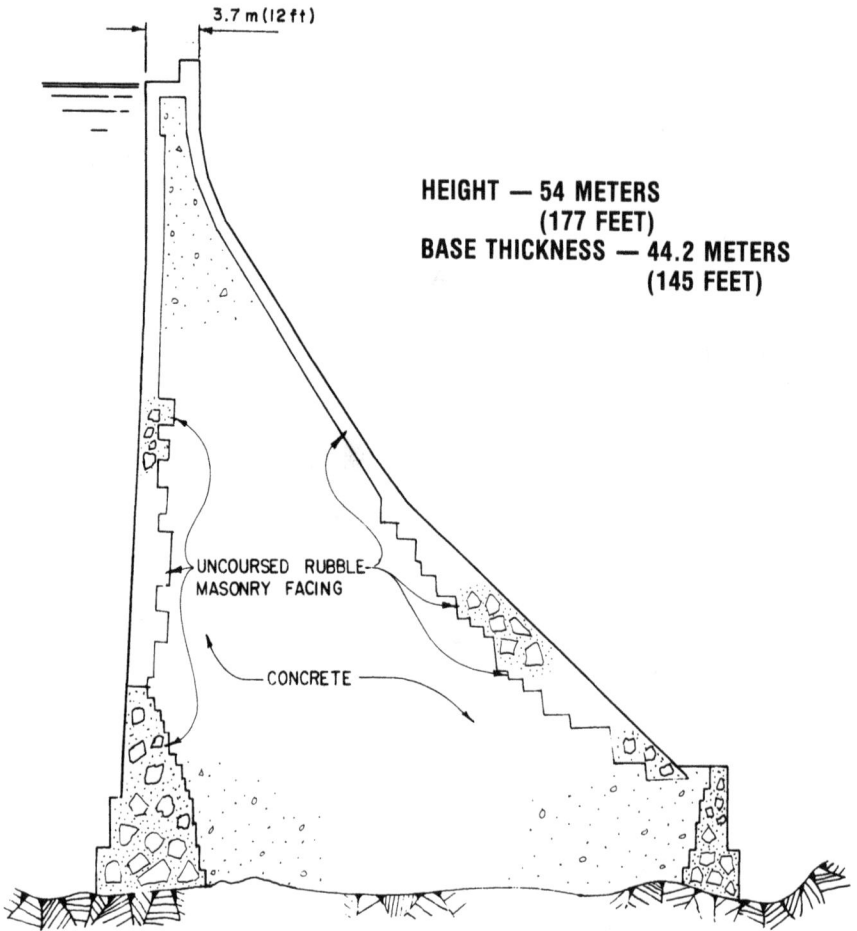

Figure 1-26.—Periyar Dam, cross section. P-801-D-79312.

The Twentieth Century

In Egypt, the Aswan Dam (fig. 1-27) was completed in 1902 on the mainstream of the Nile. The quarried-granite mass was 20 meters (65.5 feet) high and 1951 meters (6400 feet) long. Enlargement of this gravity dam to a height of 27 meters (88.5 feet) was completed in 1912, followed by a further increase in 1933. Its height above foundation is now 53 meters (174 feet). Successful operation of the Aswan storage facility was attributable in part to provision of sluiceways in the structure which allowed silt to flow through to the irrigated lands of the lower Nile.

In the United States, early gravity dams generally had conservative proportions. Cheesman Dam (fig. 1-28), completed in Colorado in 1904, was 72 meters (236 feet) in height and curved in plan on a radius of 122 meters (400 feet) even though it was a full gravity section. This established an American precedent for the arch-gravity barrier. Engineers recognized that the joints in these structures should be filled so as to resist loads. Otherwise the arch function could not develop until deflection under gravity action had closed the joints.

Design concepts for gravity dams were beginning to change. The middle-third criterion for dimensioning these structures was being questioned. It had been generally accepted as assurance against overturning of moderately loaded dams. But several failures demonstrated that uplift and sliding could be of greater concern. Designers began to consider these factors in engineering new projects.

As early as 1882, a drain network to reduce uplift had been incorporated into the design of the Vyrnwy Dam (figs. 1-29 and 1-30) for the water system of Liverpool, England. Engineers in the United States gave first recognition to uplift in design of the Wachusetts Dam in Massachusetts (1900-1906). A cutoff was built under the dam downstream from its heel, but no drains were provided. Olive Bridge Dam in New York State (1908-14) was constructed with drains in the structure itself but with none in the foundation. Among the first dams with both masonry and rock drainage were Medina in Texas (1911-12), Arrowrock in Idaho (1914-15), and Elephant Butte in New Mexico (1914-15). The foundations at these sites were drilled to control seepage. Since then, drilling has been common practice for large gravity dams.

The New Croton Dam (fig. 1-31), a gravity structure completed in 1905, was one of the last major American dams of cut-stone masonry. While natural cement was used on this job, it was also one of the first applications of American portland cement. [In later projects portland cement found increasing acceptance, but natural cement was used as a blend with portland cement (up to 25 to 30 percent) in several structures, including Bull Shoals, Clark Hill, Wolf Creek (all built around 1945-50) and the Robert Moses (Barnhart Island) Dam (1959).] New Croton was the highest dam in the world — 90.5 meters (297 feet) above its foundation. However, this new record was soon surpassed.

Within the next few years, three important arched barriers were completed in the West. The 85-meter (280-foot) high Theodore Roosevelt Dam (1911), a thick, arch-gravity structure (figs. 1-32 and 1-33) in Arizona, and the 65-meter (214-foot) high Pathfinder Dam (fig. 1-34) (1909) in Wyoming

7.00 m
(22.97 ft)

HEIGHT — 20 METERS
(65.5 FEET)
BASE THICKNESS — 24 METERS
(79 FEET)

S = 0.67

BUTTRESS

S = 0.054

QUARRIED GRANITE

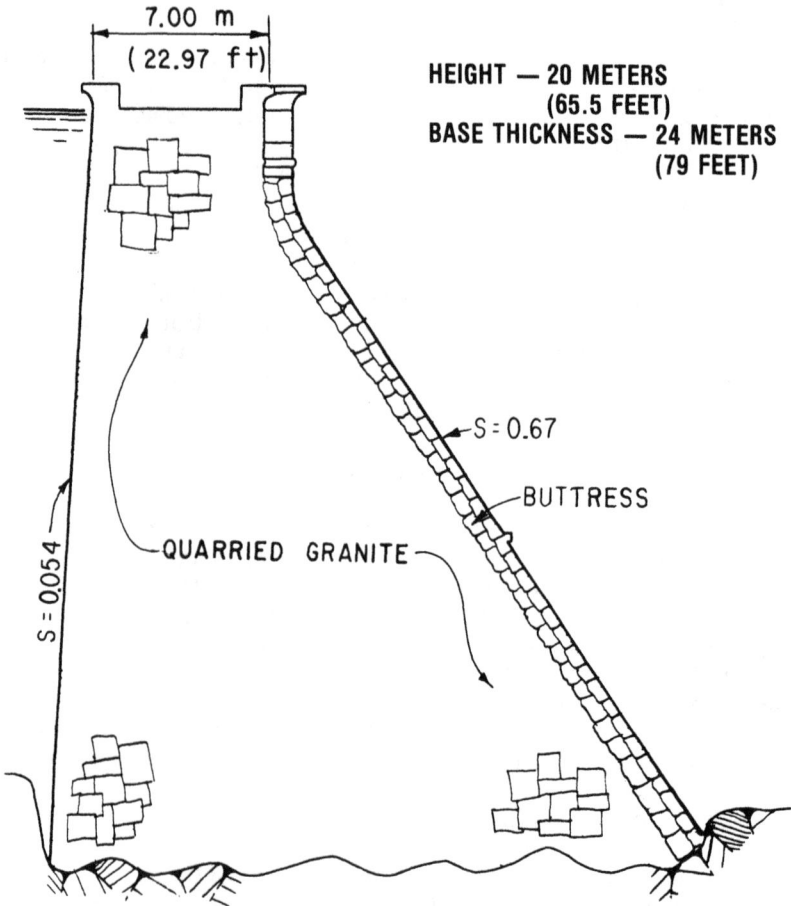

ORIGINAL DAM AS CONSTRUCTED—1902

Figure 1-27.—Aswan Dam, cross section. P-801-D-79313.

52

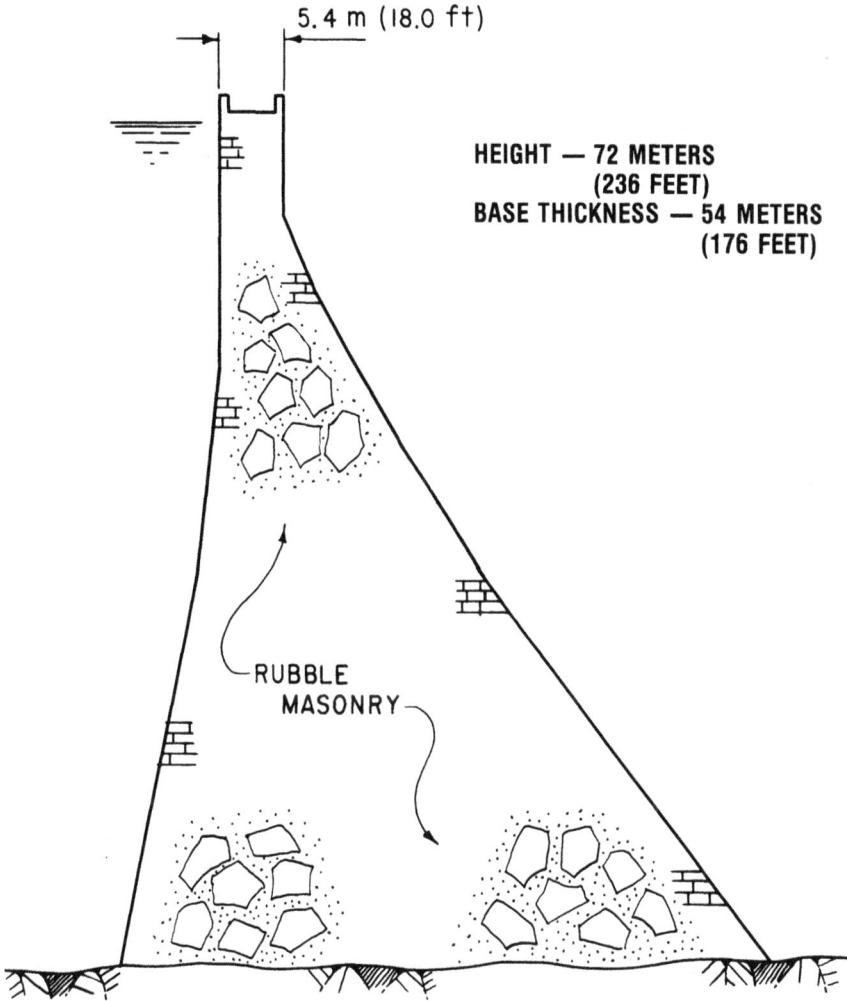

5. 4 m (18.0 ft)

HEIGHT — 72 METERS
(236 FEET)
BASE THICKNESS — 54 METERS
(176 FEET)

RUBBLE
MASONRY

Figure 1-28.—Cheesman Dam, cross section. P-801-D-79314.

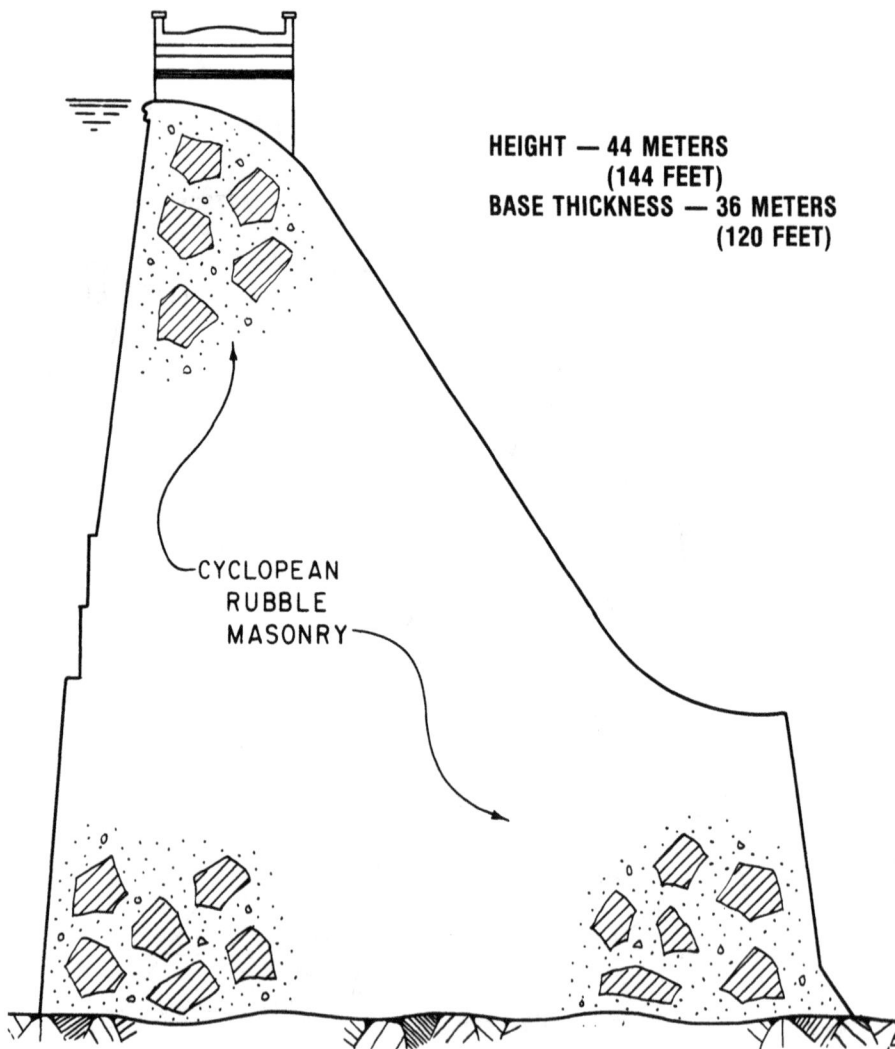

HEIGHT — 44 METERS
(144 FEET)
BASE THICKNESS — 36 METERS
(120 FEET)

CYCLOPEAN
RUBBLE
MASONRY

Figure 1-29.—Vyrnwy Dam, cross section. P-801-D-79315.

Figure 1-30.—Vyrnwy Dam (Courtesy, British National Committee, ICOLD). P-801-D-79316.

5.53m (18.0 ft)

HEIGHT — 90.5 METERS
(297 FEET)
BASE THICKNESS — 63 METERS
(207 FEET)

CUT-STONE MASONRY

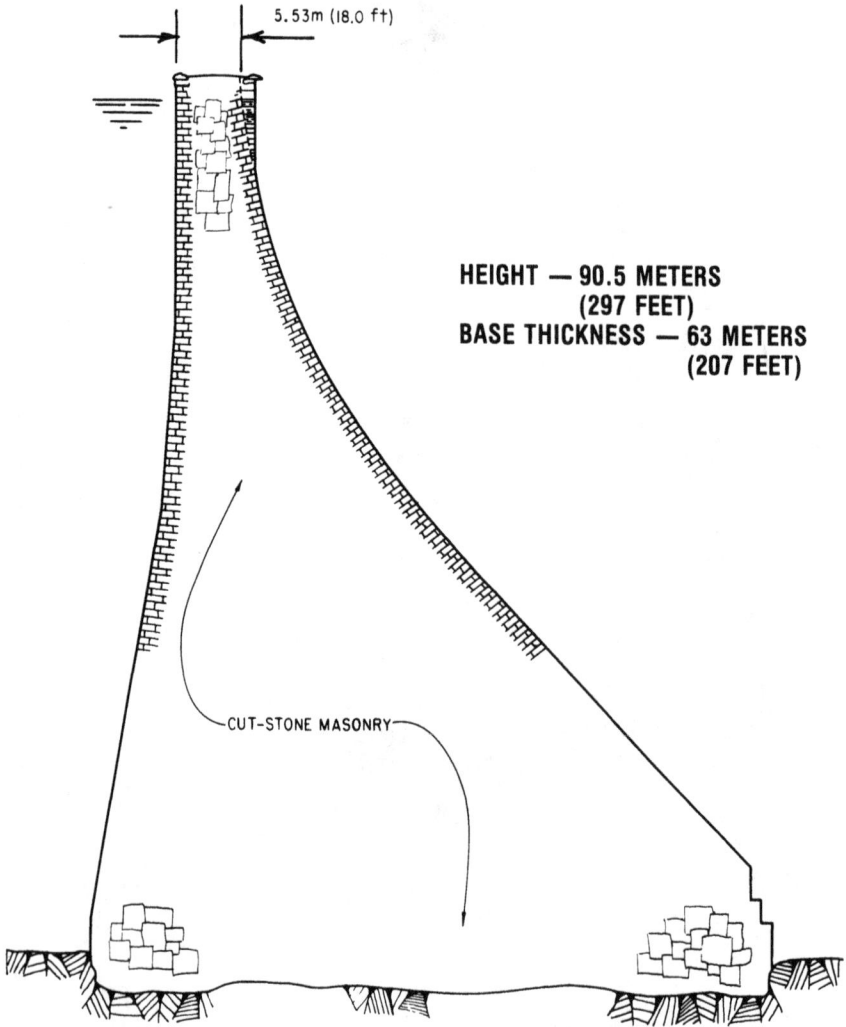

Figure 1-31.—New Croton Dam, cross section. P-801-D-79317.

were built of stone masonry. The Buffalo Bill (Shoshone) Dam (figs 1-35 and 1-36) (1910), also in Wyoming, attained a record-setting height of 99 meters (325 feet). Pathfinder and Buffalo Bill (fig. 1-37) are true arches.

New precedents were set by the 107-meter (350-foot) high Arrowrock Dam (fig. 1-38) on the Boise River in Idaho. In the construction of this concrete, thick, arch-gravity dam, completed in 1915, first use was made of spouted cobble concrete, introducing methods of placement that gained wide acceptance. Although the structure was arched on a radius of 204 meters (670 feet), the conservative section had a maximum base thickness of 68 meters (223 feet). With such mass, it must function primarily as a gravity dam, relying more on its weight than on any arch action.

The 94-meter (308-foot) high Kensico Dam, built to provide water storage for the Catskill Aqueduct, introduced a new era in United States dam construction. This straight gravity structure, erected in the period 1910 to 1915, was composed of "cyclopean concrete" produced by highly mechanized methods. Railroad trains transported the cyclopean stones and buckets of concrete to the damsite, where electriclly powered derricks hoisted and placed them in the forms. The concrete mass was finally faced with stone masonry.

Americans were also making progress in embankment construction. Some of their first major earthfills were in California, including the 67-meter (220-foot) high San Pablo Dam, completed in 1920, and the Calaveras Dam of the same height, in 1925. In terms of embankment volume, the Saluda Dam (1930) in South Carolina held the United States record of approximately 8 400 000 cubic meters (11 000 000 cubic yards) until the Fort Peck Dam (fig. 1–39) (1935–40) in Montana with nearly 96 049 000 cubic meters (125 628 000 cubic yards).

Rockfill dam technology was given new impetus in the United States. In 1924, the Dix River Dam for the water supply of Danville, Kentucky, established a height record of 84 meters (275 feet) for rockfills. It is a combination of embankment and gravity section. In 1931, the Salt Springs Dam in California raised the record to 100 meters (328 feet).

Outstanding concrete arch dams built in the United States during this period include the Pacoima Dam (1929) in California, built for flood control in Los Angeles County, to a record height of 113 meters (370 feet), and the Diablo Dam (1929) in Washington, setting a new mark of 119 meters (390 feet). The Owyhee Dam (1932), a concrete, thick, arch-gravity structure in Oregon (figs. 1-40 and 1-41), was designed as an arch. Its crest is 162 meters (530 feet) above the bottom of the cutoff trench. Officially, the height of this dam is listed as 127 meters (417 feet).

Techniques for mixing and placing concrete were undergoing significant changes. The constructors of the Diablo Dam, for example, used a dry mix placed by a belt conveyor suspended from a derrick, with a short tube known as an "elephant trunk" at the discharge end. The recognized disadvantages of segregation and voids in wet, spouted concrete spurred attempts to find methods of placing an even drier mix. In the construction of the Calderwood (Tennessee) and Chute à Caron (fig. 1-42) (Quebec, Canada) Dams, completed in 1930, use was made of bottom-dump buckets that enabled placing relatively dry concrete in the forms without segregation. This procedure became widely approved.

PLAN

4.9 m (16.0 ft)

HEIGHT — 85 METERS
(280 FEET)
BASE THICKNESS — 56 METERS
(184 FEET)

S=0.67

Cyclopean
rubble
masonry

CROSS SECTION

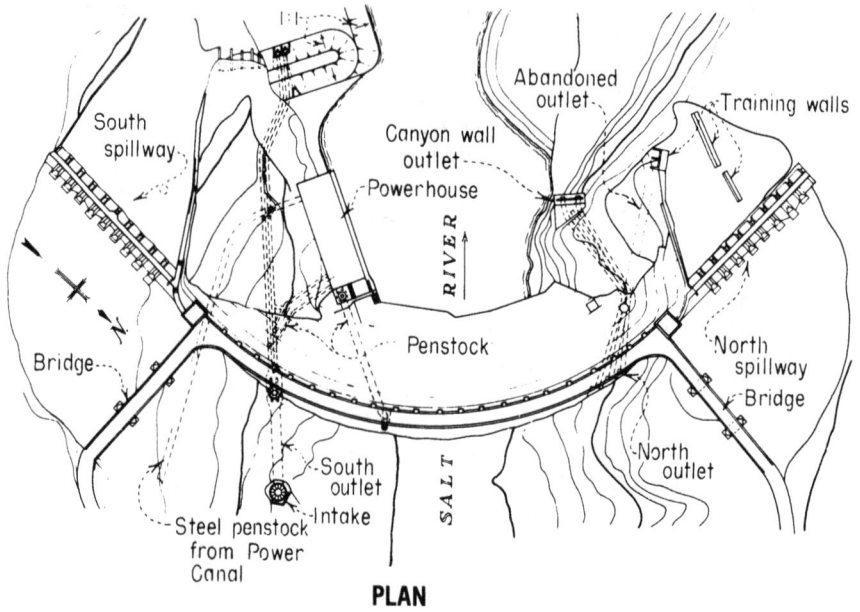

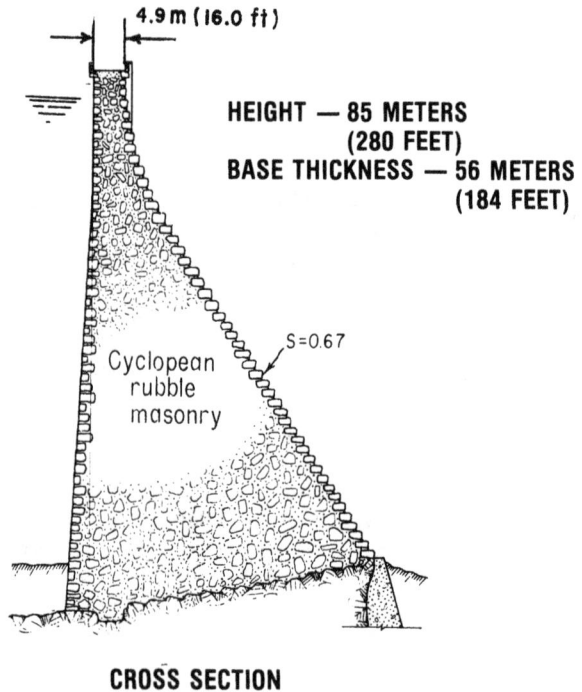

Figure 1-32.—Theodore Roosevelt Dam, plan and cross section. P-801-D-79318.

Figure 1-33.—Theodore Roosevelt Dam. P-801-D-79319.

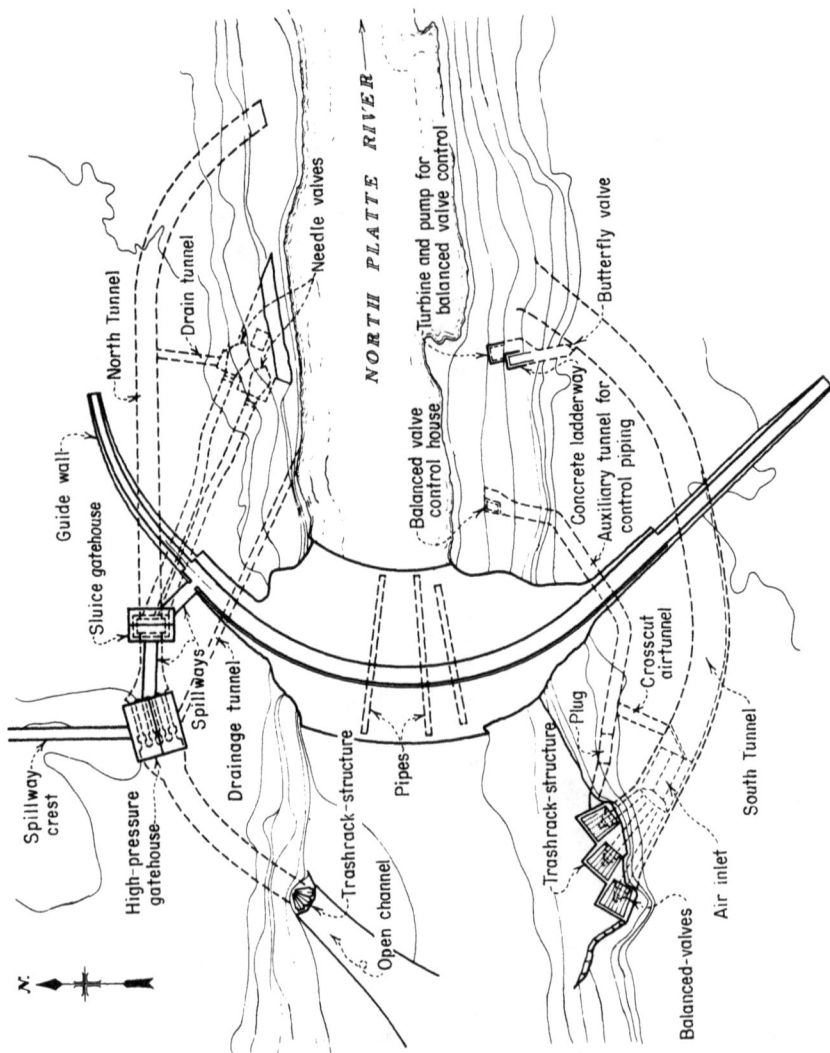

Figure 1-34.—Pathfinder Dam, plan. P-801-D-79320.

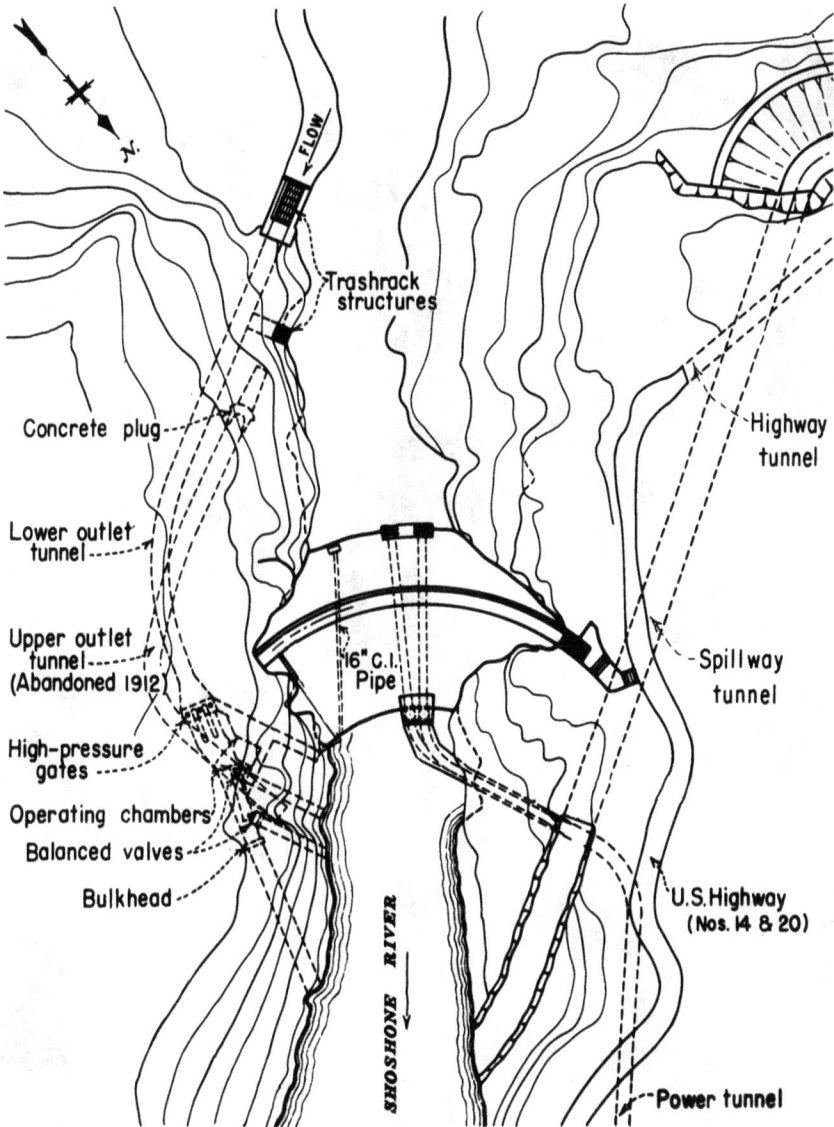

Figure 1-35.—Buffalo Bill (Shoshone) Dam, plan. P-801-D-79321.

Figure 1-36.—Buffalo Bill (Shoshone) Dam (P26-600-1353A). P-801-D-79322.

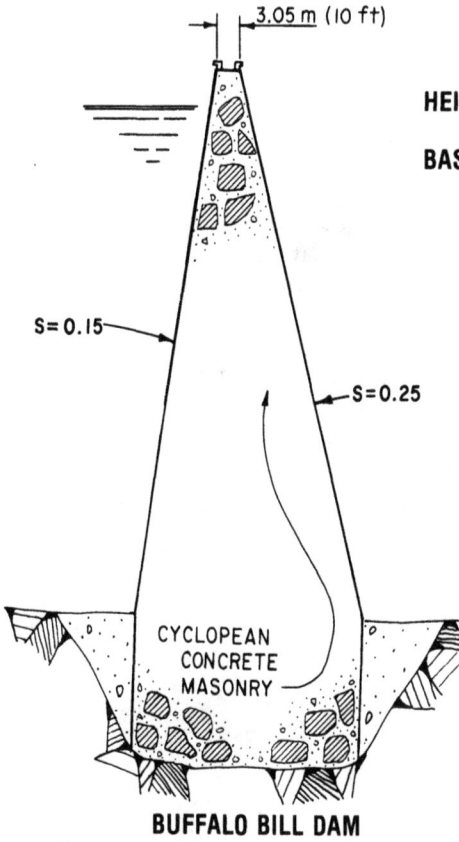

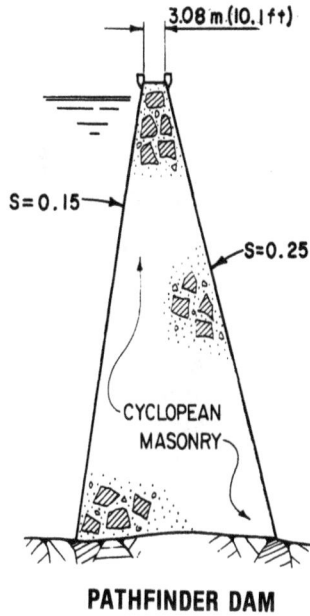

Figure 1-37.—Buffalo Bill (Shoshone) and Pathfinder Dams, cross sections. P-801-D-79323.

Figure 1-38.—Arrowrock Dam (P4-100-851I). P-801-D-79324.

Figure 1-39.—Fort Peck Dam (Courtesy, U.S. Corps of Engineers). P-801-D-79325.

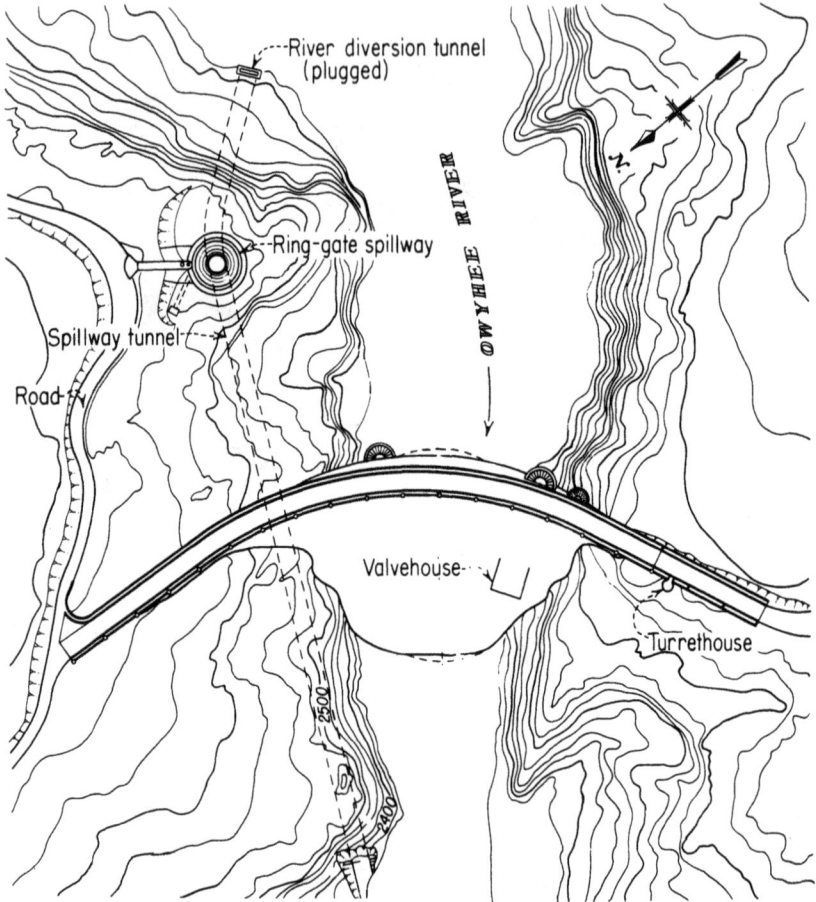

PLAN

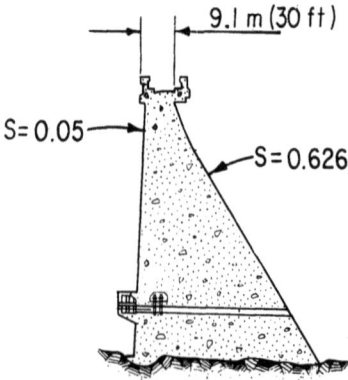

CROSS SECTION

HEIGHT — 127 METERS
(417 FEET)
BASE THICKNESS — 80.8 METERS
(265 FEET)

Figure 1-40.—Owyhee Dam, plan and cross section. P-801-D-79326.

Figure 1-41.—Owyhee Dam (P48-100-413I). P-801-D-79327.

Figure 1-42.—Chute à Caron Dam (Courtesy, Alcan Smelters and Chemicals, LTD). P-801-D-79328.

Hoover (Boulder) Dam (fig. 1-43), built during the period from 1931 to 1936 on the Colorado River, is a massive, thick, arch-gravity structure containing a total of 2 481 448 cubic meters (3 245 612 cubic yards) of concrete. It has a height of 221.4 meters (726.4 feet) above the foundation, a crest length of 379 meters (1244 feet), and a thickness varying from 13.7 meters (45 feet) at the top to 201 meters (660 feet) at the base. While this huge dam drew worldwide attention, its mass was soon exceeded by the 8 092 813 meters (10 585 000 cubic yards) of the Grand Coulee Dam (fig. 1-44) (1942) on the Columbia River in Washington, and the 6 445 156 cubic meters (8 430 000 cubic yards) of Shasta Dam (fig. 1-45) (1945) on the Sacramento River in California.

Glen Canyon Dam (fig. 1-46), an arched concrete structure, was completed in 1964. It is located on the Colorado River in Arizona and is among the highest in the United States. It is 216.4 meters (710 feet) high and 475 meters (1560 feet) long. Arch thickness varies from 7.6 meters (25 feet) at the crest to 91.4 meters (300 feet) at the base. The total concrete volume is 3 747 060 cubic meters (4 901 000 cubic yards).

Dworshak Dam (fig. 1-47) on the North Fork of the Clearwater River in Idaho was completed in 1974. It is a straight concrete gravity structure 219 meters (717 feet) high and 1002 meters (3287 feet) long, with a volume of 4 970 000 cubic meters (6 500 000 cubic yards).

Oroville Dam (figs. 1-48 and 1-49) (1968), the primary storage feature of California's State Water Project, is a 235-meter (770-foot) high zoned earthfill structure with a volume of 59 639 000 cubic meters (78 008 000 cubic yards). The selection of the embankment type dam was governed by the abundant supply of ideal pervious materials that had been produced by dredgers mining for gold in the flood plain of the feather River. Unique laboratory compaction equipment was developed for testing the large-size cobble material used in the dam. These facilities can accommodate an earth-rock specimen 914 millimeter (36 inches) in diameter and 2286 millimeters (7.5 feet) high.

Two precedent-setting Canadian structures include the Daniel Johnson (Manicouagan No. 5) Dam (fig. 1-50) (1968), a multiple arch dam 214 meters (703 feet) high, located in the bush country of Quebec 805 kilometers (500 miles) from Montreal which consists of 13 arches supported by 12 buttresses, with the central arch spanning 161.5 meters (530 feet); and Mica Dam (1972) an earthfill structure located about 128 kilometers (80 miles) north of Revelstoke on the Columbia River. Its height above lowest point of foundation is about 242 meters (794 feet), and it is 792 meters (2600 feet) long. The dam has a nearly vertical core of glacial till and outer zones of compacted sand and gravel. Slope protection is dumped rock riprap. Total volume of the embankment is 32 111 000 cubic meters (42 000 000 cubic yards).

The increase in the number of dams since 1900 has been impressive. In the United States alone, the roster of major dams — more than 15 meters (49 feet) high, or between 10 and 15 meters (33 and 49 feet) and impounding more than 100 000 cubic meters (80 acre-feet) — grew from 116 in 1900 to 2635 in 1962. Around the globe, very high dams kept up with this rapid pace. Up to 1939, only 11 dams more than 100 meters (328 feet) high were completed — 5 in western Europe and 6 in the United States. By 1960, there

Figure 1-43.—Hoover (Boulder) Dam (P45-300-10769). P-801-D-79329.

Figure 1-44.—Grand Coulee Dam (P222-117-48654). P-801-D-79330.

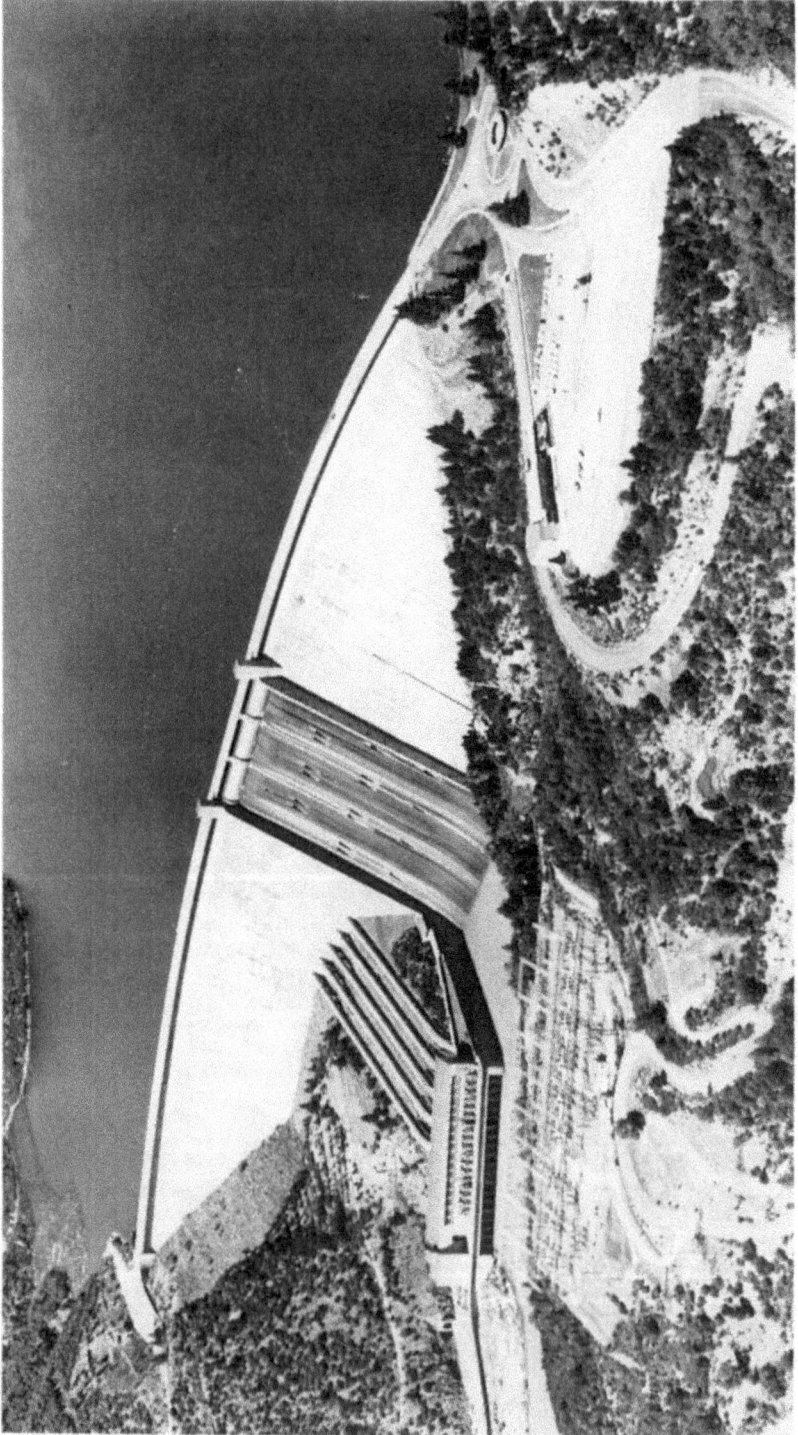

Figure 1-45.—Shasta Dam (P(S)-200-12006NA). P-801-D-79331.

Figure 1-46.—Glen Canyon Dam (P557-400-383). P-801-D-79332.

9.1 m (30 ft)

HEIGHT — 219 METERS
(717 FEET)

S = 0.8

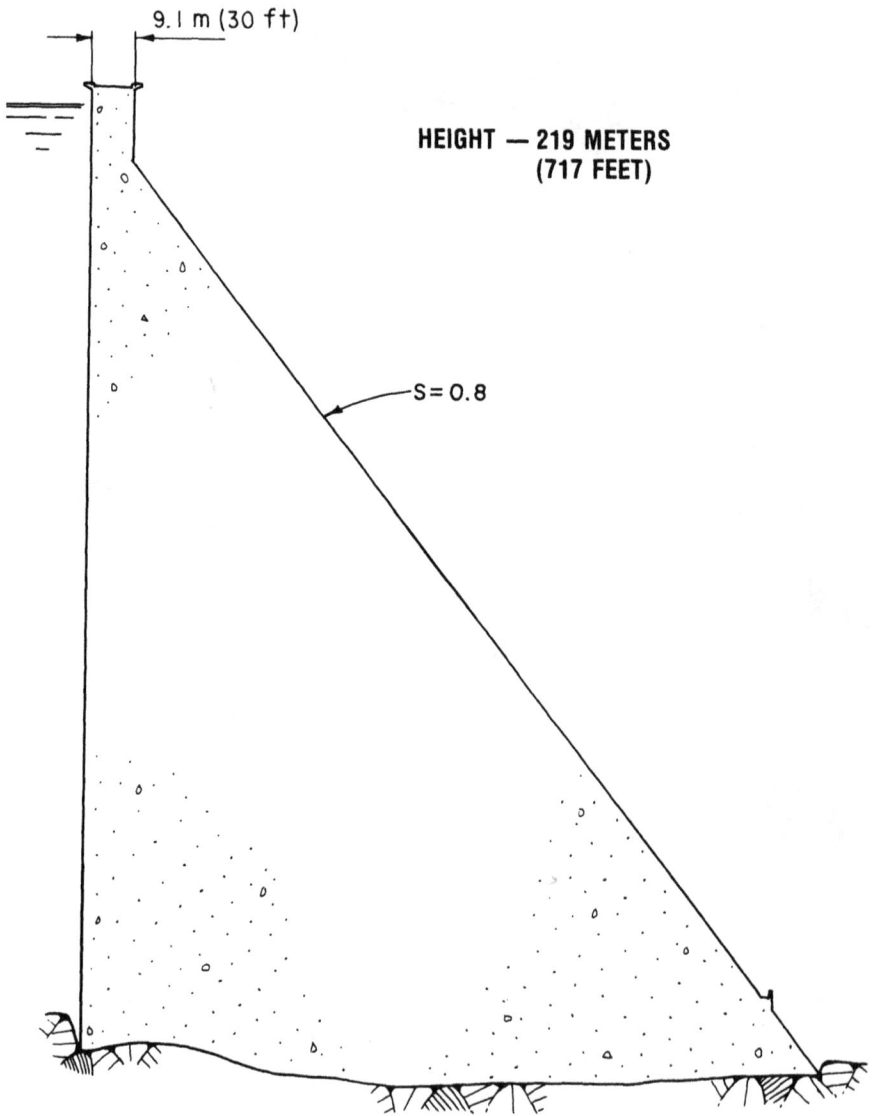

Figure 1-47.—Dworshak Dam, cross section. P-801-D-79333.

Figure 1-48.—Oroville Dam (Courtesy, Calif. Dept. of Water Resources). P-801-D-79334.

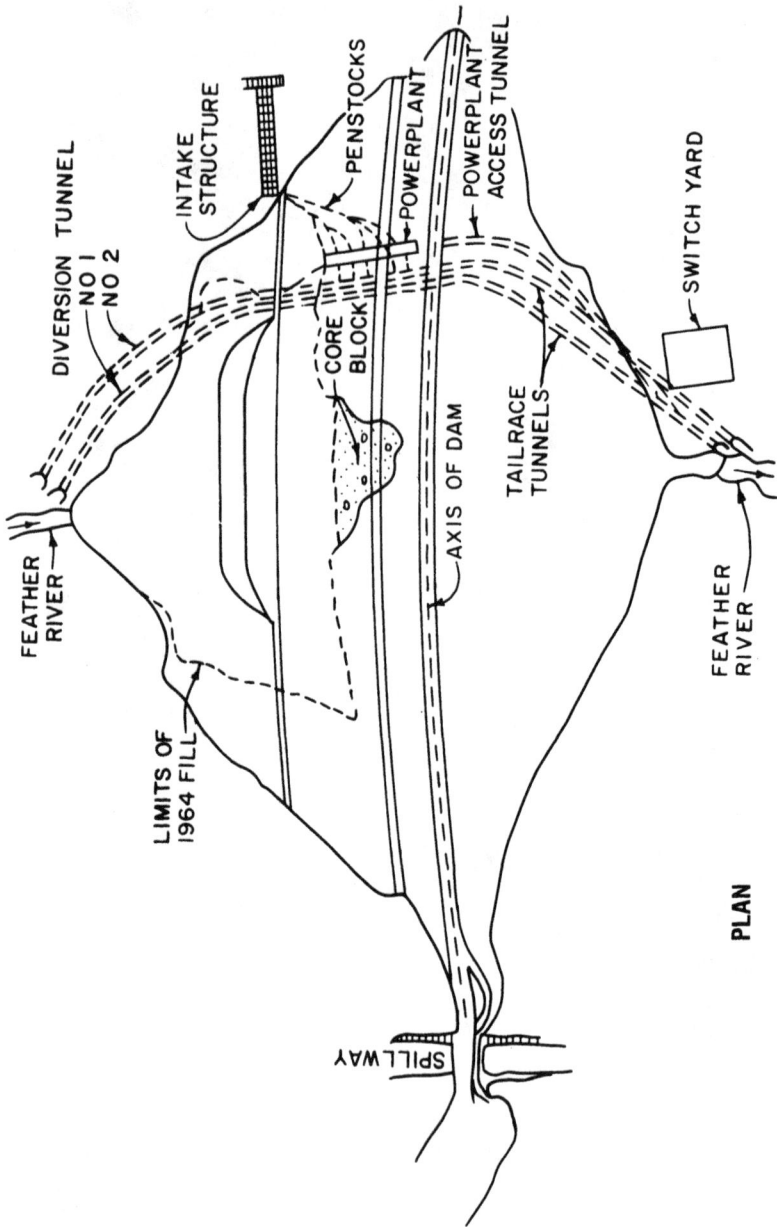

Figure 1-49.—Oroville Dam, plan and cross section (1 of 2). P-801-D-79335.

76

ZONED EARTHFILL

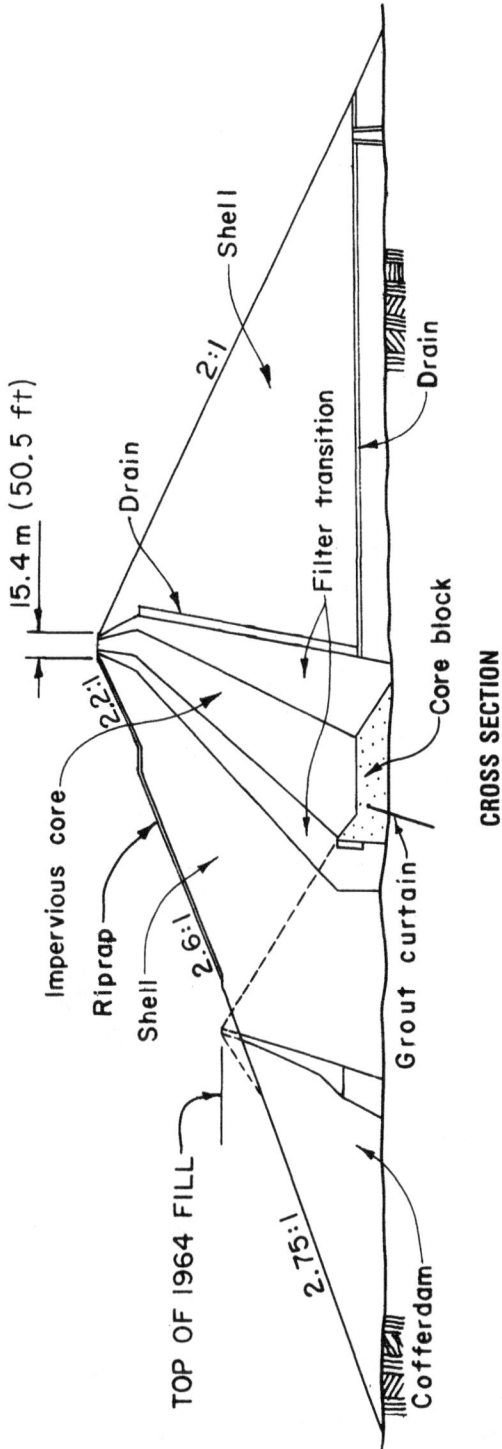

HEIGHT — 235 METERS
(770 FEET)

15.4 m (50.5 ft)

2:1

Shell

Drain

Drain

Filter transition

2.2:1

Core block

Impervious core

Riprap

Shell

2.6:1

CROSS SECTION

Grout curtain

TOP OF 1964 FILL

2.75:1

Cofferdam

Figure 1-49.—Oroville Dam, plan and cross section (2 of 2). P-801-D-79336.

Figure 1-50.—Daniel Johnson (Manicouagan No. 5) Dam (Courtesy, Canadian National Committee, ICOLD). P-801-D-79337.

were 88 such structures in operation throughout the world, and 65 more were built in just the next 5 years. New records have been set in quick succession. Outstanding height precedents for contrete structures have been achieved since midcentury by the Mauvoisin Dam (fig. 1-51) (1957) in Switzerland, 237 meters (777 feet); the Vaiont Dam (fig. 1-51) (1961) in Italy, 265 meters (869 feet); and the Grande Dixence Dam (fig. 1-52) (1962) 285 meters (935 feet) and Contra Dam (figs. 1-53 and 1-54) 220 meters (722 feet) (1965) in Switzerland.

On the Vakhsh River in the Soviet Union, the construction of the Nurek Project, with a 300-meter (984-feet) high embankment dam (fig. 1-55), is nearly complete. The total volume of embankment amounts to about 58 000 000 cubic meters (75 861 000 cubic yards).

The Soviets rate their Rogun Dam (fig. 1-56) on the Vakhsh River as the highest in the world, with a height of 325 meters (1066 feet); crest length of 660 meters (2165 feet); downstream slope of 2 to 1; upstream slope of 2.4 to 1; and a volume of 75 500 000 cubic meters (98 750 000 cubic yards).

Another large Soviet dam is the concrete arch dam of the Inguri Project. As reported in 1979 publications, the Inguri Dam (fig. 1-57) is located 7 kilometers (4.3 miles) from the Dzhvari Village in a narrow gorge of the Inguri River. The arch dam has a crest 680 meters (2231 feet) long. The dam has a projected maximum height of 272 meters (892 feet). The dam thickness is 10 meters (32.8 feet) at the crest elevation and 52 meters (170.6 feet) at an elevation 50 meters (164 feet) above its base where it rests on a concrete block which serves to plug the canyon. The dam is an arch of the double-curvature type. The estimated volume of concrete in the dam is 3 880 000 cubic meters (5 075 000 cubic yards).

The records for volume of dam are also being surpassed. The great mass of Fort Peck Dam is now overshadowed by the Tarbela Dam on the Indus River in Pakistan, with 121 720 000 cubic meters (159 203 000 cubic yards) of earth and rock.

Since the distant beginnings of human history, the engineering of dams has evolved from primitive trial-and-error ventures to increasingly sophisticated analytical approaches. Early dam building was an uncertain art resting on cumulative experience. As the centuries unfolded, the art was gradually merged with science. Mathematics and the mechanics of materials have become increasingly effective in development of safer designs. Theoretical analysis combined with the practical judgment of the experienced engineer will provide the best insurance as the search for water moves to new horizons.

A list of the highest dams in the world is shown in table 1-1.

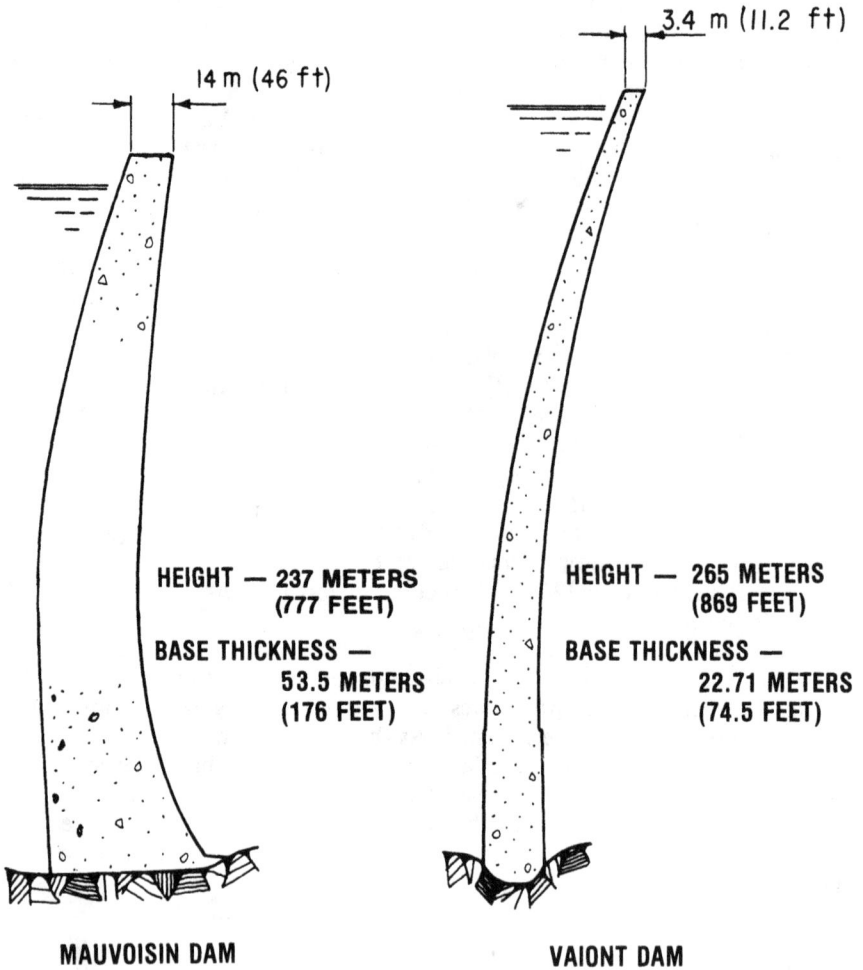

Figure 1-51.—Mauvoisin and Vaiont Dams, cross sections. P-801-D-79338.

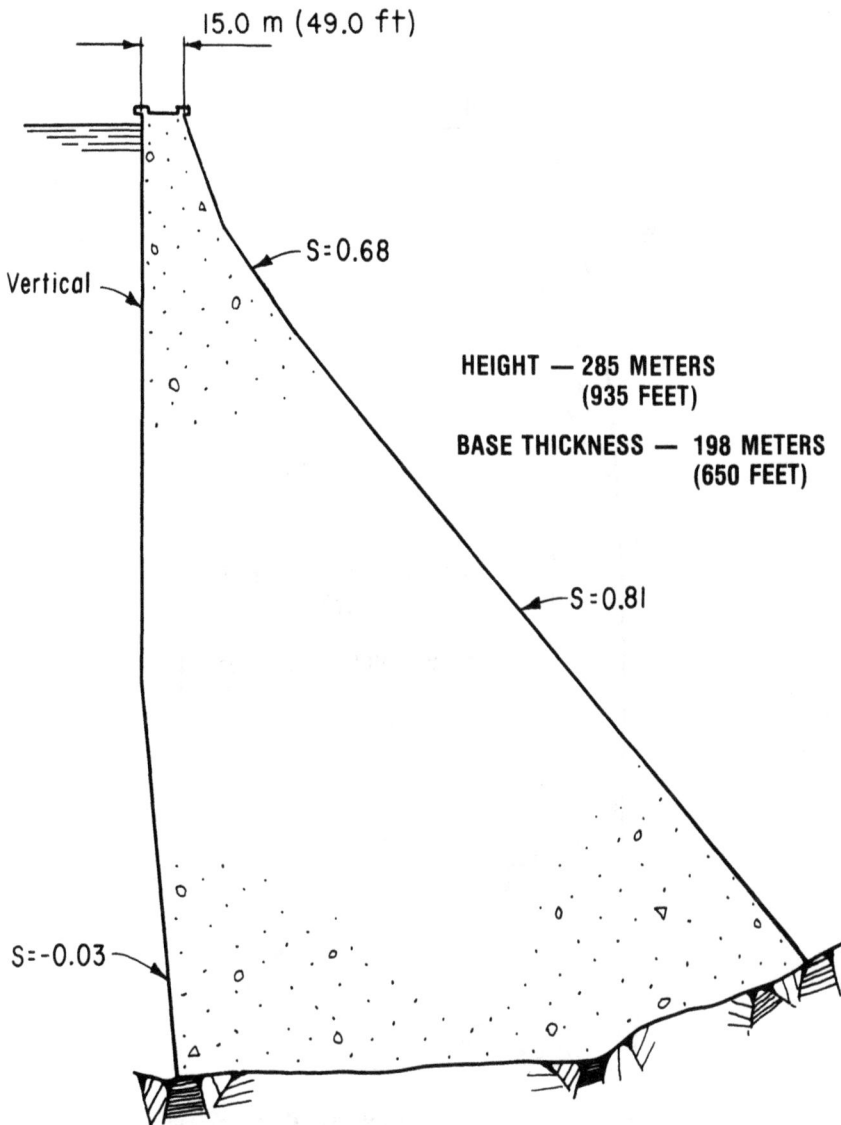

Figure 1-52.—Grande Dixence Dam, cross section. P-801-D-79339.

7 m (23 ft)

HEIGHT — 220 METERS
(722 FEET)

BASE THICKNESS — 25 METERS
(82 FEET)

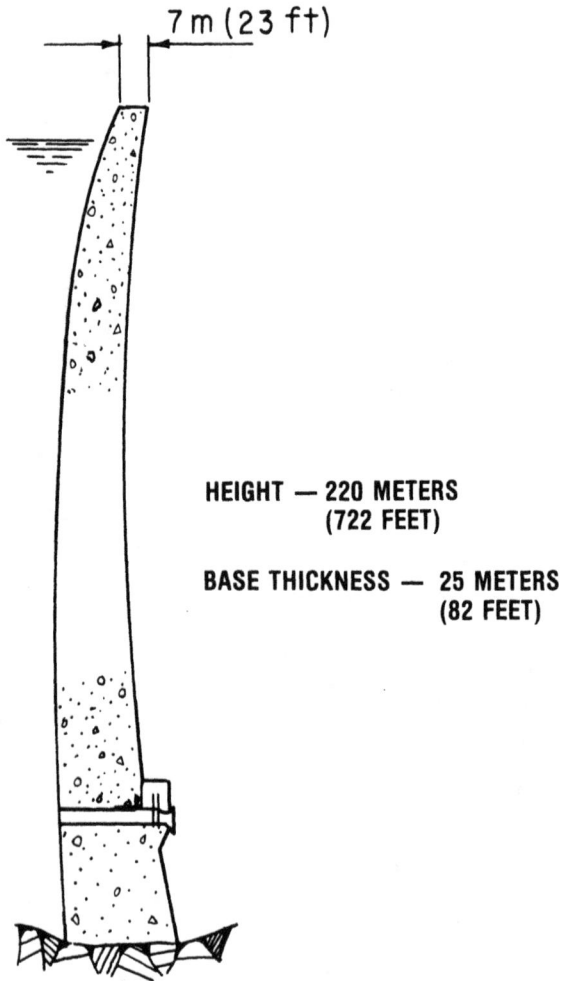

Figure 1-53.—Contra Dam, cross section. P-801-D-79340.

Figure 1-54.—Contra Dam (Courtesy, Comité National Suisse des Grands Barrages, ICOLD). P-801-D-79341.

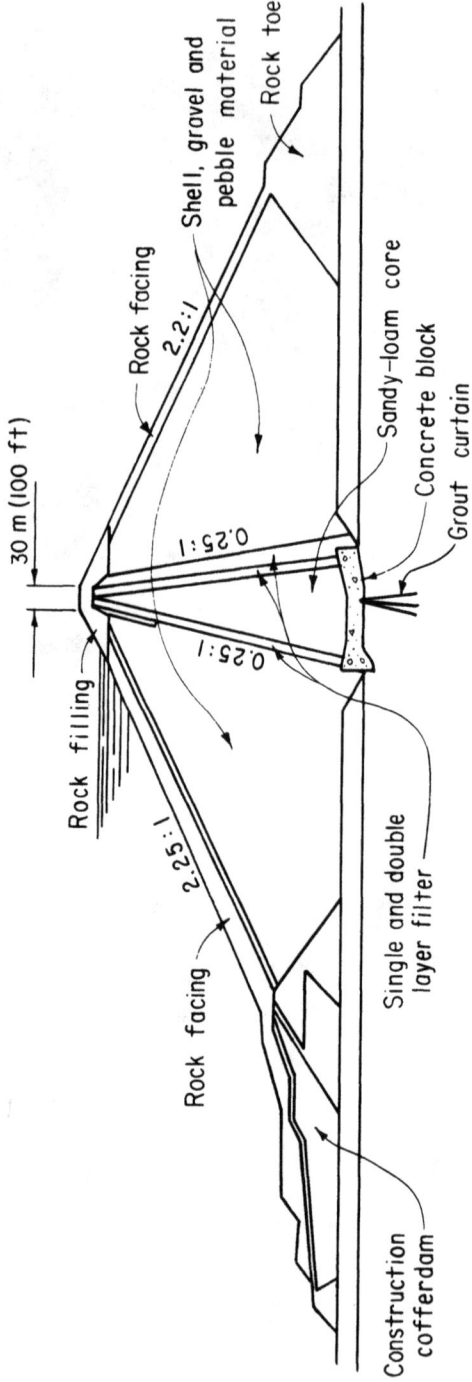

Figure 1-55.—Nurek Dam, cross section. P-801-D-79342.

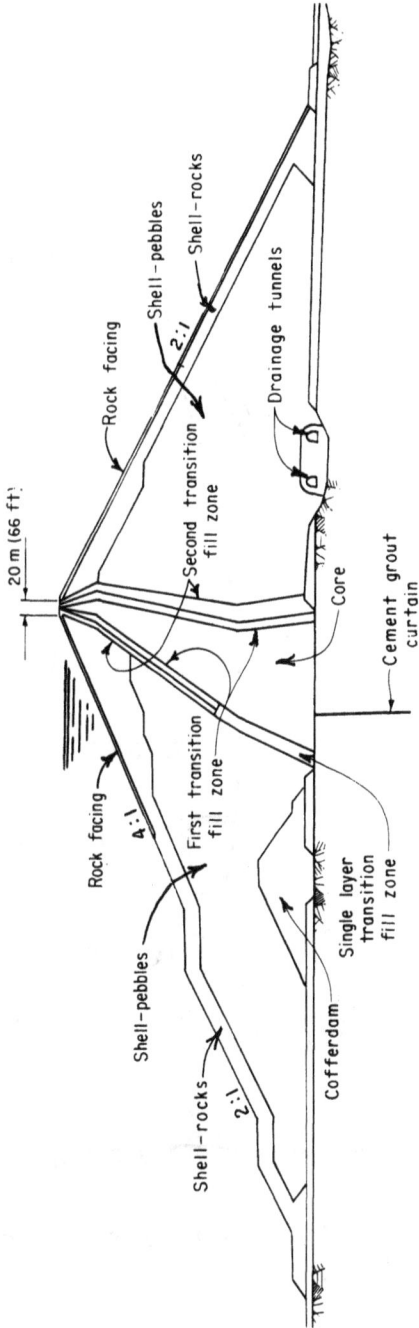

ROCK AND EARTHFILL

HEIGHT — 325 METERS
(1066 FEET)

Rock facing

Shell – pebbles

Shell – rocks

2:1

Drainage tunnels

Second transition fill zone

20 m (66 ft)

Rock facing

4:1

First transition fill zone

Shell – pebbles

Shell – rocks

2:1

Core

Cement grout curtain

Single layer transition fill zone

Cofferdam

Figure 1-56.—Rogun Dam, cross section. P-801-D-79343.

10 m (33 ft)

HEIGHT — 272 METERS
(892 FEET)

BASE THICKNESS* —52 METERS
(171 FEET)

*AT TOP OF SADDLE

TOP OF
SADDLE

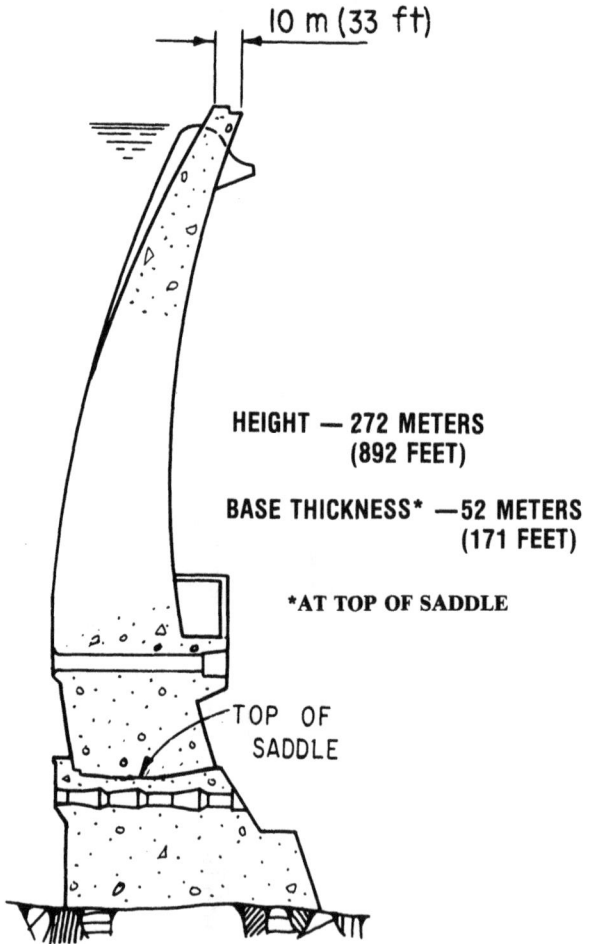

Figure 1-57.—Inguri Dam, cross section. P-801-D-79344.

Table 1-1.—*Highest dams in the world**

Name of Dam	Country	Type	Height Meters	Feet	Year Completed
Rogun	USSR	Earth	325	1066	U.C. (1985)
Nurek	USSR	Earth	300	984	U.C. (1985)
Grande Dixence	Switzerland	Gravity	285	935	1962
Inguri	USSR	Arch	272	892	U.C. (1985)
Chicoasén	Mexico	Rockfill	265	869	1981
Vaiont	Italy	Arch	265	869	1961
Tehri	India	Rockfill	261	856	U.C. (1990)
Kinshaw	India	Earth/Rockfill	253	830	U.C. (1985)
Mica	Canada	Earth/Rockfill	242	794	1972
Sayano-Shushenesk	USSR	Arch	242	794	1980
Mihoesti	Romania	Earth	242	794	U.C. (1983)
Chivor	Colombia	Earth/Rockfill	237	778	1975
Mauvoisin	Switzerland	Arch	237	777	1957
Oroville	USA (Calif.)	Earth	235	770	1968
Chirkey	USSR	Arch	233	764	1977
Bhakra	India	Gravity	226	741	1963
El Cajón	Honduras	Arch	226	741	U.C. (1985)
Hoover	USA (Ariz.-Nev.)	Arch/Gravity	221	726	1936
Contra	Switzerland	Arch	220	722	1965
Dabaklamm	Austria	Arch	220	722	U.C. (1989)
Piva (Mratinje)	Yugoslavia	Arch	220	722	1975
Dworshak	USA (Idaho)	Gravity	219	717	1974
Glen Canyon	USA (Ariz.)	Arch	216	710	1964
Toktogul	USSR	Arch	215	705	1978
Daniel Johnson	Canada	Multiple Arch	214	703	1968
San Rogue	Philippines	Earth	210	689	U.C. (- -)
Luzzone	Switzerland	Arch	208	682	1963
Keban	Turkey	Earth/Rockfill/ Gravity	207	679	1974
Dez	Iran	Arch	203	666	1963
Almendra	Spain	Arch	202	662	1970
Kölnbrein	Austria	Arch	200	656	1977
Karun	Iran	Arch	200	656	1976
Altinkaya	Turkey	Earth	195	640	U.C. (1986)
New Bullards Bar	USA (Calif.)	Arch	194	637	1968
Lakwar	India	Gravity	192	630	U.C. (1985)
New Melones	USA (Calif.)	Rockfill	191	625	1979
Itaipu	Brazil/Paraguay	Earth/Rockfill/ Gravity	190	623	U.C. (1983)
Kurobe No. 4	Japan	Arch	186	610	1964
Swift	USA (Wash.)	Rockfill	186	610	1958
Mossyrock	USA (Wash.)	Arch	185	607	1968
Oymopinar	Turkey	Arch	185	607	U.C. (1983)
Atatürk	Turkey	Rockfill	184	604	U.C. (1990)
Shasta	USA (Calif.)	Arch/Gravity	183	602	1945
Bennett W.A.C.	Canada	Earth	183	600	1967

* As reported in The World's Major Dams, Reservoirs, and Hydroelectric Plants, Bureau of Reclamation 1982, International Water Power & Dam Construction, May 1981, or The International Commission on Large Dams "World Register of Dams." U.C. = Under construction, ( ) = Estimated year of completion.

Part II

SAFETY
OF DAMS

General

The risk of the failure of a dam is one of the inevitable burdens of civilization. A primary duty of the engineer is to minimize this hazard. In no other field of engineering is the responsibility to the public heavier or more exacting.

Dam safety programs are of vital importance to all of society and call for multidisciplinary use of talent. Engineers must work closely with other professionals, including geologists and seismologists. Coordination can help to reduce the uncertainties but, due to difficult constraints, not all dam failures can be averted.

As builders are forced to use poorer sites, the job of protection becomes more difficult. Since millions of people live practically in the shadows of major dams, it is imperative that increasing attention be given to finding the best ways to ensure protection. This requires united effort by all agencies dedicated to safe water and power services.

Compensating somewhat for the increasing risk, including the greater consequences of failure as populations crowd onto the lands below reservoirs, is the growing body of knowledge of dams. More is known about how to design them, how to build them, and how to keep them safe. A significant part of this knowledge has been gained from failures. Even the best designers of any of the works of man have seen their structures in trouble. They have become even better designers by learning from these lessons.

Some critics have said that certain kinds of dams are safer than others. They recommend that we not build any of the weaker types. But there is no consensus as to which is stronger and which is weaker. The argument is largely a waste of time. Dams cannot be rated in terms of generalities. Certainly, when site specifics are considered, one type of dam may be judged preferable to another. But the selection among alternatives must follow exhaustive examination of local conditions — past, present, and future — and their possible effects on the structure.

The investment in safety should be accepted as an integral part of project cost, and not an extra item that can be eliminated if the budget is tight. This concept of safety applies throughout all phases of project development, from planning through design and construction, to operational surveillance.

Economy should not be placed ahead of doing the job right. While an engineer is trained to regard cost as a deciding factor in selecting among alternatives, each option considered must produce a safe structure. Engineers must not be concerned that good design will price them out of business. Defensive design measures have their prices, but they must not be compromised.

Striking a balance between economics and safety is not easy. Increasingly, the damsites now being considered are not of first quality, and dam designers are forced to build on foundations that would not have been selected just a few years ago. This comparative quality of sites must be recognized in budgeting for projects. Even with anticipated higher costs, we must be willing to pay the price for the necessary additional safeguards. The temptation to cut corners must be resisted. If we cannot guarantee good engineering, the project should not be built.

Yesterday's practices may not necessarily be good enough for tomorrow's needs. Future damsites will continue to present new challenges to engineers. As technology improves, we must continue to learn and to apply this new knowledge.

Existing dams and reservoirs should be reanalyzed periodically to ensure that they can still meet the test of safety by current standards. As knowledge of hydrology, seismicity, and the geological environment accumulates, and technology advances, facilities once regarded as safe may need modifications.

Safety of dams requires consideration of more than the technical factors. Looking at the organization, for example, one thing which must be assured is that all voices are heard. Ideas may come from within the organization — from nearly any level — or from outside. The latter includes, of course and particularly, consultants. One of the greatest hazards in the engineering of major structures is the exclusion of the ideas of those who may have valuable contributions to make. The management of any organization must exert special effort to assure that this does not happen.

In case histories of projects gone wrong, the dominance of single decisionmakers — sometimes authorities whose reputations for expertise were well earned — is not uncommon. Even experts can make mistakes, and probably the worst is to assume that an expert's judgment need not be questioned by those qualified to question.

Another consideration, especially in large organizations composed of many compartments, is to assure that information flows among the units. The many ideas essential to good engineering must be shared freely across the internal boundaries. The integration of separate efforts should be continuous throughout the evolution of designs, rather than simply gluing together individual final products. This means that designing must start with a general perspective and then focus on the individual parts — not vice versa.

There must be recognition of the inseparable relationship of design and construction. These functions are best considered as a single process. Design is not completed until construction is accomplished. Designers and construction engineers have to work in concert during design and while the dam is being built so that the site conditions disclosed can be weighed against design objectives. Any necessary modifications in design during this period should be a collaborative effort of designers, geologists, and construction engineers.

The vital relationship between the engineer and the geologist needs continuing emphasis. They must work as closely together as must the dam and its foundation.

The organization has to assure that its engineers, geologists, and other professionals continue to learn and apply the latest technology. This can be accomplished in various ways, including attendance at professional conferences, counseling and lecturing in-house by consultants, advanced university courses, personnel exchanges with other organizations, and recruitment periodically of new personnel with advanced education or advanced experience.

A generally competent organization must still be willing to accept — in fact it should seek — independent review of its engineering practices. The levels of technological advancement and the expertise of individual staff members will vary from unit to unit within even the best of engineering organizations. The inflow of knowledge from the outside will serve to strengthen areas of relative weakness.

Dams require defensive engineering, which means listing every imaginable force that might be imposed, examination of every possible set of circumstances, and incorporation of protective elements to cope with each and every condition. Lines of defense should be erected in succession, so that if one fails the next will take over. Each project calls for its own tailoring of defenses to meet the hazards inherent in and peculiar to the site.

To assemble the array of possible occurrences, the proposed construction site must be thoroughly known and understood. Exploration and testing must pursue all clues relating to surface and underground conditions. Those responsible for the project budget must understand that the knowledge gained is absolutely essential and worth the price. Also, the same can be said for expenditures related to instrumentation which will continue to provide information on site conditions once the project is built.

The different ways in which dams can fail are known. Gravity dams are characteristically stable. Even on poor foundations, as at St. Francis Dam (Calif.), monoliths have stood while adjacent blocks were swept away. Single-arch dams have been known to collapse quickly when their foundations failed (Malpasset in France), although arches are inherently very strong structures. Dams dependent upon buttresses, such as slabs and multiple arches, may disintegrate as the buttresses fail in succession like a row of dominoes. Embankments tend to fail more slowly, but they are obviously more susceptible to erosion than are masonry structures.

There are reliable ways to design against such tendencies. Potential causes and modes of failure must be thoroughly listed and examined. One of the greatest risks of error in the design process is to overlook any one of these possibilities. In searching for adverse combinations of occurrences, the designer has to consider the structure, the site, and the vicinity. The dam and its foundation must be designed to function together as an integral unit. We must not be in such a hurry as to overlook any aspect, however small.

These considerations are fundamental, but they need repeated emphasis. Concepts and hypotheses will often be more important than calculations, which may be worth little if founded on the wrong assumptions. Most failures can be attributed to simple, sometimes apparently insignificant, causes. Sophisticated designs have sometimes failed due to oversights that were obvious in retrospect. Designers should be encouraged to use the most advanced analytical techniques, but at the same time, they must be cautioned not to forget elementary forces.

Accidents and failures provide lessons which must be thoroughly learned and shared. A wise engineer will examine all such information that can be found, including his own mistakes as well as others. The designer must assume that any of these problems can occur on any project and must make every effort to prevent them. One of the basic guiding principles that must be followed is that extreme conditions should be averted by changing the setting. Seepage barriers or interceptors will do this.

Designs must be as foolproof as possible. This also includes auxiliary facilities. For example, even well-trained operations personnel may not perform as expected. In an emergency, they may not be where they are needed. Mechanical and electrical equipment may fail to operate. These are real possibilities or probabilities. The safest designs therefore will be those that can function despite any of these happenings. This is accomplished by minimizing dependence on operators and equipment for making emergency releases. An ungated spillway gives such assurance. Where gates or valves are used for spillways and outlets, redundant power and control systems will enhance reliability.

In imagining what can go wrong with a dam, its typical characteristics and those of its foundation must be recognized. We know, for example, that an arch depends on unyielding abutments and that a gravity structure may have to resist foundation water pressures of high magnitude. We also must accept that the zones of an embankment will not be homogeneous, despite the best of construction controls. While the designer of an embankment may assume uniform characteristics to facilitate analysis, he must remember that the materials in the as-built embankment will be variable. Borrow areas and quarries may yield soils and rock that only approximately resemble what the designer has assumed, even though the specifications are exacting. There will be further differences in the embankment from lift to lift. Moisture and density and gradation may vary considerably. We know that consolidation causes permeability to range appreciably from crest to base. Even though the design objective theoretically may be to have uniform zones — and the construction engineers must strive for such quality — this is an ideal that is never achieved. Knowing this, the prudent engineer will incorporate successive line of defense to guard against defects. Filters and drains must be incorporated into the design to control seepage that theoretical calculations may not forecast. No matter how much foundation exploration and testing have been done, the designer should be concerned about the foundation's capabilities, and should call for enough foundation treatment to compensate for the unknowns.

Let's face the probabilities: Foundation rock will never be flawless, even if grout is forced into the formation to the point of complete refusal. Dental work and rock bolting, although valuable, will not make it perfect either. Despite the best work of the best people, some seepage is inevitable at the damsite. As long as precautions have been taken for its control, the dam should be safe.

Key questions, of course, are: Where might the water go, and what damage can it do? The potential for damage will depend upon the water pressure differences along the seepage path. Consider the example of severely open-jointed foundation rock intercepted by a single grout curtain. If water can flow from the reservoir to the curtain with minimum head loss, the pressure differential through a window in the grout barrier may be high — depending on how effective the curtain is in reducing general pressures on its downstream side. Anything standing in the way of a stream through the window had better not be movable. This would suggest that a grout curtain by itself may not be totally beneficial, even if it is a zone of appreciable thickness. It may be most ineffective at its top, where grouting pressures were low. And this is where it needs to be most effective.

Unless the foundation itself is erodible, the escaping water can probably do its greatest damage where it impinges on the underside of an embankment. While grout curtains are useful to reduce foundation water losses, they should not be counted on to protect embankment materials at the contact with the foundation. The fill should be isolated from potentially detrimental flows in the foundation by one or more safeguards. These may include consolidation grouting, slush grouting, concrete dental work, or filters.

Seepage control by drainage often provides more reliable results than cutoffs, including grout curtains. But these measures are not mutually exclusive; each may have an important function. Drains should be capable of conveying water under low hydraulic gradients. In designing them, we must remember that discharge capacity can be restricted by breakdown of aggregates, by excessive fines, and by compression under embankment loads. Some engineers still rely on unscreened sands and gravels for drains and filter zones. The use of such materials transported directly from natural deposits can lead to serious problems. They cannot be condemned without reservation, however, because their characteristics vary widely from site to site. Some have enough drainage capacity and some do not.

Compared with masonry structures, embankments are generally less homogeneous, even within each zone. Differences in internal conditions may make them vulnerable. A permeable layer inadvertently placed within a zone intended to be impervious may constitute a conduit through the dam. An impermeable layer placed within a zone intended to be pervious may preclude proper drainage. Placement of a fine zone against a coarse zone without a filter may permit the movement of particles from one zone to the other. These hazards are obvious, and yet they have sometimes been disregarded, even by engineers who should have known better. Embankments and their foundations must be designed so that internal boundaries will always be maintained. The dangers may include movement or solution of the foundation rock itself, or migration of embankment material into the rock joints and fractures.

The tendency to think of averages in the engineering of dams must be resisted. Failures occur where the dam or its foundation is weakest, not where it is in average condition. The design must focus on the potential weaknesses. Exploration and testing will necessarily depend on sampling techniques, with results varying sometimes over a wide range. Natural materials available for construction may exhibit average characteristics that meet requirements; yet, they may be judged totally unacceptable when their variations are considered.

The variability of natural conditions does not encourage unreserved faith in standard guidelines or "cookbook" approaches to dam engineering. No matter how many exploratory holes are drilled or how many samples are

93

tested, the reservoir site may still contain surprises — and these may appear at any time during the life of the structure.

For example, in the interest of economy, engineers may elect to leave channel alluvium in place under an embankment. Prudent engineering will require at least that the streambed materials be thoroughly sampled and that gradations be determined. While this testing may indicate that the alluvium is compatible with overlying zones of the dam, a designer with good judgment will recognize that irregularities may still exist. Without debating the advisability of building a dam on such a base, as a minimum, the interface between foundation and superimposed materials should be protected.

Writers of manuals cannot foresee exactly what should be done in such cases. Professional judgment must be applied by the engineer on the spot. In the field of dam safety, however, those who possess such judgment are not numerous enough to match the needs. Guidelines therefore can be of value, especially if they are used with some flexibility. They do provide a way for experts to share their knowledge with those who are less experienced.

Flexibility is the key. Rigid criteria are useful only as long as conditions match the underlying assumptions. They fail when deviations are not perceived or when the latitude and the judgment are not available to make the necessary adjustments. The trouble with a "cookbook" is that some of its users may come to think that it contains all the recipes. Design by the book is especially hazardous in an organization insulated from professional interchange.

Statistics on Failures

The total number of dams in the world which represent hazards in the event of failure may exceed 150,000. Many of these structures have not performed as planned. As a rough estimate, there have been perhaps 2,000 failures, including partial collapses, since the 12th century A.D. Most of these, of course, were not major dams. There have been about 200 notable reservoir failures in the world so far in the 20th century. More than 8,000 people died in these disasters.

At the World Power Conference in Berlin in 1929, a general interest in advancement of the engineering of dams led to the founding of ICOLD (International Commission on Large Dams). This organization has universal support and is instrumental in collecting and sharing knowledge gained by professionals on dam design and construction from all countries of the world.

The following tabulation published by ICOLD indicates the numbers of known major dam failures in the historical period through 1965.

Year	Approximate number of significant failures
Prior to 1900	38
1900 to 1909	15
1910 to 1919	25
1920 to 1929	33
1930 to 1939	15
1940 to 1949	11

Year	Approximate number of significant failures
1950 to 1959	30
1960 to 1965	25
Date unknown	10
Total	202

The toll in human lives resulting from some of the major disasters throughout the world has been estimated as follows:

Dam	Country	Year of disaster	Lives lost
Machhu II	India	1979	2,000+
San Ildefonso	Bolivia	1626	Unknown*
Vaiont	Italy	1963	2,600
South Fork (Johnstown)	USA	1889	2,209
Panshet-Khadakwasla	India	1961	Unknown
Orós	Brazil	1960	Unknown**
Puentes	Spain	1802	608
Kuala Lumpur	Malaya	1961	600
Gleno	Italy	1923	600
St. Francis	USA	1928	450
Malpasset	France	1959	421
Hyokiri	South Korea	1961	250
Quebrada la Chapa	Colombia	1963	250
Bradfield (Dale Dike)	England	1864	238
El Habra	Algeria	1881	209
Sempor	Indonesia	1967	200
Walnut Grove	USA	1890	150
Babii Yar	USSR	1961	145
Vega de Tera	Spain	1959	144
Mill River	USA	1874	143
Buffalo Creek	USA	1972	125
Valparaíso	Chile	1888	Over 100
Alla Sella Zerbino	Italy	1935	Over 100
Bouzey	France	1895	Over 100
Nanaksagar	India	1967	100
Zgorigrad (Vratza)	Bulgaria	1966	96
Austin	USA	1911	80
Bila Desna	Czechoslovakia	1916	65
Frías	Argentina	1970	42+
Lower Otay	USA	1916	30
Palagnedra	Switzerland	1978	24
Eigiau-Coedty	Wales	1925	16
Teton	USA	1976	11
Baldwin Hills	USA	1963	5
Tigra	India	1917	Unknown

*Estimated as high as 4,000 but probably fewer
**Estimated as high as 1,000, but probably fewer

Governmental Supervision

Such catastrophic loss of life is of great public concern. This has led to a wide recognition of the need for governmental involvement in the supervision of dams and reservoirs. An informed citizenry is an essential element. Safety-conscious people will not only demand safer dams, but will also be willing to finance the improved protective measures.

In 1929, the year following failure of St. Francis Dam, California placed dams under an effective system of governmental supervision. The State exercises jurisdiction over the design, construction, operation, alteration, repair, and behavioral surveillance of all legally defined dams, except those owned by the Federal Government. Many States in the United States and several foreign countries have since enacted similar legislation.

In the 1920's, there was a marked increase in the number of dams constructed throughout the world. Since there were few governmental regulations to provide guidance, each builder was burdened with full responsibility for his work. Engineers were often exposed to awesome liabilities as they ventured into new frontiers. As the difficulties of erecting larger dams on poorer foundations were recognized, the engineer was joined by the geologist and other professionals in an integrated project effort.

As a result of the major disasters at Malpasset (France), Vaiont (Italy), and Baldwin Hills (United States), governments in several countries enacted new or revised laws for supervision of safety of dams and reservoirs. The Dam Inspection Act, U.S. Congressional Public Law 92-367, signed into law August 8, 1972, authorized the Secretary of the Army, acting through the Chief of Engineers, to undertake a national program of inspection of dams. Under this authority, the Corps of Engineers has (1) compiled an inventory of Federal and non-Federal dams; (2) conducted a survey of each State and Federal agency's capabilities, practices, and regulations regarding the design, construction, operation, and maintenance of dams; (3) developed guidelines for safety inspections and evaluations of dams; and (4) formulated recommendations for a comprehensive national dam safety program.

The new French regulations require annual inspections. Recognizing that the Malpasset and Vaiont tragedies happened during reservoir filling, the French rules impose especially strict inspection requirements in the initial impoundment stage. They establish and regulate the rate of reservoir filling, and they call for weekly instrumentation readings, and require inspections at daily, weekly, and monthly intervals. Regular surveillance of the reservoir's peripheral areas is included. A report on the performance of the dam and reservoir is to be ready after 6 months.

In the United Kingdom, the Reservoirs Act of 1975 was written as an updating of 45-year-old rules that were put into effect after the British dam failures at Dolgarrog and Skelmorlie, in 1925. The new law provides authority for regulators to intervene when an inspecting engineer's report has not been given adequate response. In such cases, the officials are empowered to effect the necessary corrective measures and to bill the costs to the owner. The British regulations also call for certificates which specify limits of safe reservoir operating level. An inspection is to be conducted by an independent qualified civil engineer not later than 2 years after issuance of the "final certificate," and at proper intervals thereafter.

The Bureau of Reclamation considers 2 years to be the maximum time interval between field (onsite) examinations and 6 years to be the maximum time interval between evaluations. Also, Regional personnel make periodic examinations devoted primarily to operation and maintenance.

Disaster Preparedness

The best appraisal of the hazard posed by a dam can be made by its designers and those responsible for its operational surveillance. They are especially qualified to assess the potential modes of failure, and to estimate the consequences. Such analyses, coupled with rational disaster planning, can reduce the numbers of failures and the resultant losses. Flood-plain zoning can have beneficial effects also, but economic and political factors will probably continue to limit its effectiveness.

The area subject to inundation in the event of breaching of a dam can be estimated by hydraulic engineers. They can also calculate depths and velocities of flow at any one point of the flood. In some cases, the rapidly deteriorating condition of a dam may be detected in advance so that the threatened area can be evacuated. In the Baldwin Hills (California) disaster, for example, the time between discovery of trouble and the final embankment collapse was used effectively to save many lives. About 16,500 people lived in the disaster area, and most of them were safely evacuated. Even so, 5 people died, and 27 were injured — the low casualty list was a testimony to the excellence of the warning and evacuation programs as well as the timely detection of the failure symptoms. Door-to-door warning, helicopters with loudspeakers, radio and television appeals, and effective perimeter control by the police deserve credit for the comparatively low casualty list. Of course, extensive property damage was unavoidable in the neighborhoods which had to be abandoned in the face of the flood threat.

In contrast, the Orós Dam disaster in Brazil reportedly had a high death toll with proportionally less economic impact. Here, too, alerts were issued, which helped 100,000 people to escape; even so, many human lives were reported lost.

In the Vaiont Dam disaster in Italy, telecommunication lines were torn away by the impact of the flood wave. The residents in the endangered valley were therefore unaware of the horror descending upon them until a policeman noticed the floodwater and sounded the first warning. For thousands of people, it was too late.

Involvement of the Courts

Around the world, various attitudes and policies are found regarding how the responsibility for dam safety should be shared. After the Vega de Tera Dam (Spain) failed in 1959, civil and criminal lawsuits were filed in the Spanish courts against 10 engineers. Eventually, four of them were found guilty. The charges, which led to the judgment by the court, were undoubtedly attributable in part to the public agitation stimulated by the failure.

The collapse of the Malpasset Dam (France) in the same year resulted in the indictment of its chief engineer for negligent homicide. The court, however, acquitted him since "he merely controlled and supervised so that the designer's directions were strictly observed***." In 1966, the court of appeals at Aix-en-Provence supported the verdicts of the criminal courts that the chief and four other engineers on the project were innocent.

When the Chitauni embankment dam in Uttar Pradesh, India, failed in 1968, the blame was immediately placed on the State engineers who had directed its design, construction, and operation. Without waiting for a full inquiry, the Chief Minister of the State suspended several of these engineers. While this precipitous discipline may have satisfied an aroused public, the long-range consequences must also be considered. Engineers under such penalty may act cautiously during future flood emergencies when unhesitating action is needed.

Insurance of Dams

There are other ways to share responsibility for dam safety. The financial burden, for example, can be carried by insurance. Coverage can extend to one dam or to many dams. With thousands of reservoirs in operation, comprehensive insurance responds to a common need for spreading risk over a wide area. Not all dams have high disaster potential. In remote unpopulated places, the hazard to life and property may be minimal. And in any dam failure there is an areal limit to the damage which can occur. These circumstances therefore ensure that the insurance money paid by the many can pay for the losses of the few. In its widest application, the essential spreading of risk is accomplished by reinsurance, which in international terms means that the insurance industry of one country may make reciprocal exchanges of parts of portfolios with the insurers in another country.

Residents in the potential zone of inundation might logically contend that all of the costs should be borne by the owner of the reservoir. If the dam did not exist, there would not be any threat of an uncontrolled discharge of water. And, since the owner maintains the hazard, he must be accountable for any damages, no matter what might have been the cause of failure. In many cases the owner's rebuttal can rest just as logically on the service which his dam and reservoir render the people, including those residing downstream. These views can be reconciled by having the losses underwritten by the whole community of interest through regional or national disaster funds. Any insurance program should place emphasis on the supervision of dams by experienced specialists.

Part III

KINDS OF PROBLEMS

General

The life of a dam can be threatened by natural phenomena such as floods, rockslides, earthquakes, and deterioration of the heterogeneous foundations and construction materials. In the course of time, the structure may take on anisotropic characteristics. Internal pressures and paths of seepage may develop. Usually the changes are slow and not readily discerned by visual examination.

Continuous monitoring of a dam's performance will usually ensure detection of any flaws which may lead to failure. This must be done by personnel who know the signs of distress. Knowledge of the forces which cause deterioration can be gained by studying the postmortems of failed structures. Some of these were conceived by acknowledged masters of the profession. Even they could not always foresee the potential weaknesses nor the neglect that their works might suffer. As more knowledge is accumulated, similarities are found in the malperformances of dams from site to site. These teach valuable lessons.

Analysis of the performances of the various types of dams will show their relative suitability for conditions which may be encountered at a given site. Each type can be related generally to a certain mode of failure. A gravity dam may collapse only in the section which is overstressed. A buttress dam may fall in domino fashion through the successive collapse of its buttresses. The rupture of an arch may be sudden and complete. Failure of an embankment may be relatively slow, with erosion progressing laterally and downward and then accelerating as the flood tears through the breach.

The records of dams indicate that earthfills have been involved in the largest number of failures, followed in order by gravity dams, rockfills, and multiple and single arches. That more troubles would occur among the more prevalent dam types is not surprising. Considering the number of failures compared to total numbers of dams built for each type, the multiple-arch type shows a comparatively poor record.

In 1962, the Spanish "Revista de Obras Públicas" published the results of a study of 1,620 major dams. In the period from 1799 to 1944, there were 308 accidents or failures at these structures. Actual breaching occurred in less than half the events. This list of dams with troubles consisted of 177 embankments, 70 gravity dams, 7 multiple arches, 2 arches, and 52 other types. Excepting 77 happenings for which adequate details were not given, a classification of causes of failure is presented on the following page.

Cause	Percent of failures
Foundation failure	40
Inadequate spillway	23
Poor construction	12
Uneven settlement	10
High pore pressure	5
Acts of war	3
Embankment slips	2
Defective materials	2
Incorrect operation	2
Earthquakes	1

In April 1966, the Japanese government completed a survey of impairment of dams during the period from 1950 to 1965. The total number of impaired dams was 1,046. In this number, there were 118 dams with a height of 15 meters (49 feet) or more. Impairment caused by heavy runoff affected 38 percent of the total, while problems attributable to earthquakes were reported in 6 percent of the cases. Most of the remaining impairment was due to cumulative deterioration brought on by various forces. Most studies of this kind indicate that the main dangers stem from the unpredictability of extreme floods and the uncertainties of the geologic setting.

Biswas and Chatterjee, in their 1971 article entitled "Dam Disasters — An Assessment," concluded from a study of more than 300 dam failures throughout the world that about 35 percent were a direct result of floods in excess of the spillway capacity; and 25 percent were due to foundation problems such as seepage, piping, excessive pore pressures, inadequate cutoff, fault movement, settlement, or rockslides. The remaining 40 percent of the disasters were found to result from various problems including improper design or construction, inferior materials, wave action, acts of war, or general lack of proper operation and/or maintenance.

At the Thirteenth International Congress on Large Dams in New Delhi, India, in 1979, a report was presented on 52 Spanish dams which had suffered accidents or failures during the past 200 years. Eleven fundamental causes were listed:

Cause	Number of dams affected
1. Overtopping	7 (4 concrete and 3 rockfill)
2. Erosion in the spillway	5
3. Breakage or damage of gates	3
4. Damage in outlets and conduits	5
5. Excessive seepage	33
6. Seepage with piping	9 (8 of the 33 in Cause No. 5, plus one cofferdam)
7. Freezing	6 (5 concrete and one earthfill)
8. Formation of fissures	11 (6 concrete, 2 rockfill cores, and 3 rockfill protective faces)

Cause	Number of dams affected

9. Settlements incompatible
with stability.............. 5 (4 embankments and the
abutment of one concrete
dam)
10. Stability defects, as a consequence of the project....... 8
11. Construction faults.........22

The engineers who analyzed the problems at these dams concluded that there were "two predominant causes of damage," which were closely inter-related: (1) construction defects, and (2) interstitial water, inadequately controlled.

Foundation Problems

Foundation failures may lead to the complete breaching of the dam. In other cases, inherent strength or deformability of the structure may save it from total collapse. A notable example was the failure in 1802 of the Puentes Dam in Spain wherein the rubble masonry arched over the breach caused by washout of alluvial foundation.

Foundation deficiencies may be related to the natural condition of the foundation or to its treatment during construction. Differential settlement, sliding, high piezometric pressures, and uncontrolled seepage are common evidences of foundation distress. Cracks in a dam, even relatively minor ones, may also be indicative of a foundation problem.

Concrete dams can withstand overtopping for at least a limited time without damage. The key to safety may be the ability of the foundation to bear the impact of the overflow, rather than the resistance of the dam itself, which is likely to be more than adequate.

The safety of arch dams is highly dependent upon the strength of their abutments. Failure may stem from weakness in the rock resulting from saturation or deterioration, or excessive flood loading, or from abutment shearing under hydrostatic pressures such as occurred at Malpasset Dam in France. Failures of arch dams also may be triggered by the erosion of foundation materials by overtopping. However, arch dams are inherently capable of passing floods. The Vaiont Dam in Italy, in a remarkable demonstration of this capability, withstood a sudden water surcharge of 100 meters (300 feet), in the rockslide disaster of 1963 without distress.

Potential erosion of the foundation itself must be considered. Clay or silt in weathered joints or faults cannot easily be removed by washing and therefore may preclude effective grouting. Seepage may gradually transport these materials into voids downstream. Consequent enlargement of the joint or fault conduits may threaten the integrity of the dam.

Foundation seepage can cause internal erosion or solution. The removal of foundation material may leave collapsible voids and consequently precarious support for the dam. Such potential weaknesses sometimes can be identified by examining geologic conditions in the immediate vicinity of the reservoir. Actual deterioration may be evidenced by increased seepage, by sediment in seepage water, or an increase in soluble materials disclosed by chemical analyses. The records of site exploration may yield clues of the presence of materials vulnerable to attack, such as dispersive clays, water-reactive shales, gypsum, and limestone.

Uncontrolled seepage through an erodible foundation may open voids which must be bridged by the dam. A concrete structure may have such capabilities so long as stresses are within tolerable limits and the opening is not too great. The same phenomenon in the foundation of an embankment can cause collapse of overlying fill and eventual breaching of the dam.

Subsidence of terrain caused by pumping from the underground can cause foundation settlement and distortion of the dam. Such distress can also be due to the collapse of foundation soils caused by loading and wetting. Fine sands and silts with low densities and low natural moisture contents are especially susceptible to this phenomenon. The consequent cracking of the dam can create a dangerous condition, especially in earth-fills of low cohesive strength.

General settlement of a rock foundation under the weight of dam and reservoir usually is not a cause of concern. However, differential settlement at irregular rock surfaces has not been an uncommon problem. The resultant cracking of the embankment is one of the most threatening conditions to be encountered. Preparation of rock foundations therefore should include shaping of projections and overhangs by removal and/or filling with concrete or shotcrete.

Foundation failures may occur due to saturation of foundation material and consequent washout or sliding. Foundation erosion may progress slowly, but major slides may occur suddenly.

Foundations with generally low shear strength or with seams of weak material such as clay or bentonite may be vulnerable to sliding. Seams of pervious material also may contribute to sliding if seepage through them is not controlled to preclude detrimental uplift.

Shear zones frequently have caused problems at damsites and therefore warrant close examination. Two common types that have been troublesome are bedding-plane zones in sedimentary rocks and foliation zones in metamorphics. Shales and schists, respectively, are prime suspects in such cases. Meticulous work must be done to identify and evaluate the potentially hazardous interbeds or foliations, which may be deceptively thin. Because of their inherent weakness, drill core recovery may be difficult. Where they pose significant threat to the dam or reservoir, exploratory excavations by trenching, tunneling, or shafts may be justified.

Once the dimensions, orientation, and materials in the shear zones are known, preventive or remedial engineering may call for drainage, rock reinforcement, and/or buttressing to reduce sliding potential. Water pressures on the suspect shear surfaces may be lowered by vertical or horizontal drain holes, drainage adits, or toe drains.

Seepage

Water movement through a dam or through its foundation is one of the important indicators of the condition of the structure and may be a serious source of trouble. No one can be sure what effect the construction of a dam may have on its foundation. The impoundment usually can be expected to increase — substantially in the case of deep reservoirs — the percolation and the pore pressures in the underlying formations, unless seepage control facilities are installed. The consequences may be important not only at the damsite but elsewhere on the reservoir rim, particularly where the natural barrier is thin.

Seeping water naturally tends to carry away constituents that may be vital to the integrity of the dam. Turbid flow issuing from a dam or its foundation may be an indication of internal erosion. Such removal of material is typically progressive, so that the structure is gradually weakened. A sometimes more subtle attack may be launched through chemical solution of foundation rock. Some damsites, for example, have large quantities of gypsum and other soluble minerals in the foundation. Appreciable volumes of such material may dissolve as water percolates.

The Lower Van Norman (San Fernando) Dam, an embankment structure in California, rested on sedimentary rock which had been subjected to high rates of solution. During the 60 years of operation, there were numerous observations of caving and erosion in the foundation. Apparently most of the leaching occurred at one abutment which was composed of fractured shales and lightweight siltstones. During a grouting program, more than 850 cubic meters (30 000 cubic feet) of cement was injected into this abutment area in a curtain less than 213 meters (700 feet) long. Although most of this grouting was done at low pressures, the average cement requirement was more than 0.26 cubic meters per meter (3 cubic feet per foot) of drilled hole.

Since seepage at this dam did not increase appreciably over the years, some investigators have assumed that the weight of the embankment upon the weak abutment tended to close the foundation voids as solution progressed. Since the fill over this foundation had been subjected to much settlement, some of this possibly could be attributed to consolidation of the foundation as the solution openings were closed. This foundation area had been grouted 30 years previously, but the later drilling program revealed that seepage had been continuing through the fractured and weathered zone via passages where gypsum and other soluble minerals had been leached out, primarily along bedding planes and, to a lesser extent, along joints and cracks.

Such solution of rock can jeopardize the safety of a dam by enlarging underground passages, weakening the foundation, and making seepage control more difficult. Gradual development of a honeycombed foundation structure could lead to collapse during a severe earthquake.

Investigators are not always able to pinpoint quickly and accurately the location or to détermine all the characteristics of this kind of foundation deterioration. Correction of such conditions should begin with the establishment of a thorough monitoring system to ascertain the quantity, the composition, and the sources of seepage. Adequate measurements must be taken of the piezometric surface within the foundation and the embankment, as well as any horizontal or vertical distortion of the abutments and the fill. Constant attention must be focused on any changes such as in the rate of seepage, settlement, or in the character of the escaping water. Generally, differentials caused by dissolving of solid material develop slowly enough to provide advance warning of the need for any remedies.

The presence of gap-graded materials such as openwork gravels or segregated nests of materials in a foundation or in drains or filters may be conducive to internal erosion.

Any leakage at an earth embankment may be potentially dangerous, since rapid erosion may quickly enlarge an initially minor defect.

Seepage paths may be opened by settlement cracks caused by weak material in the embankment or foundation and by shrinkage cracks in highly plastic clays in the embankment. Other dangerous water passages may be created by burrowing animals, decaying tree roots, and leakage along conduits improperly placed in an embankment.

Uncontrolled seepage may be accompanied by excessive embankment pore pressures and consequent weakening of the soil mass. High pore pressures can result from the placement of embankment too rapidly or too wet, or because of seepage through pervious materials in the embankment or along foundation joints and cracks.

Erosion

Embankments may be susceptible to erosion unless protected from wave action on the upstream face and surface runoff on the downstream face. Groins are especially vulnerable to such damage. The downstream toe of the fill may also be subject to erosion if outlet or spillway flows are not kept at safe distances.

Riprap armors the upstream slope of an earthfill structure against wave erosion. Rockfill or gravel is also sometimes used on the downstream slope to protect from rain and wind attack. Seeding with grasses may be an acceptable alternative. Berms on the downstream face may also serve to control erosion by intercepting and diverting runoff.

The dislodging of riprap by wave action may leave the embankment exposed to erosion, but this deficiency can usually be detected and corrected before serious damage has developed.

Embankment Movement

The deformation characteristics of embankment materials are usually not precisely predictable, and the effects of weather and poor construction effort may also be uncertain.

In an older embankment dam, the condition of materials may vary considerably. There may be small or extensive areas of low strength. Location of these weaknesses must be a key objective of the evaluation of such dams.

Instability of an embankment may be caused by deleterious materials used in its construction. Soluble minerals such as gypsum may be carried away, leaving solution channels or cause general settlement due to loss of volume. Erosion of dispersive clays may result from seepage of water with a low salt content. Decomposition of wood or other organic material in an embankment can leave voids and cause settlement cracks.

Adverse conditions which have been well known at embankment dams, and which deserve attention are listed on the following page.

- Poorly sealed foundations
- Cracking in the core zone
- Cracking at zonal interfaces
- Soluble foundation rock
- Deteriorating impervious structural membranes
- Inadequate foundation cutoffs
- Desiccation of clay fill
- Steep slopes vulnerable to sliding
- Blocky foundation rock susceptible to differential settlement
- Ineffective contact at adjoining structures and at abutments
- Pervious embankment strata
- Vulnerability to "quick" conditions during an earthquake.

Embankment dams may be damaged to a dangerous degree by distortions at critical points. If, when placed, the embankment materials are poorly compacted or their moisture content is too low, excessive or uneven settlement may result, especially if quickly saturated upon rapid initial filling of the reservoir. Differential settlement may be most severe at steep abutments and at buried structures where effective compaction is difficult to obtain. At such locations the fill may crack or slump and arch, opening paths of seepage which may be dangerous. For this reason, many failures have occurred along outlet pipes. The fill materials used in contact with rock foundations or concrete abutments should possess plastic properties which will allow them to accommodate any movements that may occur.

An embankment may be most vulnerable at its interface with rock abutments. Especially during first impoundment of the reservoir, saturation of granular materials in the upstream shell may result in substantial settlement. The crest tends to develop extension strains near the abutments and increased compression in its central sections. At this critical stage, the embankment may be susceptible to transverse cracking.

Deformations of an embankment or its foundation may have especially adverse consequences at structures in or adjoining the dam, such as:

- Thin concrete cutoff walls projecting from abutments into the fill may be cracked or sheared,
- Conduits constructed through or under the embankment may be subjected to tension that tends to pull joints apart,
- Vertical towers within the embankment may be bent or tilted.

Dumped and sluiced rockfills were built for many years and generally have given good service. However, such dams usually undergo appreciable settlement. This may cause cracking of thin, sloping cores. A rockfill compacted by heavy rolling equipment and constructed to slopes flatter than the angle of repose should have an inherent resistance to failure greater than a rockfill which has been dumped and sluiced to natural slopes. Major difficulties have also been experienced with concrete face slabs placed on dumped rockfills. Characteristically, adjustment of the rockfill results in horizontal compression at the middle of the slab and tension at the abutments. Settlement can be significantly reduced if the entire rockfill is mechanically compacted rather than dumped and sluiced. Also, it is preferable that placement of the concrete slab on the rockfill be delayed long enough to permit the maximum compaction or settlement.

In some ways, a compacted earth core is superior to a concrete slab as the impervious element of a rockfill dam. If constructed of materials of sufficient plasticity, the core should be flexible enough to adjust without significant damage. During settlement, it should tend to mold itself to the abutments more readily than a relatively rigid concrete slab. A well-graded impervious earth core also has the important advantage of healing minor cracks which may develop during adjustment of the fill.

Because of the cited difficulties at earlier rockfills with concrete slabs, the engineering of this type dam has undergone important changes since the 1960's. Improvements in zoning, compaction, and cutoff and slab details have produced superior embankments with much less deformation, and have led to construction of such dams to greater heights.

Liquefaction

Improved methods for analyzing the stability of dams subjected to seismic loading provide reliable indications that many old dams may be vulnerable to earthquakes. Hydraulic fill dams especially have become suspect. The potential for development of "quick" conditions in such embankments is generally recognized. The possibility is acknowledged that such weaknesses may also exist in loose cohesionless soils in the mass or the foundation of other kinds of dam. Reevaluation of the stability of any embankment incorporating or founded upon such materials should be given high priority.

An historic accident occurred on February 9, 1971, with the liquefaction of a large part of the Lower Van Norman (San Fernando) Dam in California during an earthquake of Richter Magnitude 6.4. A major disaster was averted by the very narrowest of margins. This event emphasized the inherent instability of certain embankments constructed of uniformly fine-grained soils.

The Lower Van Norman Dam was a hydraulic fill. Experience in California, where more than 30 relatively large earthfill structures of this type were built in the period 1850 to 1940, has clearly demonstrated their flaws. The sluiced embankment is not as stable or as free-draining as its advocates supposed. The near disaster at San Fernando confirmed its susceptibility to liquefaction under seismic vibration.

A true hydraulic fill dam (fig. 3-1) was built by conveying earth materials from borrow to embankment as a liquid mixture and placing them in the embankment by water. This entailed continuous ponding. The foundation cutoffs for some of these structures were made by dumping fine-grained soils into a water-filled trench. In some cases the cutoffs were narrow and extended 30 meters (100 feet) or more into alluvial foundations.

An alternative to the true hydraulic fill was the semihydraulic fill, which was adopted at some sites where there was not enough water to transport materials over longer distances. In this type of construction, hauling from the borrow area was accomplished by other means, and the material then was moved into place in the embankment by water.

In dams built by the semihydraulic method, the outer zones of the embankment usually consisted of car-dumped fills. Material was sluiced from the inner slopes of these fills by water jets. The finer material was washed into a central pool, forming the core. The coarser particles tended to be

Figure 3-1.—A hydraulic fill under construction (San Pablo Dam in Calif.) (Courtesy, Calif. Dept. of Water Resources). P-801-D-79345.

segregated out before traveling very far from the outer zone toward the core.

There was some agreement that fine materials which were very uniform in size should not be used in hydraulic fills. Such materials were recognized as susceptible to flow slides. The most desirable materials in the shells of the embankment would have been those of nonuniform gradation. These objectives apply as well to other kinds of earth dams.

In general, the deficiencies of hydraulic fills became of concern to some engineers in the 1920's and 30's when special remedial measures were found to be necessary. Because of the importance of some of these dams, much publicity was focused on several slides and construction accidents.

Hydraulic fills are now known to be characteristically vulnerable during an earthquake. Two of these dams suffered serious damage during the 1952 Arvin-Tehachapi shock in California, although both were far from the epicenter. Typical damage consisted of cracks extending longitudinally in the embankment. The same manifestations were observed at the Van Norman reservoir complex in 1971.

Judging by these and other experiences, the most pronounced effect of severe seismic activity at a hydraulic fill is likely to be distortion of the embankment in response to low-frequency vibrations of comparatively long duration. This would be manifested by settlement and lateral spreading. Such effects can be intensified by liquefaction. Susceptibility to this is highest in saturated low-density soils with uniform gradation and fine-grain size. Liquefaction is a potential problem in any embankment, such as a hydraulic fill, which may have continuous layers of such materials.

After the massive slide, 3 820 000 cubic meters (5 000 000 cubic yards), in the upstream portion near the right abutment of Fort Peck Dam in 1938, the hydraulic fill concept came under a cloud of suspicion. Even though investigation of that accident finally focused the blame on an incompetent foundation, the hydraulic fill lost popularity after the period of the 1930's. This was probably as attributable to economics as it was to structural inadequacy. In the 1940's, the advent of heavy compaction equipment brought the rolled embankment to the fore as a competitive alternative. Since then, the hydraulic fill has been given little consideration for new construction in the United States, but has been of continuing interest to engineers responsible for operation and maintenance and remedial programs.

Concrete Deterioration

Aging of concrete dams can be attributed to both physical and chemical factors. The former relate to changes in forces acting on the structure, including those caused by temperature variations. The latter are associated with infiltration into the dam of aggressive waters containing inorganic acids, sulfates, and certain other salts. Chemical reactions of these substances with constituents of concrete can result in leaching of the concrete. Soft water, for example, may attack concrete causing serious deterioration in a few years. Defective or inferior materials used in the construction of a concrete dam can result in deterioration and possible failure of the structure. Poorly bonded cement, weak aggregates, or mineral-laden water can produce low-strength concrete. Highly absorptive aggregates may be susceptible to freeze-thaw damage. Aggregate contaminated by soils, salts, mica, or organic material, also may produce substandard concrete.

Concrete mixes for massive structures usually contain air-entraining agents. This appreciably improves the durability of the concrete and increases resistance to freezing and thawing. However, such distress still can occur where entrainment is insufficient or when the aggregate itself is vulnerable to freeze-thaw action. Closely spaced parallel cracks at edges of concrete blocks may be symptomatic of freeze-thaw expansion. Entrance of water into the cracks and subsequent freezing are likely to further the deterioration.

Disintegration of concrete may be caused by freezing and thawing, thermal expansion and contraction, or wetting and drying. Freeze-thaw effects are most likely to be found in parapets, cantilever beams, slabs, and walls of appurtenant structures.

Many gravity dams constructed in the 19th century were of stone masonry with lime mortar. This is susceptible to deterioration and loss of strength over long periods of exposure to seeping water. Once its bond has been broken, water pressure in the joints may actuate a sliding or overturning failure. Failure of the Bouzey Dam in France in 1895 was attributed to this.

Several concrete dams have suffered alkali-aggregate reaction. Typically, this chemical process is evidenced by upstream movement of an arch crown, by spalling of the concrete at extremities, and by characteristic pattern cracking and crazing of the dam faces.

The strength of the concrete mass may be reduced by alkali-aggregate reaction. Visible clues to the deterioration include: (1) expansion, (2) cracking of random pattern, (3) gelatinous discharge, and (4) chalky surfaces.

Petrographic examination of the concrete cores taken from affected structures has revealed severe fracturing. Core tests have shown a strength regression as high as 25 percent or more. Expansion in the decomposing concrete can be substantial. Total upstream deflection of the arch crown at one 61-meter (200-foot) high dam in California was about 127 millimeters (5 inches) in the first 10 years after completion of construction. Rates of movement usually appear to decrease as the dam increases in age.

Alkali-aggregate reaction sometimes causes the disbonding of blocks at lift surfaces. Loss of strength by disbonding, and the accompanying increase in hydrostatic pressure along the lift surfaces, will reduce resistance to sliding and overturning. Alkali-aggregate reaction can cause expansion of a concrete dam with consequent cracking and deterioration, and possible binding of gates, valves, and metalwork. Once alkali-aggregate reactivity has developed in a relatively thin concrete dam, it cannot be stopped practically by any means now known. Where deterioration has progressed to a dangerously advanced stage, the effective remedies are to remove and replace the defective concrete or to build a new dam to replace the old one.

Settlement and cracking of concrete structures may be attributable to uplift, foundation displacement, ice thrust, or seismic forces. In spillways or outlet works conveying high velocity flows, offsets in the conduit surfaces may cause cavitation.

Vibration of structures by earthquake, water surges, or equipment operation may damage concrete.

Damage due to the overstressing of a concrete dam often may be identified by examination. Clues include cracking, opening at joints or lift surfaces, seepage variations, and displacement.

Erosion of concrete may be caused by flowing ice, rocks, logs, wind,
traffic, or cavitation.

One of the most common problems reported at concrete gravity dams is
clogging of drainage systems. The need for regular maintenance of drains is
well recognized.

Obstruction of dam and foundation drains may be attributable to various
causes, including displacement, soil or rock deposits, biological growth,
and leaching and deposition of chemicals.

Spillways

Overtopping of the dam may result from failure to make timely and
adequate releases through the spillways and outlets.

Overtopping has been the most common cause of failure of embankment
dams. Several failures attributable to overtopping have occurred while the
dams were still under construction. The Orós Dam in Brazil, which was
destroyed by flooding in 1960; and the Sempor Dam in Indonesia, which
failed in 1967, are notable examples.

The value of adequately sized and readily operable spillways has been
convincingly demonstrated. However, the determination of the proper
capacity may be difficult. Voluminous records of precipitation and runoff
on watersheds have been collected since most dams were constructed. Also,
more reliable methods for analysis of hydrologic data have been developed.
Inevitably, some existing spillways have been shown to be unable to pass the
maximum floods that can now be forecast from new data. Where the risk of
dam failure is unacceptable, the total discharge capacity should be increased.

The comparatively limited time span of most meteorological and hydrological data suggests the probability that historical extremes eventually will
be surpassed. Notable disasters stemming from inaccurate forecasting of
flood conveyance requirements include South Fork (Johnstown) in the
United States, Orós in Brazil, and Panshet and Machhu II in India.

In all these events the largest anticipated floodflows were proven to be
unrealistically low. However, even with the benefit of hindsight, the next
forecaster of flows in those watersheds may not be assured of better
immunity from error.

Aside from the very important consideration of discharge capacity,
spillways must be checked also for such common flaws as slides or debris
obstructing channels, erosion and undermining, broken linings, and
inoperable mechanical equipment. Since the life of the dam depends upon
safe functioning of the spillway during emergency, regular examinations
and thorough maintenance are essential. Maintenance of the facilities for
conveying water past a dam is seldom difficult, yet sometimes receives too
little attention. Even carefully designed and expensive equipment and
structures have been known to suffer from neglect, especially where frequent operation is not required.

Malfunctioning gates, valves, or hoisting equipment may result from:
(1) displacement of the structure; (2) corroded, worn, broken, or loose

parts; (3) misalinement of parts; (4) binding due to infrequent operation; (5) insufficient lubrication; (6) improper operating procedures; (7) power outage; (8) electrical circuit failures; (9) icing; and (10) silt or debris. Inadequate maintenance of electrical and mechanical equipment may lead to operational failure at a crucial time.

The design criteria for most effective operation of conveyance works must be strictly followed. Some spillways and outlets require symmetrical operation. In others, waterhammer, equipment vibration, and flow velocities must be carefully controlled. Guidelines for operators must be kept permanently at the dam. In the Bureau of Reclamation, these guidelines are designated as SOP's (Standing Operating Procedures) and DOC's (Designers' Operating Criteria).

Outlets

One of the most prevalent adverse conditions at reservoirs, particularly where small or medium-sized dams are involved, is a poorly constructed outlet or one that has deteriorated through lack of maintenance. And yet, the capability of rapid lowering or emptying of a reservoir during a crisis can be extremely important. In some cases, distress in a dam has been alleviated by reducing the pool elevation just a few feet. On such occasions, a properly functioning outlet works is essential. The value of control at the upstream end of an outlet, to limit conduit water pressures within or under the dam, is generally recognized.

Demolition

Deliberate efforts have been made to destroy dams, including bombing, sabotage, and demolition for public safety. Of course, the number of dams which have failed from other causes is probably much greater than the number destroyed intentionally. The potential for hostile action, though, does warrant some examination. Military strategists can be expected to see the advantages of attacking any conspicuously vulnerable structure that may be of value to the enemy. In past wars, commanders have launched assaults on dams to flood out enemy forces or to cut off routes crossing rivers.

Two gravity dams, the 91-meter (298-foot) high Burguillo near Avila, and the 56-meter (184-foot) high Ordunte near Bilbao, were attacked and damaged in 1937 during the Spanish Civil War. General Franco's forces reportedly set off a 15-ton charge in an inspection gallery at Ordunte; however, the dam, just completed in 1934, was not permanently impaired.

Soviet troops withdrawing under German attack in September 1941 reportedly detonated 90 tons of explosives in a tunnel in the Dnjeprostroj Dam on the Dnjepr (Dnieper) River. The blast disintegrated an upper section of the structure about 200 meters (660 feet) long. The resultant discharge through the breach was reported to have reached a maximum of 35 000 cubic meters (1 240 000 cubic feet) per second.

Two gravity structures in Germany, the 40.3-meter (132-foot) high Möhne Dam and the 48-meter (157-foot) high Eder Dam, were bombed in 1943 by the British Air Force. The deluge released on the downstream valleys caused an estimated loss of about 1,200 human lives.

The 69-meter (226-foot) high Sorpe Dam near Sorpe, Germany, an earth-fill structure, was bombed at the same time as the attack on the Möhne Dam. Two direct hits on its crest produced craters about 12 meters (40 feet) deep, but the dam did not fail. The Sorpe Dam was bombed several more times during the war, and suffered a total of 11 hits, but remained in service. Temporary repairs of the broken concrete core and the cratered embankment were made in 1945 and 1946.

In January 1951, strong surges of water and clay suddenly occurred in the drainlines of the Sorpe Dam. A cement grouting program eliminated about 75 percent of the muddy leakage. However, by 1956 the cumulative loss of material had increased to several thousand cubic meters, causing settlement of as much as 1.4 meters (4.6 feet). Extensive repairs were then started. The extent of wartime damage was revealed during this work. The outlet was found to be broken upstream from the core, allowing water under reservoir pressure to escape into the dam and its foundation. This had caused much erosion of the material surrounding the conduit. The remedial program in 1956 included the thorough grouting of the voids at the concrete core and the outlet tunnel. Also, the bomb craters remaining in the lower slopes of the embankment were filled and the upstream face of the dam was paved. The repair was completed in 1962. The grouting required a total of 53 000 meters (174 000 feet) of drilling, 3950 metric tons (4350 tons) of cement, and 1500 metric tons (1700 tons) of clay.

The 81-meter (266-foot) high Hwa Cheon Dam, a concrete gravity struc-ture on the North Han River just above the 38th Parallel was attacked and extensively damaged by both sides during the Korean War. The structure had been built by the Japanese during World War II; and, due to difficult conditions at that time, the work was substandard. A hurried construction schedule combined with shortages of materials caused design changes at the site. The height of the 18 spillway gates was reduced by 1.5 meters (4.92 feet), and the elevation of the overflow crest was increased the same amount. This modification of the original design to substitute 6000- by 12 000-millimeter (19.67- by 39.33-foot) gates for 7500- by 12 000-milli-meter (24.58- by 39.33-foot) gates resulted in an ogee shape that induced severe erosion, to depths of more than 2 meters (6 feet) in some places.

When repair was undertaken after the Korean conflict, the dam was found in very poor condition, not only from the spillway scouring but also from the blowing out of six gates by the North Koreans and hits by three 907-kilogram (2000-pound) torpedoes launched by the forces from the south. In rehabilitation of the structure, the outside spillway gate on each end was eliminated, the four middle gates were enlarged, and the other gates were rebuilt. The spillway was repaired by placement of new concrete with a minimum thickness of about 1 meter (3 feet). Total cost of the remedial work on the dam was approximately $13 million.

In 1966, sabotage was suspected as a possible cause of the breaching of a dike impounding a sediment basin for a lead and zinc plant near Vratza in Bulgaria. The collapse of the earthfill created a 4.6-meter (15-foot) high floodwave through the towns of Zgorigrad and Vratza. Reports indicated that as many as 600 people perished, but the accepted record shows a toll of 96.

Although the consequences of hostile action against dams have been severe in some cases, the historical frequency of such events has been comparatively low. This is not necessarily reassuring, however. Looking to the future, the increasing potential of damaging attack cannot be disregarded. Both the numbers and sizes of dams have expanded rapidly in the 20th century. The record height doubled approximately in the period between the two world wars and has increased more than half again since World War II. This rapid growth has been experienced in both embankment and concrete dam construction. Even with the significant advances which have been achieved in the technology, these phenomenal statistics warrant serious analysis as parameters in the hazard equation.

Sliding

The possibility of sliding on the reservoir slopes or on the dam abutments or of the dam itself must be taken into account in assessing safety at a water storage site.

The consequences of landslides may include blockage or rupture of essential appurtenances or overtopping of the dam by waves, as occurred at Vaiont Reservoir in Italy in 1963. While the potential for landslides may exist in nearly any kind of rock, some slates and schists are notoriously susceptible to movement. Shales and claystones have also caused problems.

Where such rocks are present in the foundations of concrete dams, special precautions may have to be taken. Many dams have failed where the sliding hazard was ignored or given inadequate attention. The Ohio River Dam No. 26 can be cited as an example. The dam was completed in 1911 and failed on August 8, 1912. A long section of this Chanoine wicket structure about 6 meters (20 feet) high slid on a shale bedding plane immediately under the concrete base. The shale was described by investigators as soft, poorly cemented, and "greasy." Another concrete dam, near Austin, Pennsylvania, suffered a similar fate. It was founded on weak sandstone with shale bedding seams, one of which was the plane of failure. Austin Dam, on the Colorado River in Texas, was built on limestone which had shale seams. The concrete barrier, along with a layer of rock just under its base, slid on one of these seams. Rock under the toe of the dam had been scoured, undermining the structure at that point and thus further lowering the resistance to sliding.

Induced Earthquakes

The filling of a large reservoir behind a high dam may actuate an earthquake. Various factors can be contributory to such movement, including the superimposed water weight, reduction of frictional resistance in the underlying rock due to pore pressures, and decline in rock strength caused by chemical alteration. Infiltration of water into the foundation under high pressure can trigger the release of cumulative tectonic strain. Intensified pore pressures tend to diminish friction by reducing normal stresses on the planes of fracture. Consequent movement will extend until irregularities at the interface again exert sufficient restraint.

Damaging earthquakes of Richter Magnitude greater than 6 have occurred in the regions of Kariba Dam (Zambia), Koyna Dam (India), Kremasta Dam (Greece), and Hsinfengkiang Dam (China).

Kariba Lake is located on the Rhodesia-Zambia border in a vicinity which had been reported to be quiescent seismically prior to the impoundment in 1958. The first earthquakes were recorded near the site in 1961. Thereafter, seismic activity intensified, culminating in a series of strong shocks between August 14 and November 8, 1963. In this period, beginning just after full reservoir level had been reached, nine earthquakes ranging between 5.1 and 6.1 in magnitude were recorded, with all epicenters near the dam. Thereafter, seismic activity declined.

The experience at Kariba suggests that frequency and magnitude of tremors may increase during and following first impoundment but then may tend to diminish.

Koyna Dam (India) is a rubble-concrete gravity structure 103 meters (338 feet) high and 853 meters (2800 feet) long, located south of Poona on the Indian Precambrian Shield, which had been regarded as a region of relatively low seismicity. A search of historical records, however, has revealed that about 20 earthquakes of moderate intensity occurred in the period 1594-1967 on the western edge of the Indian peninsula which includes the Bombay-Poona-Koyna area. Soon after the filling of the reservoir, tremors were felt near the damsite. These were the beginning of a succession of thousands of shocks within a radius of 25 kilometers (15.5 miles) of the dam.

Construction at the site was started in 1954 and completed in 1963; impoundment was begun in 1962. First tremors were recorded in 1963. The reservoir reached capacity in 1965. The first two significant tremors, with magnitudes of 5.0 and 5.5, were recorded on September 13, 1967 causing mild local damage. Soon thereafter, on December 11, 1967 (December 10, G.m.t.) a quake of magnitude 6.5 occurred, with epicenter in the vicinity of the dam. The peak acceleration recorded at the damsite was 0.63g. The toll from the shock was about 180 dead and 2,200 injured, mostly in the village of Koyna Nagar, where most of the buildings were damaged or destroyed. Koyna Dam suffered extensive horizontal cracking but did not fail. Seepage through the structure increased appreciably. Repairs and strengthening were required.

Additional quakes were felt in the following several months. Records of the Hyderabad seismographic station, about 490 kilometers (304 miles) east of Koyna Dam, show that the frequency of damaging earthquakes tended generally to decline after 1967. However, the region did not return to its preproject quiescence. In the period 1968-73, about 30 shocks of magnitude 4 or greater were recorded. Most of the epicenters were either under the lake or in the general vicinity of the dam.

The Koyna Reservoir site is geologically similar to the vicinity of Grand Coulee Dam in the United States, as both are on basaltic plateaus. No earthquakes have been generated by the reservoir at Grand Coulee. Also, several other reservoirs not far from Koyna have been quiescent, even though the geologic environment is similar.

At Kremasta Reservoir (Greece), water storage began in July 1965. Tremors first occurred about a month later. Both the reservoir filling and the seismicity increased rapidly between November 1965 and January 1966. In February 1966, as storage was approaching the maximum, a shock of 6.3 magnitude occurred. Destruction of villages was reported, but the dam was not damaged. This earthquake was succeeded by a series of lesser quakes later in 1966, at least six of which had magnitudes exceeding 5.0.

Kremasta Reservoir is in an earthquake zone, although the vicinity of the reservoir itself had been comparatively inactive before construction of the project. The tremors which occurred in the 14 years immediately preceding reservoir operation were focused about 40 kilometers (25 miles) from the damsite, while subsequent seismic disturbances have been at or near the lake.

The Greek Ministry of Industry arranged for an investigation through an International Committee on Kremasta. The report of the Committee, published by the International Commission on Large Dams in 1974, pointed out that although a significant fault crossed the Kremasta Reservoir, geological observations and leveling measurements did not indicate any change at the reservoir that could be related to the earthquakes.

Earthquakes were stimulated by filling of the 11.5-billion-cubic-meter (9 300 000-acre-foot) reservoir at the Hsinfengkiang Dam about 160 kilometers (99 miles) northeast of Canton, China. Impoundment behind the 105-meter (344-foot) high concrete buttress dam was begun in 1959 and completed in the fall of 1961. Although there was no prior record of destructive earthquakes in the area, filling of the reservoir was followed by a series of earthquakes. By 1972, over 250,000 earthquakes of Richter Magnitudes greater than 0.2 had occurred. In the 6 months immediately after full reservoir was attained in 1961, tremors increased both in magnitude and frequency. On March 19, 1962, an earthquake of magnitude 6.1 caused cracking over a length of 82 meters (269 feet) in the upper part of the 440-meter (1443-foot) long dam. In a subsequent remedial program, the structure was strengthened and equipped with strong-motion instruments at various levels. These recorded several aftershocks.

One of the important conclusions reached as a result of the Hsinfengkiang experience was that the high dynamic magnification at the crest of the dam resulting from the vibration mode at the site required special design emphasis. In such circumstances, the top of a dam may be the most vulnerable part during an earthquake.

Part IV

SIGNIFICANT ACCIDENTS
AND FAILURES

Alla Sella Zerbino Dam

On August 13, 1935, a concrete gravity dam, near Genoa in northern Italy, collapsed under the pressure of floodwater produced by intense rainstorms. Destruction in the vicinity of Ovada was extensive. The toll of human lives exceeded 100.

The reservoir, constructed in 1923, is about 32 kilometers (20 miles) northwest of the city of Genoa, on the Orba River, a tributary of the Tanaro River in the Po River watershed. Two dams, the 42-meter (138-foot) high Zerbino and the 12-meter (39-foot) high Alla Sella Zerbino, formed the Ortiglieto Reservoir. The smaller of the two, a saddle dam, was the one that failed.

The larger structure, on the river, is of a curved gravity design and was constructed using cyclopean concrete. It has a crest thickness of 6.00 meters (19.7 feet). The slope of the upstream face is 0.05 to 1, and the downstream face is 0.80 to 1 near the top, with varying slopes averaging about the same for the remainder of the face down to the toe. The radius of curvature in plan is 200 meters (656 feet).

The smaller dam, with a straight axis, had a crest thickness of 3.30 meters (10.8 feet). The faces were sloped 0.05 to 1 upstream and 0.55 to 1 downstream.

Discharge capacity at the reservoir was provided by: (1) 12 automatic siphons 2 by 3 meters (6.56 by 9.84 feet), built into the crest of the main dam, capable of discharging 500 cubic meters (17 700 cubic feet) per second; (2) a bottom gate intended to pass 55 cubic meters (1940 cubic feet) per second; (3) a valve with a capacity of 150 cubic meters (5300 cubic feet) per second; (4) a spillway adjoining the main dam, with a crest length of 68 meters (223 feet), capable of discharging 150 cubic meters per second; and (5) powerplant turbines with a design discharge capacity of about 25 cubic meters (880 cubic feet) per second. Total discharge capacity was therefore estimated as 880 cubic meters (31 000 cubic feet) per second. The siphon spillways were originally thought to be capable of handling the expected floods. But after the failure of the Gleno Dam in that same year (1923), a valve was installed at the Zerbino Dam as a precaution.

A pressure tunnel, 3 kilometers (1.9 miles) long, leads from the reservoir to the penstock of a powerplant connected to the system which serves Genoa. The tunnel intake is near the Alla Sella Zerbino Dam.

At the damsite after the failure, little remained of the broken structure or its foundation. Extensive erosion of the rock testified to the low quality of the geologic formation. The penstock below the tunnel was not damaged, but the powerplant was torn away down to the machinery. Bridges were destroyed as far as 14 kilometers (9 miles) downstream. The concrete mass was gone along with an appreciable part of its foundation. Except for the walled approach road embankment, little structural evidence remained at the site of the smaller dam to guide investigators.

The drawings of Alla Sella Zerbino Dam show conventional details such as vertical joints, an inspection gallery, and a drainage system extending both vertically and horizontally. They suggest that reasonably careful consideration was given to provisions for structural stability, including a cutoff into rock at the heel, a line of grout holes, foundation shaping to enhance sliding resistance, and a concrete toe block at the maximum section. The dam was reported to have been constructed in a manner, and of materials, similar to those used at the main dam. The slope of the downstream face of the smaller gravity structure was obviously steeper than traditional, but the total mass including the toe block may have been regarded as ample for such a low dam.

The characteristics of the geological formation were probably a primary factor. The rock at the sites of the two dams for the Ortiglieto Reservoir was reported to be a serpentine schist. In the Apennines of Liguria, schistose formations are common and have served as foundations for several other dams in that region. However, the local rock at the Alla Sella Zerbino site was reportedly not the best for concrete aggregate nor for the foundation of an important structure. Relating this to the failure, the lack of resistance to erosion is clearly suspect. There appears to be little doubt that the toe of the small barrier was undermined by the heavy overflows, which precipitated sliding and overturning.

Accounts of the failure indicated that erosion had occurred also at the larger dam as result of its overtopping. However, the design of that structure had provided toe protection to withstand the effects of discharges through the siphon spillways. Also, some reports suggested that the foundation at the main dam was as sound as could be found in the area.

The small dam closed a gap in the reservoir rim on a narrow ridge. This alone necessitated preparation of the foundation so that the gravity monoliths would be solidly based to resist lateral forces. The designers apparently respected this requirement and even undertook extensive grouting of the rock. However, the extreme flood surcharge and the lack of safeguards against erosion led quickly to collapse. The spilling and the resulting high water pressures alone would not have caused the dam to break, even though it would have been severely tested by such conditions. But considering the erodibility of the formation and the absence of works for effective energy dissipation, the instability during the 1935 flood is explained.

After the failure, analysis was made of the runoff from the drainage area of about 141 square kilometers (54 square miles). Flood frequency calculations prepared by the designers had shown that an inflow of 800 cubic meters (28 200 cubic feet) per second was to be expected once in 20 years on the average. The consequences of such occurrences were apparently accepted as controllable by maintenance.

The normal water level at elevation 322 meters (1056.4 feet) provided a storage volume of 18 000 000 cubic meters (14 600 acre-feet). But the reservoir impoundment immediately before the disaster was substantially greater as the flood overtopped the crest of the dam at elevation 325 meters (1066.3 feet). Assuming the mean storage level during the overflow at elevation 327 meters (1072.8 feet), there would be an additional discharge of about 1350 cubic meters (47 600 cubic feet) per second for the two dams, assuming a crest length of 160 meters (525 feet) for the large dam and 80 meters (262.5 feet) for the small dam.

With storage to elevation 327 meters (1072.8 feet) compared with the normal level of 322 meters (1056.4 feet), there was a surcharge of about 5 meters (16 feet) on the Zerbino Dam, which in relation to its height of 42 meters (138 feet) was not significantly excessive. However, at the small dam, which was only 12 meters (39 feet) high, this overload amounted to nearly 50 percent.

Alla Sella Zerbino Dam held back almost half the total water stored even at normal water level, i.e., 8 000 000 cubic meters (6500 acre-feet) out of 18 000 000 cubic meters (14 600 acre-feet). Assuming a 5-meter (16.4-foot) rise above normal water level at time of failure, the total storage would have been increased to about 24 000 000 cubic meters (19 400 acre-feet), of which the smaller dam upon failure would release 14 000 000 cubic meters (11 300 acre-feet). Erosion of the foundation at the site released additional amounts of water. The hazard represented by the small dam was therefore much greater than its dimensions would suggest.

The calculations are only approximations, since there were uncertainties about the conditions at the reservoir during the flood. Reportedly, the siphons may have been partially blocked and the bottom gate did not operate well. Some difficulty was experienced in hoisting the gate. It may have been only partially opened. This is not an uncommon disadvantage of such gates in times of emergency, especially where the facility is infrequently operated and perhaps poorly maintained. Under such circumstances, the gate may tend to bind because of rust or silt.

Austin Dam

A concrete gravity dam at Freeman's Run near Austin, Penn., failed suddenly on September 30, 1911, with the loss of 80 human lives. The dam was 15 meters (50 feet) high, 9 meters (30 feet) thick at the base, 0.76 meters (2.5 feet) thick at the top, and 166 meters (544 feet) long, with a 1.2- by 1.2-meter (4- by 4-foot) cutoff wall excavated into rock. The structure, completed in 1909, was of cyclopean concrete buttressed by a rolled earthfill. The foundation consisted of interbedded shale and sandstone.

The initial trouble occurred during the dam's first year of operation. In January 1910, when the reservoir had reached full capacity for the first time, the dam began to give way. Disaster was reportedly averted by blasting holes in the structure. Evidently, during the initial introduction of water into the reservoir, the dam was loaded before the concrete had set sufficiently. This caused the opening of cracks and the development of excessive pressures under the dam. As a consequence, in the 1910 accident the dam dropped about 150 millimeters (6 inches) at the toe and slid out about 450 millimeters (18 inches) at the spillway.

The paper company which owned the dam allegedly did not strengthen the dam, and allowed the reservoir to fill again. Presumably, the holes were plugged, but adequate repairs were not made; and the structure remained in this hazardous condition until its sliding collapse on September 30, 1911.

Failure of the dam evidently was due to weakness of the foundation, or of the bond between the foundation and the concrete. A witness to the collapse stated that part of the west end of the dam first gave way near the base. He saw it begin to move and had time to enter his house, telephone an alarm to Austin, and go back outside in time to see the dam break.

His report was confirmed by the condition of the western extremity of the structure and the fragments just downstream from it. Approximately 23 meters (75 feet) of the west end of the dam survived, with a regular, nearly vertical break about 11 meters (35 feet) high. The adjoining section had fallen on its downstream face about 30 meters (100 feet) downstream, with its top toward the west. The smooth condition of its base, with a number of twisted 32 millimeter (1¼-inch) steel rods protruding and some signs of laitance, showed that there had been a horizontal joint in that plane.

The eastern end of the dam failed in about the same way and probably at nearly the same time. A center section, approximately 60 meters (200 feet) in length, was shifted from 1.5 to 4.6 meters (5 to 15 feet) downstream as a single block, the eastern end having moved farther. This monolith evidently slid on its base without rotating about a horizontal axis.

The foundation rock at the site is a sandstone, with essentially horizontal beds from 0.3 to 1.0 meter (1 to 3 feet) thick interlayered with shale and earth material. The concrete in the dam contained large pieces of the sandstone, and postfailure examination found those which were exposed in the remains to be generally sheared — an effective demonstration of the relative weakness of the aggregate.

The dam had sloping faces down to the natural ground. Below this point the faces were vertical. On the downstream side this face averaged about 1.2 meters (4 feet) high. Apparently the bond at the foundation contact was assumed to be strong enough to resist sliding, even with such a comparatively narrow dam base. A more prudent designer would have continued the slopes of the structure all the way to the foundation and might have excavated even further to ensure enough structural mass, base area, and anchorage to preclude movement.

Babii Yar Dam

News dispatches from Moscow, U.S.S.R., on March 25, 1961, confirmed reports of a recent disaster in the Ukrainian capital city of Kiev. Floodwaters pouring through a gorge on the edge of the city reportedly swept away a number of apartment houses on the bank. Numerous casualties were mentioned in the news accounts, but there was no indication of how many of these were fatalities.

Subsequent information confirmed that in March 1961, the rolled earth dam failed in the Babii Yar Gorge of the Dneiper River in the Ukraine. The failure was attributed to overtopping by waves driven by a windstorm. In Kiev, extensive damage was done as the disgorging water from the broken reservoir inundated the lower levels of the city. The data now available show that about 145 people died in the disaster.

Baldwin Hills Dam

The failure of the 12-year-old Baldwin Hills Reservoir (fig. 4-1) on a sandy hilltop in Los Angeles, Calif., on Saturday, December 14, 1963, focused attention on the subtle changes which can occur at dams to threaten their safety. This reservoir failed suddenly following displacement in its foundation. Although the activity which precipitated the collapse was not attributed to earthquake, movement was concentrated at faults which were planes of foundation weakness. As a consequence, the reservoir's lining and underdrains were ruptured; and water under pressure entered the pervious and highly erodible soft sandstones and siltstones in the foundation. Destruction was rapid once the uncontrolled leakage began.

The main dam had a height of 71 meters (232 feet) and a crest length of 198 meters (650 feet). The reservoir consisted of compacted earth dikes on three sides, the fourth or north side being closed by the dam.

At about 11:15 a.m. on December 14, 1963, an unusual sound of running water was detected in the spillway discharge pipe at the reservoir. This was the first of a series of observations at the scene of a catastrophe in the making. The caretaker then noted that water was running freely from the drains under the asphalt-paved bottom of the reservoir. He summoned the operating system engineer. Discharge lines were then opened to lower the reservoir level. The time was about 12:20 p.m. Nearly 24 hours would have been required to empty the reservoir.

As the draining was started, police were asked to evacuate the area below the dam. Motorcycle and patrol-car officers were dispatched throughout the danger zone to sound sirens, to call at doors to warn occupants to leave their homes, and to close streets to traffic.

At about 1:00 p.m., muddy water was discovered emerging downstream from the east abutment of the dam. Men repeatedly risked their lives while the leakage was ominously worsening. They worked below the dam clearing debris from the inlets to the storm drain system; they entered and examined the inspection chamber under the reservoir; and they hung by ropes on the upstream face of the dam looking for a way to control the outflow.

Transmission of signal alerts via radio and television began at 2:20 p.m. Helicopters equipped with loudspeakers flew over the threatened neighborhood warning residents. By 3:20 p.m., about 1,600 people had left the area.

At about 2:20 p.m., the receding of the water level revealed a 0.9-meter (3-foot) wide rupture in the lining of the reservoir opposite the downstream break in the dam. A futile effort was made to plug the hole with sandbags, but the bags disappeared without noticeable effect. The men were ordered to safety as the crack widened below them.

At 3:38 p.m., a huge gush of water blew mud and debris through the lower face of the dam, and poured down the steep ravine leading to a residential street 275 meters (900 feet) away. A witness described the collapse: "(The) sloping earth wall was leaking muddy water from a top to bottom crack. It gushed in a 10-foot-wide brown stream from a fissure at the bottom ***. Next, there was a roar like a great cannon, then a rumbling and shuddering of the ground as the face of the slope erupted. A mighty jet of water and mud and fragments of earth shot out ***. The explosion point

Figure 4-1.—Baldwin Hills Reservoir after failure (Courtesy, Calif. Dept. of Water Resources). P-801-D-79346.

was at the bottom, opening a huge hole into which upper wall portions fell.''

By 4:55 p.m., the reservoir was empty and five persons had drowned, several of them trapped in automobiles. The flood destroyed 41 homes and damaged 986 others. Its waters poured into nearly 100 apartment buildings; gouged out streets; and tore away water pipes, sewers, and gaslines. Houses immediately below the dam disintegrated and disappeared as the full impact of the raging water hit them.

When the reservoir was empty, a crack could be seen in the asphaltic lining extending all the way across the bottom in line with the cut through the dam. A fault runs directly under the reservoir floor and the dam along this alinement. It is one of three faults discovered during construction of the facility.

The Baldwin Hills Reservoir was located on the highest hills in the southwest part of Los Angeles, a logical location to serve the south and southwest sections of the city. At the time of site selection, the hills were relatively undeveloped, except for an oil field lying generally south and west of the proposed reservoir site.

Painstaking care had been taken by the Los Angeles Department of Water and Power in its design and construction of this hilltop reservoir. From the time it was placed into service in 1951, it had been kept under exceptionally close surveillance.

The State Engineering Board of Inquiry which investigated the failure concluded that earth movement occurred at the reservoir on December 14, 1963, following an apparent long-term development of stress and displacement in the foundation.

Evidence indicated that the movement which triggered the reservoir failure could be associated with land subsidence which had continued in the vicinity for many years. This relatively fast settlement over a limited area was superimposed on the very slow tectonic deformations along the nearby Newport-Inglewood fault system.

The Inglewood fault strikes prominently through the Baldwin Hills, roughly parallel to and about 150 meters (500 feet) from the west rim of the reservoir. It would appear that significant fault movements had not occurred there during the past 10,000 years.

The sedimentary formations that comprise the foundation for the Baldwin Hills Reservoir consist of sands, silts, and gravel which are in part loose and in part moderately consolidated.

Several minor faults were mapped in the reservoir vicinity during its construction. Two of these passed through Baldwin Hills Reservoir. As much as 180 millimeters (7 inches) of vertical displacement occurred at the trace of one fault on the east side of the reservoir during the failure. The breach in the reservoir developed along the plane of this fault.

The records of the Seismological Laboratory at the California Institute of Technology, located 24 kilometers (15 miles) from the reservoir, disclose that, in the months immediately preceding and during the failure, the reservoir was not subject to any shocks believed to be of sufficient strength to cause damage.

The embankment design concept required an impervious reservoir lining and an underdrainage system to ensure that the phreatic line would not rise into the embankments and impair their safety. The uppermost member of the lining was a 75-millimeter (3-inch) thickness of porous asphaltic paving. Immediately below the paving was the main impervious member, a compacted earth lining, which was designed to be 3 meters (10 feet) thick in the bottom of the reservoir, decreasing to 1.5 meters (5 feet) in thickness normal to the slope at the top. Below the compacted earth lining was a 100-millimeter (4-inch) cemented pea-gravel drain designed to act as a collector for all seepage finding its way through the compacted earth lining. A system of 100-millimeter clay tile pipes was to convey this water into the drainage inspection chamber. The pea-gravel drain was capped with a 6-millimeter (1/4-inch) porous sand-gunite layer to prevent infiltration of compacted earth. Immediately below the pea gravel was an asphaltic membrane about 6 millimeters thick, applied directly to the foundation subgrade. The function of the membrane was to prevent all water from reaching the foundation. The security of the reservoir was dependent upon the impermeability of the asphaltic membrane.

The reservoir was kept under strict surveillance by means of a complex system of safeguards. The caretaker made daily inspections of the seepage from drain networks underlying the entire reservoir and its embankments. Monthly surveys were conducted to detect movements in the reservoir and in the surrounding area. A squad of maintenance specialists inspected the reservoir once each month, on the alert for factors related to the safety of the facility. Instrumentation at the site, such as strain gages, seismoscopes, and tiltmeters, was carefully planned and closely watched.

Beginning in the spring of 1963, there was a slight but detectable and consistent uptrend in the measured reservoir seepage. All the discharging horizontal drains under the main dam began to experience rapid variation whereby the drainage reduced to zero and in some cases then increased to its former amount, followed by continued fluctuations.

During the first year of operation, a crack was discovered in the drainage inspection chamber under the reservoir near the location of a known fault in the foundation. This crack, and others in the structure, continued to develop over the years of operation. Various other evidences of movement were observed at the reservoir from time to time during the 12-year operation period.

Over the period of several decades, starting long before the reservoir was built, a shifting and subsiding of survey stations had been noted in the Baldwin Hills. Since 1957, conspicuous earth cracks developed, particularly in an area southeast of the reservoir.

Periodic surveys delimited a substantial area of subsidence generally elliptical in shape and including Baldwin Hills Reservoir. With these survey data it was estimated that a maximum settlement of about 2.9 meters (9.7 feet) had occurred over about a 40-year period at a point approximately 0.8 kilometer (1/2 mile) west of the reservoir. During this period, subsidence at the reservoir aggregated about 0.9 meter (3 feet), the southwest corner dropping more than the northeast corner.

Triangulation surveys revealed stations in Baldwin Hills to be moving laterally in the general direction of the axis of the subsidence bowl. Furthermore, measurements of the reservoir dimensions indicated a progressive elongation of the northeast-southwest diagonal between 1950 and 1963 of about 122-millimeter (0.4 foot).

The Inglewood Oil Field lies under the Baldwin Hills, with its long axis approximately parallel with the Inglewood fault, which cuts through the oil field. The greatest subsidence in the area had occurred over the most productive oil zones. Occasional earth rebound had been measured since 1957 when substantial repressuring of the field was started by saltwater injection.

The foundation of the reservoir had been subjected to progressive horizontal stretching, concentrated at the steep fault planes in the soft rock. The loosely articulated fault blocks were very sensitive to the concentrated pressures caused by the injection program. The foundation blocks under the reservoir literally tended to pull apart and drop down in a staircase descending toward the center of the subsidence bowl. This movement was intermittently accelerated by rebound triggered by repressurization. The gaps opened between the fault blocks became ready conduits for leakage once the integrity of the protective lining and drain system had been destroyed.

Bila Desna Dam

In September 1916, the 17-meter (56-foot) high Bila Desna Dam, a 1-year-old earthfill structure near Jablonec nad Nisou in Czechoslovakia, failed under the pressure of floodwaters. The resulting deluge destroyed the village of Desna and claimed 65 victims.

Following completion of construction in 1915, reservoir impoundment proceeded during the spring of 1916. When the water depth had been increased to 11.5 meters (38 feet), leakage of about 3.8 liters per second (60 gallons) per minute was observed. At that time, fissures were discovered at the outlet conduit, indicative of inadequate compaction of the backfill around that structure. This reportedly was especially evident at the outlet tower.

In the month of September 1916, runoff from heavy rainstorms rapidly raised the level of reservoir storage to full capacity. Leakage increased substantially, and a muddy discharge was observed near the downstream end of the outlet works. Recognizing the threat to the dam, the operators began to open the control gates in an attempt to evacuate the reservoir as rapidly as possible. Their efforts were too late; the dam broke suddenly.

People living downstream from the reservoir did not have advance warning of the failure. Some were able to escape as the river rose around them. The flood wave from the broken embankment swept through the densely forested valley, downing trees and washing out brush. This debris accumulated in the gorge downstream to form a new barrier which partially retained the waters for a short time. When this obstruction collapsed, a second wave rushed down the valley, causing devastation as severe as that of the first surge.

The Bila Desna embankment had slopes of 1.5 to 1 on the upstream and downstream faces. Its width was 4 meters (13 feet) at the crest and about 55 meters (180 feet) at the base. The construction materials were primarily decomposed granite and sandy clay. An inclined core was reportedly incorporated into the fill. A "wooden trench constructed as far as rock" was provided for passing water in case of flood.

The concrete outlet works, with gate tower built into the embankment, were the focal point of the dam's troubles. During its short life, the fill was subjected to severe distortion at the outlet. The tower, founded on a timber grillwork, undoubtedly complicated the construction of the dam. Backfill at the tower and the conduit was not uniformly compacted. At the intersection of the embankment core and the outlet, differential movement disrupted the already vulnerable seal against seepage. A cavern was created over the conduit as leakage developed. Progressive deterioration in this zone eventually led to failure. There were reports of difficulty in opening the gates during the emergency.

Bouzey Dam

The Bouzey Dam (fig. 4-2) was located on L'Avière, a tributary of the Moselle River near Epinal in Vosges Province in France. It failed on April 27, 1895. The designers of the dam evidently disregarded the "middle-third" principle, which states that on any horizontal plane within a gravity dam, the resultant of forces should act within the middle third. This was to ensure that the stress would be compressive at all points on that plane. The rule assumed a linear variation of stress, which is now recognized as only a crude approximation.

The Bouzey Dam was a straight masonry gravity structure with a length of 528 meters (1732 feet). Apparently the original design would have provided a height of about 20 meters (66 feet) with a crest thickness of about 4 meters (13 feet). This thickness was to be carried down to a level about 4.3 meters (14 feet) below the top of the dam where the downstream face was to begin a concave circular curve extending down to the toe. The resulting base thickness was approximately 11.3 meters (37 feet).

Construction was started in 1878. Subsequently, a decision was carried out to increase the height about 2 meters (6.5 feet) without altering the other dimensions of the dam. Late in 1881, the work was complete and the impoundment was started. Leakage soon appeared, totaling approximately 57 liters per second (900 gallons per minute). About a year later, two cracks were discovered in the dam. As a safeguard, the allowable operational levels of the reservoir were lowered. However, on March 14, 1884, with the reservoir level about 2.7 meters (9 feet) below the maximum, a 137-meter (450-foot) long section of the mass slipped abruptly on its base and moved downstream as much as 380 millimeters (15 inches). This was accompanied by a rapid increase in leakage to about 108 liters per second (1700 gallons per minute). Despite this conspicuous hazard, no immediate remedial measures were taken. In fact, the operational regime was unchanged, and the reservoir level rose to exceed the preaccident elevation by a few inches.

In the fall of 1885, about 1½ years after the mishap, the reservoir was drained. Inspectors then learned that there were many cracks in the upstream face, with one break extending 91 meters (300 feet) horizontally. Evidently, cracking had also severed the dam from its cutoff wall, which extended into the foundation at the heel. This crack at the base was then covered by a longitudinal block of masonry, which in turn was sealed with puddled clay. The downstream toe was extended outward and downward by addition of a masonry mass which roughly doubled the dam's base thickness. It gave a flatter slope to the lower downstream face, keying into the original mass at about midheight. The cracks in the old masonry were

4.0 m
(13.12 ft)

HEIGHT* —22 METERS
(72 FEET)
BASE THICKNESS* —11.3 METERS
(37 FEET)

NOTE: SHADED AREA INDICATES
CONCRETE ADDED AFTER FAILURE.

*ORIGINAL DAM

BACKFILL

CLAY

MASONRY

BACKFILL

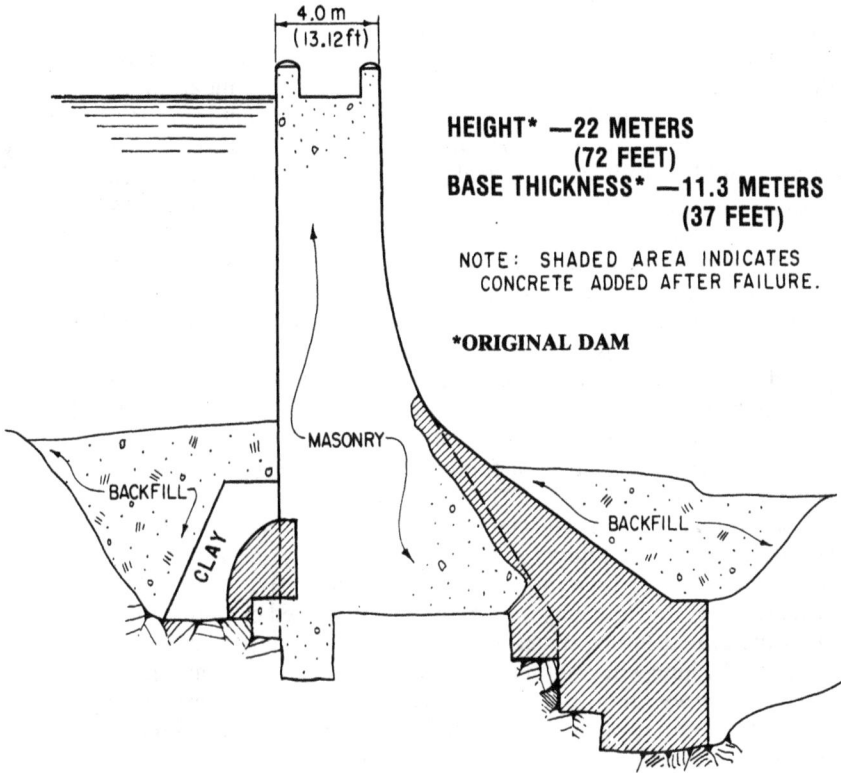

Figure 4-2.—Bouzey Dam, cross section. P-801-D-79347.

grouted, and drains were installed. The remedial program was finished in September 1889.

In November 1889, reservoir impoundment was begun. When the water level reached maximum, the crest deflection was nearly 25 millimeters (1 inch) at some points. This evidently did not cause any concern, and the reservoir remained in operation.

On April 27, 1895, the entire top part of the dam — a mass roughly 10 meters (33 feet) high and 183 meters (600 feet) long — broke away and loosed a torrent of water upon the village of Bouzey. Continuing its rush down the valley of L'Avière, the flood left several other villages in ruins and caused the death of more than 100 people.

Among factors related to the failure, water pressures in and under the dam were undoubtedly contributory. The sandstone foundation was cracked, and the cutoff at the heel was not extended adequately into the rock. Both of these conditions could allow water under the structure. Faulty masonry joints could permit similar leakage at higher elevations. Detrimental hydrostatic pressures, therefore, could have developed easily. Such pressures, if applied in the upper parts of the dam where structural thickness was marginal, may have been a primary source of trouble.

Yet the weakening of the dam appeared to be progressive. Failure did not occur as soon as the reservoir was full, but instead, many months passed before the signs of distress were discovered. After the repairs had been completed in 1889, an even longer period ensued before the collapse, even though the storage levels were kept high. Investigators of the disaster wondered whether other, more subtle, forces of destruction were at work.

In the search for a cause of failure, the mortar used to bond the masonry was an outstanding suspect. There is apparently no question that it was inferior. Instead of using the cement and clean sand required by the contract specifications, the builders mixed lime with dirty sand of poor quality. If the preparation of lime mortar was done carelessly, as was alleged in this case, some of the lime may have remained unslaked or only partially slaked. If incorporated into a structure in this unstable state and then exposed to wetting, the lime will complete its slaking and tend to expand in the process. Used in Bouzey Dam, this could have resulted in weak mortar joints constituting potential planes of separation. Whether the cracking in the dam followed such planes is not clearly evident.

Various possible causative factors may have acted in unison. The comparatively thin masonry section would have been susceptible to tilting away from the thrust of the reservoir water, accompanied by opening of cracks at the base and in the masonry and by gradual deterioration of mortar. Even acknowledging that the design dimensions were marginal, including the last-minute increase in height, one unavoidable conclusion is that defective materials and poor workmanship deserve much of the blame for the Bouzey Dam disaster.

Bradfield Dam
(Dale Dike)

Disaster struck the Bradfield Dam (fig. 4-3) — or Dale Dike — near Sheffield, England, on the evening of March 11, 1864. The breaching of this

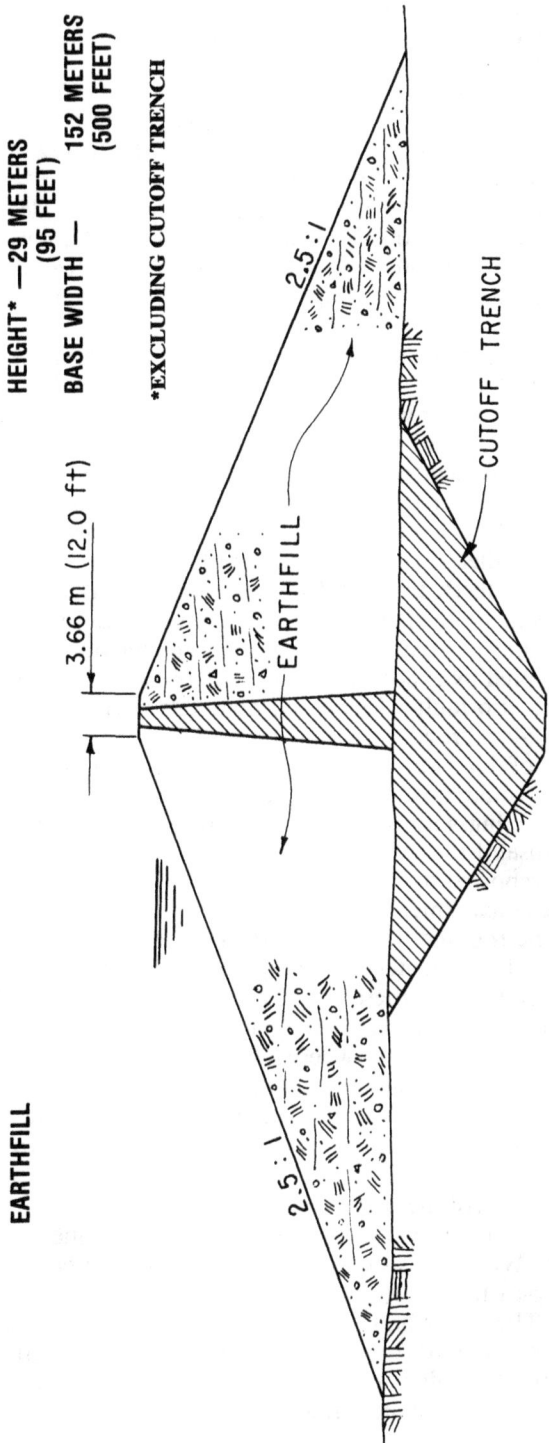

Figure 4-3.—Bradfield Dam, cross section. P-801-D-79348.

embankment loosed a flood which swept into the valley with such velocity that any effective warning of the populace was impossible. As a consequence, 238 people died.

The Bradfield Dam was built to supply water to Sheffield. It was a 29-meter (95-foot) high, 382-meter (1254-foot) long earthfill structure. The crest was 3.66 meters (12 feet) wide, and the base width was 152 meters (500 feet). Outer embankment slopes were 2.5 to 1, both upstream and downstream. A rough masonry lining was placed on the upstream side. The total volume of fill was nearly 306 000 cubic meters (400 000 cubic yards). There was a spillway 7.3 meters (24 feet) wide and 3.4 meters (11 feet) deep. Maximum reservoir impoundment was approximately 3 200 000 cubic meters (2600 acre-feet).

The construction contract was awarded in 1858. However, exploratory holes at the first damsite selected indicated a disturbed foundation. Therefore, a location farther upstream was approved, and project construction was initiated at the new site in January 1859.

Most of the embankment contained a mixture of shale and other rock excavated from the reservoir floor. This was dumped loosely and insufficiently compacted. The specifications required these zones of the fill, on either side of the puddled core, to be built up of successive layers not more than 0.9 meter (3 feet) thick before compaction. The actual layer thicknesses averaged more than twice that specified. Even the specified layer thickness would not have allowed proper compaction, and this fact was generally accepted by experienced engineers of that time. The compaction methods were also faulty. Evidently, materials were transported onto the embankment by railroad cars, and the only compactive effort came from the impact of load on the rails. A contemporary and more conventional practice would have employed carts, traversing the fill in random pattern and compacting with their wheels. Bradfield Dam did not receive this benefit. Because of the inferior placement and compaction, the dam's outer zones were reported to be quite porous.

The core of the embankment was composed of puddled clay, 1.2 meters (4 feet) thick at the top and 4.9 meters (16 feet) thick at the base. Initial planning envisioned a 3-meter (10-foot) deep core trench, but upon exposure of the foundation, there was a clear need to excavate deeper to reach the impervious rock strata. Trenching was then extended to a maximum depth of 18 meters (60 feet), where a satisfactory cutoff was achieved.

During the excavation and puddling of the trench, there was troublesome inflow of water from the sides. This was kept under control by two steam pumps until the puddled core had filled the excavation enough to block the seepage.

Two 450-millimeter (18-inch) cast-iron outlet pipes were laid diagonally under the embankment in a trench which was about 3 meters (10 feet) square in section. Within this excavation, the pipes were embedded 0.76 meter (2.5 feet) apart in puddled clay to levels 0.46 meter (1.5 feet) above and below them. The upper half of the trench was backfilled with gravel, which presumably was interrupted by puddle at the intersection of the trench and the core. Beginning about 30 meters (100 feet) on each side of the core, the outlet trench was sloped down to the bottom of the core trench.

This was intended to distribute deflection over several pipe lengths in case of settlement of the puddled core.

The outlet pipes were laid in lengths of 2.7 meters (9 feet), with spigot-and-socket joints filled with lead. The valves were installed at the downstream ends of the pipes, thus subjecting the whole outlet to full reservoir pressure. To counter the resulting internal forces tending to separate the joints, the sockets were made with conical shapes so that the lead would act as an anchoring wedge. The pipes were not provided with the conventional exterior cutoff collars to inhibit seepage.

Reservoir impoundment was commenced in June 1863 and the reservoir level increased gradually. On March 11, 1864, just before the failure, storage was essentially at the normal maximum. Since the end of February the weather had been wet and windy, and large areas of the lowlands near the North Sea coast northeast of Sheffield had been flooded. On the afternoon of March 11, the resident engineer inspected the dam to see if there had been any storm damage. Although wave spray dissuaded him from venturing too far onto the crest, he could not discern any distress in the embankment. Feeling reassured, he went home. But, within a short time he was notified that a crack had been found in the dam. He returned with a lantern and inspected the crack, which extended longitudinally at a slope distance of about 3 meters (10 feet) down from the top. It appeared to be open approximately 13 millimeters (1/2 inch), but its length was uncertain. Although the engineer judged it to be shallow, he still directed that the water level be reduced to lower than the elevation of the crack by blasting out an emergency spillway. The dam broke before this could be done.

Very few people could be warned of the sudden discharge. The reservoir was emptied within about 45 minutes. Estimates of the flow rate were as high as 1130 cubic meters (40 000 cubic feet) per second. The flood wave, moving with a velocity of about 29 kilometers (18 miles) per hour, swept toward Sheffield, 11.2 kilometers (7 miles) away. It wiped clean the narrow valley below the dam and then spread out before entering Sheffield. Sections of the city were inundated to depths as great as 2.7 meters (9 feet). In its violent path the water destroyed about 800 homes and damaged at least 4,000 others. Industrial facilities were also ruined.

The verdict of the coroner's jury was that "there has not been that engineering skill and that attention to the construction of the works which their magnitude and importance demanded" and that "the Legislature ought to take such action as will result in a governmental inspection of all works of this character and that such inspection ought to be frequent, regular, and sufficient."

A committee of engineers was retained by the dam's owner, the Sheffield Waterworks Company, to investigate the failure and file a report. This group attributed the disaster to ground movement at the damsite. Unwilling to accept this verdict, the government obtained the opinions of other experts in dam engineering. They were unanimous in concluding that faulty design and construction should be blamed for the failure. Their disapproval focused on the pipelaying and embankment compaction methods.

There is not much doubt that the work on the dam was inferior. The fill was inadequately compacted, the spillway was too small, and the outlet pipes were vulnerable to displacement. Evidently, as the embankment consolidated, the pipes deflected and seepage began to erode the puddled clay. The core could have been undermined where it crossed the outlet. This would have precipitated destructive leakage through the fill. Washing of materials from the interior of the dam probably was accompanied by substantial settlement, leading to overpouring of the erodible crest.

Buffalo Creek Dam

On February 26, 1972, the Buffalo Creek (Tailings) Dam in West Virginia failed, causing a flood wave that killed 125 people. The embankment, which consisted of a pile of coal mine waste, impounded the reservoir but lacked the features of an engineered dam. It was part of a system of spoil embankments and sediment basins on a tributary of Buffalo Creek.

Waste had been accumulating for about 25 years before the failure. The piles consisted of shale, sandstone, low-grade coal, and various kinds of timber and metal scrap. Unit weight of the material ranged from about 1360 to 1600 kilograms per cubic meter (85 to 100 pounds per cubic foot), contrasting with native soils of the area which had a unit weight of about 2000 kilograms per cubic meter (125 pounds per cubic foot). By 1960, the first embankment had been extended to a length of approximately 366 meters (1200 feet), a width of roughly 152 meters (500 feet), and a height of 46 meters (150 feet). This embankment had evidently burned for many years.

In about 1960, the mining company, to reduce stream pollution, began to run waste water from its plant into pondage behind the embankment. Clarification was accomplished by settling in the pool and filtration through the comparatively loose fill. Then the plant began to process coal from strip mining, and the waste from this operation also was dumped in the basin. However, the material from this source was naturally finer and tended to seal the embankment. The seepage slowed, and the reservoir level rose.

Federal inspectors visited the complex in 1966 and reportedly called attention to the precarious condition of the embankment. In 1967, a new embankment was constructed 183 meters (600 feet) upstream from the first barrier. Then, in 1970, a third fill was placed 183 meters (600 feet) upstream from the second. The result was a staircase of poorly built embankments, with the upper two founded on the soft sediment in the settling basins.

By 1972, the newest of the three embankments was roughly 152 meters (500 feet) in length and had risen about 13 meters (44 feet) above the sediment in the middle pool. Its broad crest was nearly as wide as it was long. A 610-millimeter (24-inch) steel overflow pipe was reportedly installed in July 1971, which extended diagonally through the fill from one side toward the center. Aside from this, the reservoir had neither spillway nor outlet. The pipe evidently did not have an inlet structure or any cutoff collars.

Occasional slips and breaks had occurred during the lifetime of the embankments. In 1971, a mining company worker said that he had seen black water issuing from the floor of the middle pool, indicating leakage through or under the uppermost dam.

In the three days preceding the failure, about 94 millimeters (3.7 inches) of rain fell in the area. Storm runoff caused the reservoir behind the third dam to rise. The water level reportedly was within 0.3 meter (1 foot) of the crest 4 hours before the collapse. Between 6 and 8 a.m. on that day, the water rose onto the graded crest and washed through dumped waste that stood as high as 2 meters (7 feet) above the crest. A mining company employee reportedly was dispatched at 6:30 a.m. to find bulldozers for excavation of an emergency spillway, but the equipment never reached the site. Longitudinal cracks appeared in the soggy fill. Slumping of the downstream face dropped the crest and accelerated the overflow. The dam broke at about 8 a.m.

There were no known witnesses to the initial collapse, but several people were nearby. One observer of the dam just after the first phase of the failure said that water was cutting away the dam from one side. Then a larger section broke out near the center. A man in the village immediately downstream from the pools reported that there was a power outage at 8 a.m. and the floodwaters started pouring from the lower embankment. He and his family escaped to higher ground, and he watched his house float off its foundation. Then he saw the burning refuse bank explode violently, loosing a deluge of water and debris. Observers at the site of the upper dam, a few minutes after its rupture, saw a broad overpouring which concealed whatever remained of the embankment.

The upper pool had contained approximately 500 000 cubic meters (400 acre-feet) of sludge and water, which was completely discharged within a quarter of an hour. During the next 3 hours, a flood wave estimated as high as 6 meters (20 feet) moved down the 24 kilometers (15 miles) of the Buffalo Creek valley at about 8 kilometers (5 miles) per hour. The village of Saunders at the upper end of this reach was washed out, and extensive damage was done to several other settlements downstream. The torrent left 4,000 people homeless.

Canyon Lake Dam
(Rapid City)

One of the most intense floods in American history struck South Dakota's Black Hills on June 9, 1972, and destroyed much of Rapid City, a community of 43,000 people. The cause of the disaster was a violent rainstorm which developed suddenly. The weather forecast for that Friday was "partly cloudy, with scattered thundershowers, with some possibly reaching severe proportions." This was not regarded as unusual enough to excite alarm. However, by late afternoon a moisture-laden southeast wind was driving hard against the hills and deflecting upward. Air currents at high elevation normally would have swept the moisture away, but on June 9, the movement aloft was negligible. The damp air concentrated over the eastern slope of the hills.

Beginning early in the evening and continuing into the night, as much as 250 millimeters (10 inches) of rain fell on a watershed where the normal annual precipitation was only about 355 millimeters (14 inches). Runoff accumulated rapidly on the steep rock slopes and gained velocity in the narrow canyons on its way to the populated areas to the east. The first indication of hazard came just before 6 p.m. when a highway patrolman radioed that he had encountered a foot of water flowing over a road in the Black Hills about 64 kilometers (40 miles) northwest of Rapid City. Ten minutes later the highway patrol reported a cloudburst 16 kilometers (10 miles) west of Rapid City. Farther south, Battle Creek reportedly rose 2.4 meters (8 feet) in just a few minutes and rushed through the old gold mining town of Keystone. Eight people camping along the stream were drowned.

Although intense rain had begun in Rapid City, few of its citizens were apprehensive. Radio and television broadcasts were interrupted periodically by news reports of storm conditions in the hills, but there was little recognition of a threat to the city itself. Despite the apparent calm, a disaster was already in the making. Rapid Creek was discharging an estimated 850 cubic meters (30 000 cubic feet) per second into Canyon Lake, a 16-hectare (40-acre) reservoir west of town. Peak inflow was recorded at 884 cubic meters (31 200 cubic feet) per second later in the night. Floodwaters were rising fast against the 6-meter (20-foot) high earth dam, constructed by the Works Progress Administration in 1938. At approximately 8:30 p.m., spillway releases were made in an attempt to control the lake level.

Beginning at about 9 p.m., a cloudburst brought as much as 150 millimeters (6 inches) of rain in 2 hours. Rapid Creek broke out of its banks. Dark Canyon, a section of homes in the foothills, was flooded.

The mayor and the city engineer of Rapid City inspected the Canyon Lake Dam just before 10 p.m. Men from the police and fire departments were dispatched to warn people downstream from the reservoir. Many residents underestimated the danger at first and remained in their homes. Water was surging down the streets. Three firemen were swept away as they tried to evacuate citizens. Another fireman went into an unoccupied dwelling and soon discovered that it was being torn loose from its foundation. He clambered onto its roof and held on as the house floated downstream. Then it came apart, leaving him clinging to part of the roof as it washed across Canyon Lake and snagged in trees on the dam crest.

Near 10:30 p.m., there was a report of a wave several feet high coming down Rapid Creek from the Black Hills. At 10:39 p.m., the order for general evacuation was given, but by that time the Canyon Lake Dam was in precarious condition. Its spillway was obstructed by debris and the embankment was on the verge of overtopping. The dam was able to resist the first flood wave, but the next surge went over the 152-meter (500-foot) long dam and began to scour the embankment. The muddy torrent pouring from the reservoir overwhelmed the winding channel of Rapid Creek all the way through the city. At about 10:45 p.m., the dam washed out. The fireman, still on the broken section of roof, witnessed the failure. As the water and debris disgorged through the breach, the roof dropped onto a part of the dam that remained intact for the moment. He was able to work his way up the abutment.

The cloudburst that sent the killer waves down the creek was limited to a relatively small area. Only about 24 kilometers (15 miles) upstream from Canyon Lake, the Pactola Reservoir was untaxed by the storm. The maximum inflow there was reported to be only about 62 cubic meters (2200 cubic feet) per second, and there was a 2 cubic meters (74 cubic feet) per second discharge from the reservoir. The Pactola Dam is a 70-meter (230-foot) high earthfill structure completed in 1956 by the Bureau of Reclamation. The capacity of its reservoir is 122 000 000 cubic meters (99 000 acre-feet), much larger than Canyon Lake. However, since most of the disastrous runoff in 1972 came from the watershed between the two dams, the Pactola flood storage capacity offered little protection.

When the Canyon Lake Dam collapsed, the surge of debris-laden water struck Rapid City with full force. Buildings near the creek were shattered. Many of the occupants were unable to escape. Mobile homes and trailers were washed away. Powerlines were knocked down and propane tanks were ruptured. There were many fires and explosions. Natural gas escaped from broken pipelines and burst into flames from the sparks of the downed powerlines.

At about midnight, the floodwaters began to recede. In the flood's swath through Rapid City, more than a thousand people had been forced to take refuge on roofs and in trees. Rescue efforts were hampered by broken communications. Due to the power outage, radio stations went off the air at about 2 a.m. Four hours later, over a Civil Defense band and using emergency power, official bulletins were again broadcast.

At daybreak the rainstorm was weakening, giving way to fog. About 5,000 rescue workers, half of them National Guardsmen, combed the city in search of survivors. They found the victims at every turn. In one 10-block section more than 85 dead reportedly were counted. The final toll was 237 fatalities, 5 persons missing, and 5,000 homeless in the path of the flood. More than half of Rapid City was said to be devastated. Twelve hundred houses were demolished, and 2,500 others were extensively damaged. About 100 commercial and industrial buildings had been ruined. Approximately 5,000 wrecked automobiles were scattered throughout the city. Seven of the nine bridges which had spanned Rapid Creek, 80 blocks of street, and 8.9 kilometers (5.5 miles) of railroad trackage were reported to have been destroyed. Total property damage was estimated at $60 million.

Canyon Lake Dam was small and — as at the South Fork Dam near Johnstown, Penn. — its contribution to the disaster toll cannot be calculated precisely. In each case, the flood was already of alarming proportions at the instant of dam failure. The sudden release of reservoir waters superimposed one disaster upon another.

Dnjeprostroj Dam

During their retreat from the German Army in September 1941, Soviet troops destroyed part of the Dnjeprostroj Dam on the Dnjepr (Dnieper) River near the industrial city of Saporoshje, in the southwest part of the Soviet Union approximately 150 kilometers (90 miles) from the Azov Sea. While this concrete gravity dam was still occupied by fleeing soldiers, about 30 trucks loaded with 3 tons of dynamite each were driven into a tunnel in the dam and exploded. The resulting breach in the structure was approximately 200 meters (660 feet) wide.

Water surged through this opening under a head of about 20 meters (66 feet). The flow of 30 000 to 35 000 cubic meters (1 060 000 to 1 240 000 cubic feet) per second was nearly 50 percent greater than the design flood. Concrete pieces weighing as much as 180 metric tons (200 tons) were later found 200 meters (660 feet) downstream. The pressure of the blast transmitted through the galleries and shafts of the structure threw parts of machinery and other objects about and caused much damage to equipment. The adjoining navigation locks were undamaged.

At the damsite, the Dnjepr River has a normal flow of about 350 cubic meters (12 400 cubic feet) per second and a high flow of 23 500 cubic meters (830 000 cubic feet) per second. The water is impounded by a curved concrete gravity dam approximately 40 meters (131 feet) high and 800 meters (2600 feet) long. There are 47 slide gates (Stoney or roller-mounted type) on the spillway crest, each with a clear span of 13 meters (42.65 feet) and an opening height of 9.7 meters (31.8 feet). Spillway discharge capacity is 23 500 cubic meters (830 000 cubic feet) per second. Each gate weighs approximately 54 430 kilograms (60 tons) and requires a lifting force of about 890 kilonewtons (100 tons) which is provided by two gantry cranes.

There are two longitudinal tunnels, or galleries, in the dam, with floor levels about 12 meters (39 feet) and 28 meters (92 feet), respectively, below the spillway crest. The upper tunnel, in which the explosion was set off, is large enough to pass heavy vehicles, while the lower tunnel is smaller.

The dam did not have low-level outlet works. The powerplant intake is about 11 meters (36 feet) below the crest. This is a large run-of-the-river plant with a 5-meter (16-foot) maximum range in operating head.

In addition to the service bridge for the cranes, a highway bridge extends the entire length of the dam on the downstream side. The powerhouse, at the right end, is approximately 230 meters (755 feet) long and has nine turbines. Eighteen intake slide gates (Stoney or roller-mounted type) with a clear span of 6.5 meters (21.3 feet) and an opening height of 9.9 meters (32.5 feet) control flow into the penstocks, which are embedded in the structure.

On the left abutment is the river lock with three stages of about 12 to 13 meters (40 feet) each. The chambers are 180 meters (590 feet) long and 18 meters (59 feet) wide. This structure has large bulkhead gates as well as intakes over the entire length of the chamber. These are supplied from a conduit under the chambers.

Immediately after the breaching of the dam, the Germans began to study alternatives for reconstruction. As a first step the service cranes, one of which was in a hazardous position at the edge of the breach, had to be secured. The next consideration was the dewatering of the damaged part of the dam. This required diversion of the flow during the repair period, estimated as 1 year. Floods usually occur during April, May, and June. The normal flow in other months would not exceed about 2400 cubic meters (85 000 cubic feet) per second. Only 12 of the spillway gates had been destroyed, leaving 35 undamaged openings for control of the floodwaters during reconstruction.

One of several proposals considered was the erection of a rock cofferdam to isolate the worksite. This was rejected as too time-consuming. The possibility of diversion through the powerplant was studied. Six turbines had to be taken out for repairs anyway. However, this plan was also dismissed, because of their low discharge capacity. Another idea was to drive a diversion tunnel in the right abutment. This was judged to be too slow. The navigation lock was considered as a bypass. Since the reservoir water level was 15 meters (49 feet) below the sill of the highest chamber, the two upper lock stages had to be removed to provide enough discharge capacity. This plan was also discarded.

The alternative finally selected was to install low-level outlets in the undamaged part of the dam. This required ten tunnels, each 5 by 5 meters (16.4 by 16.4 feet) in section. The beginning of the work was scheduled for December 1941 but was delayed by the cold winter until February 1942. The tunnels were driven from the downstream face, starting high enough to be above floodwater level during spring flow through the breach in the dam. Blasting was used to within about 4.5 meters (15 feet) of the upstream face. Pneumatic equipment was used to make the final breakthrough.

To close each tunnel intake, hemispherical reinforced concrete covers were built on the site. These were to be replaced later by roller gates. The work was delayed from January to March 1942 because cement and steel arrived late. Each cover had a thickness of 0.4 meter (1.3 feet) and measured approximately 7 by 7 meters (23 by 23 feet). The weight of each was approximately 80 metric tons (88 tons). Their placement over the tunnel inlets just before holing through was accomplished without incident. Breakthrough was accomplished with a small detonation in the center of the tunnel heading. The water pressure then forced the covers against the dam face. Sealing was more difficult than expected. Extensive use of divers was required. Final sealing was accomplished with hemp rope and rags. The job was slow and dangerous due to floating debris in the river and the suction at the leaks between the concrete covers and the dam face. This operation took the life of one diver.

While the river was still high, the damaged piers on both sides of the breach were removed by blasting. As water flowed through the new outlet tunnels, the reservoir level was lowered so that the damaged section was accessible in July 1942. The break had occurred almost entirely above the concrete joint at the elevation of the floor of the upper tunnel. The rebuilding of the piers and the dam crest was started without any difficulties. During the summer and fall, the flow in the river was considerably less than normal so the capacity of the new outlets was not taxed. In fact, since not all of the outlets were needed, four of them were closed with concrete plugs. Roller gates were installed on the remaining six tunnels.

After the low-level outlets had been closed, the reservoir level rose rapidly. Work of rehabilitation in the powerhouse had kept pace so that in January 1943 the first repaired unit was ready and generation at the plant was resumed. Construction had taken 10 months, from February through December 1942.

Nine months later, as the tide of war turned, the dam was sabotaged again, but this time by the Germans. Reportedly, aerial bombs w re detonated in the lower gallery. Although the structure was not breacned, there presumably was enough damage to necessitate some rehabilitation by the Soviet forces.

Eder Dam

The Eder Dam is a concrete gravity structure, completed in 1914, near Waldeck in the vicinity of Kassel, Germany. Its height is 48 meters (157 feet), and the crest length is 399 meters (1309 feet). The dam thickness ranges from about 6 meters (20 feet) at the crest to 35 meters (115 feet) at the base.

During an air raid on the dam on May 17, 1943, a breach of about 50 meters (164 feet) was opened in the structure. This was reported to be a somewhat smaller break than at the Möhne Dam, which was bombed at the same time. The resultant maximum discharge rate from Lake Eder was estimated at 8500 cubic meters (300 000 cubic feet) per second, and the volume of water lost was 154 000 000 cubic meters (125 000 acre-feet), compared with the reservoir's total capacity of 202 000 000 cubic meters (164 000 acre-feet). These losses continued for a day and a half.

The bombing also cracked the structure extensively in areas other than the breach, and caused severe damage to the powerplants below the dam. Soon after the 1943 attack, and continuing for about 13 months, repairs were made so that the reservoir could be put into limited operation. This work required a total of 13 000 cubic meters (17 000 cubic yards) of masonry and 12 000 meters (39 000 feet) of drilling for grouting, which took 136 metric tons (150 tons) of cement.

In the period of 1943 to 1946, reservoir levels were cautiously raised in stages while structural movement and seepage were kept under close surveillance. Leakage increased substantially with the higher pressures. Therefore, further rehabilitation was done in 1946 and 1947, including a grouting effort that required an additional 4000 meters (13 000 feet) of drill holes and 27.2 metric tons (30 tons) of cement.

Eigiau and Coedty Dams

On the evening of November 2, 1925, the town of Dolgarrog in North Wales was hit by a disaster resulting from a blowout under the base of the Eigiau Dam, a 17-year-old concrete structure. The water escaping through the hole caused the overtopping and destruction of the Coedty Dam, a small embankment dam about 4.0 kilometers (2.5 miles) downstream.

The Eigiau Dam, a concrete gravity structure, was built in 1908 and 3 years later enlarged to create a dam 992 meters (3253 feet) in length and 10.7 meters (35 feet) high above the streambed. The reservoir stored 4 500 000 cubic meters (3670 acre-feet) of water for generation at the Dolgarrog powerplant.

In 1924, to provide more water storage for the powerplant, the Coedty Dam was constructed downstream from the Eigiau Dam. The new dam was an earthfill structure, composed of local moraine material, with a concrete core wall. Its height was 11 meters (36 feet) and its length 262 meters (860 feet). Storage capacity was 311 000 cubic meters (252 acre-feet). The dam had a side channel spillway at one end.

Evidence indicates that Eigiau Dam had been poorly constructed. The concrete was inferior, and the foundation was unsound. Evidently, there had been no effort to excavate to sound rock. The entire dam was founded on a thick stratum of blue glacial clay. Excavation for the dam had been extended only a few feet below the natural ground surface and in some places only a few inches into the clay. At the location of the blowout, the base was only about 0.6 meter (2 feet) below the top of the clay.

Seepage under the dam evidently induced piping. Part of the foundation was washed out, opening a hole 21 meters (70 feet) wide and 3 meters (10 feet) deep. The dam bridged the gap without apparent distress in the structure itself, even though the breach was large enough to discharge 397 cubic meters (14 000 cubic feet) per second.

The torrent from the blowout rushed into the Coedty Reservoir, but its spillway was incapable of passing such an unexpected flow. The dam was quickly overtopped. In rapid succession the downstream slope was eroded away, the core wall broke, and a breach about 67 meters (220 feet) long was cut in the crest. The resulting flood wave carried the worst destructive capacity since, in contrast with the comparatively controlled flow through the foundation orifice at Eigiau, essentially the entire storage at Coedty was dumped in a very short time. The flood hit the powerplant first and then inundated the village of Dolgarrog 1.6 kilometers (1 mile) farther downstream. Sixteen people died.

An employee at the powerplant provided the following account:

"The disaster occurred in the evening about 9:15 p.m. Providentially, however, it was the one evening of the week when a film show was held in the Assembly Hall. This place was situated on high ground, away from the flooded area. The cinema was well attended at the time of the disaster and the audience was unharmed by the flood. If the dam had burst an hour earlier or later, there would have been heavy casualties, because the people would have been in their homes.

"When the dam burst, the water washed away the communication circuit from the reservoir to the power station. It was impossible for the reservoir attendant to race the flood, and so no warning of the approaching danger could be given. The shift engineer on duty in the power station was about to read the instruments for the 9:30 p.m. log when the water entered the power station. The operating staff stuck to their posts and succeeded in getting the plant shut down, and then managed to escape from the building. Happily, the lights in the station did not fail, because they were supplied from a small Pelton-driven generator above flood level."

Later, the Coedty Dam was reconstructed to essentially the same dimensions and continues in operation. The Eigiau Dam was abandoned.

El Habra Dam

Among the reservoirs built by the French in Algeria, the largest was on the Habra River. It had the distinction of failing on three separate occasions.

The construction of El Habra Dam was undertaken in November 1865 and completed in the winter of 1871-72. It was a gravity rubble-masonry structure. The main section of the dam was straight, with a length of 325 meters (1066 feet) and a height of about 33.5 meters (110 feet). A 125-meter (410-foot) long overflow wall connected with the dam at an angle of 35 degrees, giving a total length of 450 meters (1476 feet). The overflow crest was 1.6 meters (5.25 feet) lower than the top of the main structure. This masonry barrier was higher than the Bouzey Dam (in France), but evidently the design was intended to be more conservative by avoiding tensile stresses in the masonry. At least that was the objective of the designers. However, the construction of the masonry structure was deficient in several respects.

At the outset of impoundment, the dam leaked extensively. This gradually diminished. But on March 10, 1872, a section of the overflow wall failed during a flood.

After repairs and an 8-year period of safe operation, the dam was again ruptured in December 1881. The breach was 100 meters (328 feet) long and 35 meters (115 feet) deep, extending into the foundation. This disaster struck after a severe storm in which 165 millimeters (6.5 inches) of rain fell in a short time. Due to an inadequate spillway, the reservoir level rose in excess of the design maximum. The overloaded dam collapsed, disgorging a flood which destroyed several villages, damaged sections of the city of Perregaux, about 10 kilometers (6 miles) from the dam, and caused the loss of 209 human lives.

The failure could be attributed partly to inferior aggregates. The mortar used to bond the masonry contained fine, clean sand and a poor lime from a local source. Over the years of operation, there was an accumulation of leached calcium carbonate on the downstream face of the dam. This indicated that the slaking of the lime in the mortar had been incomplete and suggested therefore that expansion of mortar may have been a cause of disintegration of the dam. El Habra Dam had been conspicuously pervious from the beginning of impoundment. The increasing hydrostatic pressures evidently produced tension cracks in the upstream part of the dam and dangerously high compressive and shear stresses across horizontal sections.

The dam was repaired in the period 1883 to 1887, during which the section was reportedly enlarged and strengthened. But on November 26, 1927, it failed again when a flood caused the reservoir to rise 4 meters (13 feet) above the normal maximum. A government commission attributed the collapse to the weight of the muddy floodwaters and to uplift pressures on horizontal joints in which the mortar had deteriorated. Fortunately, this third failure of the dam reportedly did not result in a loss of human lives because of adequate advance warning.

Fontenelle Dam

In September 1965, during filling of the reservoir at the Bureau of Reclamation's Fontenelle Dam (fig. 4-4), a 39.6-meter (130-foot) high earthfill structure completed in 1964 on the Green River in Wyoming, seepage through joints in the shales and sandstones of the right abutment caused deep erosion which seriously endangered the dam. Quick action by the operations crews narrowly prevented a disaster. This event demonstrated the hazards of relying on single rather than redundant seepage defenses. In the case of Fontenelle Dam, the lone defense in the abutment was a single-line grout curtain. Remedial work included broadening and extending the grout curtain by additional lines of grout holes.

The Fontenelle Dam accident had several contributing factors. When the embankment remnant in the distressed area was excavated to foundation, cracks were found in the sedimentary rock (sandstone, shale, siltstone). There also was reported to be sandy core material at the foundation contact. The postemergency investigation also produced evidence that in some places the grout injected during construction had not traveled far enough to seal foundation cracks completely.

Figure 4-4.—Fontenelle Dam (P154-400-1501). P-801-D-79349.

The records of the Bureau of Reclamation provide an instructive case history of this embankment. As reported by Chief Engineer B. P. Bellport, initial foundation excavation in 1961 revealed that the sedimentary rock was much more fractured than had been expected. The cutoff trench was therefore deepened about 2 meters (6.6 feet). Grout takes in the upper 20 meters (66 feet) of the foundation were large. At the outlet location and on the right abutment, an additional line of holes was drilled and grouted. At the site of the spillway intake, on the right abutment, there were several open cracks that required treatment. More grouting was done to improve the foundation in this area. To provide additional seepage control, an earth blanket was placed on the lower part of the abutment in that vicinity. The complete foundation treatment at the damsite included about 14 000 meters (46 000 feet) of drill holes and over 4000 cubic meters (140 000 sacks) of cement for grouting.

Reservoir impoundment was begun in the summer of 1964. Seepage soon appeared but it was not considered dangerous. However, on May 6, 1965, when the reservoir water depth reached 26 meters (85 feet), water was seen leaking from the rock cut of the spillway discharge channel on the right abutment and from a cliff on the left side of the river approximately 1 kilometer (0.6 mile) downstream.

Reservoir spill began on June 15, 1965. Total leakage at maximum level was estimated to be 2000 liters (70 cubic feet) per second. On June 29, 1965, a small slough was discovered to the left of the spillway chute about halfway up the slope from the toe of the embankment. Approximately 30 liters (1 cubic foot) per second of water was flowing from the rock in that location.

On the morning of September 3, 1965, a wet area was seen on the downstream slope of the dam about 30 meters (98 feet) to the left of the earlier slough. During that day, leakage increased to 150 liters (5 cubic feet) per second and was accompanied by sloughing and erosion of the embankment. Early the next day leakage had enlarged to about 600 liters (21 cubic feet) per second and roughly 8000 cubic meters (10 500 cubic yards) of embankment had been removed by erosion extending nearly to the crest. An emergency effort was started to fill the hole with rockfill. This temporarily stabilized conditions. In the meantime, work was underway to open the main outlet and to complete discharge channels for the abutment outlets.

On the afternoon of September 5, the leakage was surging violently and was carrying large amounts of earth material. Evidently the dumped rockfill had forced the flow toward other exits. The estimated head on the discharge was 14 meters (46 feet). Dumping of rockfill was terminated and the flow tended to stabilize again. However, the next afternoon, according to Chief Engineer Bellport, "an area on the dam crest about 6 meters (20 feet) in diameter with its center near the upstream edge of the dam suddenly collapsed with a drop of about 10 meters (33 feet). Bedrock was exposed on the abutment side of this cavity and water appeared to be issuing from cracks in the rock. At this time, the reservoir level was about 4 meters (13 feet) above the base of the cavity. Heavy rock was dumped into the hole to stabilize the area against further collapse."

The emergency passed as the reservoir level continued to drop at a rate of approximately 1.2 meters (4 feet) per day. The leakage decreased to safe rates.

In appraising the accident, Bureau of Reclamation investigators concluded that grouting of the foundation joints was not fully effective because materials filling the rock openings precluded or inhibited grout flow, or possibly because soluble salts reacted with the grout to cause premature set or ultimate softening. They noted that to avoid rock movement in the steep abutments, grout pressures used in shallow grouting had to be low and therefore not as effective as desirable. The Chief Engineer suggested the need to reemphasize the avoidance of steep abutments, especially where there are stratified materials of variable strength and permeability. He also recommended grouting in multiple lines to assure a broad impervious foundation barrier in such cases. This was the remedy selected at Fontenelle. Eight lines of holes were grouted in the right abutment in the vicinity of the damage. After the embankment breach was repaired, more holes were drilled and grouted to seal further the fractured rock near the foundation surface.

In its final review of the experience at Fontenelle Dam, the Bureau made three important observations: "(1) Over-consolidated materials may develop stress relief cracking in the bottoms of valleys as well as abutments, (2) the degree of stress relief jointing will vary according to the speed of unloading and so consequently will the relative permeability of the mass vary, and (3) soluble solids in even small quantities can have a substantial effect on the quality of the barrier produced by grouting."

Frías Dam

On the Sunday evening of January 4, 1970, a dam on the Río Seco Frías (figs. 4-5 and 4-6) near Mendoza, Argentina, in the Andean foothills 966 kilometers (600 miles) west of Buenos Aires, failed after a day of intense storm. A torrential rain had swollen the stream to overflowing, causing flooding in the environs of Mendoza. Runoff from the steep local watershed smashed the Frías Dam in the adjoining Godoy Cruz township, releasing a 2-meter (6-foot) wall of muddy water through the city of 300,000 population, which was crowded with summer visitors.

The flood raced through 20.7 square kilometers (8 square miles) of the urban area, trapping motorists in their vehicles. People drowned in cars as they were washed down the streets. Houses were demolished, trees were uprooted, and outdoor cafes were swept away. A mother and her two children were knocked down by the torrent. The woman survived but the children died. A 7-year-old running from an automobile was overtaken by the flood and drowned. A funeral procession was hit by the waters; one mourner died. Hundreds of other people in the city were injured. At least 500 were left homeless. Power facilities were put out of service, hampering the efforts of emergency crews. Two months after the disaster, the official casualty count was still uncertain — 42 people dead and 60 still missing.

Mendoza is one of the important centers of the wine industry in Argentina. Some vineyards were severely damaged by the floodwaters. Officials reported that the general storm had also destroyed much of the year's apple and pear crops in Neuquen and Río Negro Provinces, south of Mendoza.

The reservoir on the Río Seco Frías had a storage capacity of only 200 000 cubic meters (161 acre-feet), whereas the flood volume was many

Figure 4-5.—Frías Dam after failure, downstream view (Courtesy of C. J. Cortright). P-801-D-79350.

Figure 4-6.—Frías Dam after failure, upstream view (Courtesy of C. J. Cortright). P-801-D-79351.

times greater than this. As a consequence, the embankment crest reportedly was overtopped by a 1-meter (3-foot) depth of water lasting for about 15 minutes.

The dam was one of several built on local streams 30 years previously after a similar disaster. Its crest length was 62 meters (204 feet), its height was 15 meters (49 feet), and its crest width was 3.85 meters (12.6 feet). Both the upstream and downstream faces had 1 to 1 slopes. This homogeneous rockfill structure was faced on the upstream side with a reinforced concrete slab 300 millimeters (12 inches) thick, and on the downstream side by mortared rubble masonry of roughly the same thickness. The top of the dam was paved in a similar manner using mortared stones. Weathering of the crest over the years may have reduced this protection. Some patching of the surface had been done. There was some question whether the crest was adequately sealed to prevent the entry of water into the interior of the embankment during overtopping. In view of the fact that the faces of the dam were effectively sealed, the pressure of confined water in the fill could have been destructive. Corrective measures recommended for the similarly designed surviving embankments in the vicinity were the sealing of the crest and the providing of drains in the downstream facing.

The outlet works of the dam on the Frías had a vertical 10-meter (33-foot) high concrete tower 800 millimeters (2.6 feet) in inside diameter, with five levels of ungated ports. This discharged into a 30-meter (98-foot) long concrete conduit of 1000 millimeters (3.28 feet) inside diameter from floor to crown, which was placed on the conglomerate foundation immediately under the embankment near its maximum section. The conduit was designed with cutoffs at its upstream and downstream ends, but evidently did not have them at intermediate points.

A spillway with a capacity of 40 cubic meters (1400 cubic feet) per second, said to be equal to the maximum recorded flood, flanked the dam in natural terrain beyond the left end. It was a steep chute constructed of concrete with a channel about 3.5 meters (11 feet) wide. There was a concrete cutoff extending approximately 2.0 meters (6.5 feet) into the foundation at the downstream end of the spillway. Evidently the design did not incorporate any provision for dissipation of the energy of the falling water. The facility therefore not only had insufficient discharge capacity but potentially was vulnerable to undermining. However, it remained essentially intact during the failure of the dam, with some limited erosion of the foundation at the lower end.

At the damsite, the sandstone and conglomerate beds of the Mogotes Formation vary considerably in thickness, consolidation, and particle size. Generally the conglomerates are poorly cemented. Both the sandstones and the conglomerates probably have constituents in some of the beds which would tend to soften and disintegrate under pressure when saturated. For this reason, concrete dams have not been regarded as suitable for construction on this type of foundation. Although a concrete dam would have provided a spillway over the river section, the Mogotes Formation is recognized as highly vulnerable to scour.

The investigators found no reason to blame the foundation, after analyzing the washout of the Frías embankment. The dam itself and its appurtenant facilities were judged inadequate. An official of Agua y Energía

146

in Buenos Aires theorized that water overflowing the crest undermined the downstream facing and eroded the fill, precipitating the sudden collapse. The construction supervisor for the dam stated that insufficient maintenance was contributory to the dam's failure. Sediment accumulation had encroached upon the capacity of the reservoir and blocked two of the five openings in the outlet tower. This, he asserted, forced more of the floodwater over the crest.

Three steep adjoining watersheds discharge flood runoff into the Mendoza vicinity. The streams, in order from north to south, are the Río Seco Papagayos, the Río Seco Frías, and the Río Seco Los Pardos (Maure). The direction of flow from these drainage areas is from west to east, dropping in a distance of about 22 kilometers (14 miles) from extreme elevations of 2500 meters (8200 feet) above sea level to 760 meters (2500 feet) at Mendoza. The slopes lying immediately above the Mendoza metropolitan area are subject to flash floods from rainstorms of short duration and high intensity. These cause extensive erosion and sedimentation. The terrain has minimal natural vegetal cover.

A difficult design problem was presented by the flash floods and attendant erosion and sedimentation. Adequate spillway capacity was a factor vital to the safety of the embankment dams in the watershed. Spillway designs at the three detention basins in the vicinity — Frías, Maure, and Papagayos — were all based on hydrologic data and criteria developed in the midthirties and found to be deficient when analyzed with the benefit of advanced procedures and more than 30 years of additional precipitation and runoff records.

The reservoirs store water infrequently, and only for short periods — usually just a few hours. During the flood inflow, hydrostatic pressures were suddenly imposed on the dams and their foundations after they had been dry and unloaded for an extended period. This could make them susceptible to cracking and internal erosion.

Apparently, siltation was a continuing maintenance problem in the Frías reservoir, as well as in its sister reservoirs to the north and south. Periodic sediment removal was necessary to maintain flood detention space and to keep the ungated tower ports free of obstruction. The project records show that in 1961, silt had encroached upon the tower at Frías Dam, leaving only the upper two tiers of openings exposed. In 1970, apparently three of the five sets of ports were open.

Three years after the failure, a new embankment dam was completed on the Río Seco Frías at a site near the old one. The new dam is a 38-meter (125-feet) high, zoned earthfill structure, with a crest length of 495 meters (1624 feet) and a crest thickness of 10 meters (33 feet). Volume of the fill is 600 000 cubic meters (785 000 cubic yards). Slopes of the dam are 2.5 to 1 downstream and 3.0 to 1 to 3.5 to 1 upstream.

The new outlet and spillway are of considerably greater capacity than at the original structure. The outlet was designed for a maximum discharge of 32 cubic meters (1130 cubic feet) per second. Capacity of the spillway, based on a storm of 1,000-year frequency, is 390 cubic meters (13 800 cubic feet) per second. It has a side channel entrance, and a diverging chute terminating in a slotted-bucket energy dissipator combined with a stilling basin. The new reservoir has a storage capacity of 2 330 000 cubic meters (1890 acre-feet).

Gleno Dam

On December 1, 1923, a dam near Gleno, about 48 kilometers (30 miles) northeast of Bergamo in the Alps of north-central Italy, failed suddenly following heavy rains. The dam had been completed in that same year, and met disaster only 30 days after the first filling of the reservoir. Destruction was widespread along the Dezzo River in the 19-kilometer (12-mile) reach to its confluence with the Oglio River at Darfo. Lesser flood damage occurred in the Oglio Valley in the remaining 8 to 10 kilometers (5 to 6 miles) to Lake Iseo at Pisogne. The height of the flood wave in the narrow Dezzo Valley was as great as 30 meters (100 feet). It destroyed five powerplants between Gleno and Darfo. Several factories and many homes were ruined. Loss of human life was estimated at 600. The severity of damage was due partly to the high velocity of the torrent as it rushed down from the 5 400 000-cubic-meter (4400-acre-foot) reservoir at about 1585 meters (5200 feet) above sea level, to Darvo at 250 meters (820 feet) and Lake Iseo at 185 meters (607 feet). The toll would have been even greater except for some detention of the floodwaters in Lake Iseo.

The Gleno Dam, erected soon after World War I, was a composite of a reinforced concrete multiple-arch structure founded on a gravity stone-masonry base. It was 43.6 meters (143 feet) high and 263 meters (863 feet) long, comprising a curved central portion about 76 meters (250 feet) long adjoining straight end sections.

The original design of Gleno Dam provided for a gravity structure. However, the contractor proposed that a multiple-arch dam be substituted. While this alternative idea was under review by the authorities, and apparently without submitting the complete plans which had been requested, the construction company proceeded with its scheme.

This change in concept resulted in a dam with a gravity masonry base in the lower part of the canyon and a multiple-arch structure in the upper part. The base was 16 meters (52 feet) high and about 76 meters (250 feet) long. It supported the curved central section of the multiple arch, while the wings were built directly on rock. The buttresses at the points of tangency were made thicker than the others. Semicylindrical arches spanned 8 meters (26.25 feet) between centers of buttresses and were sloped 53 degrees with the horizontal. Nine of the 25 arches were in the central section above the gravity base, while 12 were in the right wing and 4 were in the left.

Nine arches collapsed, along with their supporting buttresses. These were eight on the masonry base and the adjoining arch of the left wing. The gravity mass remained essentially intact. Just before the failure, at about 7 a.m., a caretaker had gone down to operate a valve and was walking along the masonry base when, at the 12th buttress from the left abutment, he saw small pieces of concrete falling. A crack began to open in that buttress. Then, as the caretaker hurried to safety, the buttress broke along with the two arches which it had supported. Other arches and buttresses went out in quick succession.

Evidently, most of the blame could be placed on inferior materials and poor workmanship. Dirty aggregates were used in the concrete of the superstructure, and the buttresses were inadequately reinforced with war-surplus antigrenade mesh. The foundation had a natural slope downstream; and the entire dam, including the wing sections, had been built directly on the inclined rock surface without trenching or other shaping to provide anchorage. The masonry of the gravity base was set in lime mortar even though cement mortar had been specified. The superstructure concrete was porous and had been hand-mixed and placed in the forms without vibration. One of the arches reportedly had leaked where a timber was embedded in the concrete.

Postfailure analysis of the design of Gleno Dam disclosed that high shear stresses had existed in the superstructure. The collapse undoubtedly was triggered by concentration of high-intensity forces on precariously weak structural elements. In retrospect, at this natural damsite with its presumably sound rock formation, just a little more care in workmanship could have produced a safe reservoir.

Grenoble Dam

The earliest documentation of a dam failure is given in some histories as the collapse of an "earth embankment" near Grenoble, France, in the year 1219. Although this was undoubtedly a major disaster, the embankment which failed was not manmade.

This catastrophe was caused by the rupture of an earth barrier which formed a lake over a period of about 28 years in the plain of Bourg-d' Oisans. Around 1191, an enormous landslide which fell on Vaudaine, opposite l'Infernet in the valley of Livet, blocked the course of the Romanche River and transformed the plain of Bourg-d'Oisans into a vast lake which reached a depth of 20 meters (65 feet). Under the pressure of these waters, the barrier collapsed during the night of September 14, 1219.

In that year, the September Fair at Grenoble, which lasted 21 days, had been well attended. Numerous merchants had come from throughout the world, and the hotels were full. During the night of September 14, the day of the Feast of the Holy Cross, when visitors and inhabitants were just getting to sleep, a terrible noise woke them. The Drac River, which joins the Isère River a short distance downstream from Grenoble, had overflowed its banks and poured water and debris into the city.

The floodwaters released from the lake had followed the course of the Romanche River, carrying away all the bridges across that river and the Drac River, arriving finally, around 10 p.m., to batter the walls of Grenoble. The waters of the Isère River, augmented by the sudden inflow, spread over the countryside several miles above Grenoble. The inhabitants of the city fled in disorder from their homes. Some took refuge in high places such as the bell tower of the cathedral and the towers of the enclosure. Others tried to cross the Isère to reach the heights of Chalemont. Unfortunately, the gate to the bridge was closed, and before it could be opened many people became terrified by the waters that battered the parapets and returned to the city, where they met death. A few succeeded in breaking open the bridge gate and finding refuge on the hillside, where they witnessed the destruction of their homes. When the waters from the lake had passed, the Isère River returned violently to its channel, carrying away that which the Drac River had spared and destroying the stone bridge constructed a century before by Saint Hugues.

The deluge left a very high toll. Grenoble's population, doubled by the presence of visitors that the fair had attracted, mourned a great number of victims. The material damages would not be repaired for a long time. The Old Bishop of Grenoble, Jean de Sassenage, then more than 80 years old, addressed an eloquent letter to his parishioners calling for donations to finance reconstruction. This work took many years.

Hyokiri Dam

At least 114 Koreans were killed and 13 were missing after floodwaters burst through a dam at Namwon, about 240 kilometers (150 miles) south of Seoul, South Korea, on the night of July 12, 1961. The structure was identified as the Hyokiri Dam.

There were estimates that 425 houses had been destroyed in the disaster, which followed several days of torrential rains. In addition to Namwon, which had a population of 38,000, two villages were also reported damaged when the nearby reservoir burst.

Khadakwasla (Poona) Dam

On July 12, 1961, tragedy struck the populous city of Poona on the Deccan plateau behind Bombay, India. Large areas of the city were destroyed, and many people died in the waters released when the Panshet and Khadakwasla Dams in the Mutha River watershed failed after about 1780 millimeters (70 inches) of rain had fallen in a 23-day period.

The breach of Panshet Dam a few miles upstream released about 247 000 000 cubic meters (200 000 acre-feet) of water into the Khadakwasla Reservoir and overtopped the dam. The inflow was far in excess of the design flood, which was calculated as 2100 cubic meters (75 000 cubic feet) per second with 1.8 meters (6 feet) of freeboard. A flood peak of about 2800 cubic meters (98 000 cubic feet) per second had occurred in 1958.

Construction of Panshet Dam was in its final stage as the storm waters filled its reservoir for the first time. It failed as a result of subsidence followed by overtopping in the morning of July 12, 1961. The released mass of water surged into the old Khadakwasla Reservoir at a moment when it was already full, and the gates discharging at nearly full capacity. The flood wave swept through the lake and spilled over the entire length of Khadakwasla (Poona) dam to a depth of 2.7 meters (9 feet) over the top. There were reports of vibration of the structure as the waters battered it. The overtopping washed away a large volume of material at the downstream toe of the dam. The thin dam resisted the forces imposed by the spilling for almost 4 hours; then broke at about 2 p.m. The failure came as the water level was dropping after the peak but while it was still estimated to be as much as 1.8 meters (6 feet) above the crest of the dam.

The breach was not at the highest cross section of the dam but rather at points where the structure was appreciably lower. Evidently, sharp angles in the foundation had caused a severe concentration of stresses. Failure occurred in two stages, the first being the breaking of the weir section. An hour later another part failed, apparently at an abrupt step of about 6 meters (20 feet) in the foundation. At this point a section of the dam rotated ''like a door'' and fell over.

This is said to be the oldest major dam of the British era in India. When the project was being planned, the design of masonry dams was not well known to Indian and British engineers. However, such structures had been constructed on the European continent and provided guidelines for the Khadakwasla design.

Construction of the dam was started in 1869 and completed in 1879. It was a gravity-type structure about 40 meters (131 feet) high, composed of rubble masonry in clay-lime mortar. Its maximum base thickness was about 18.5 meters (61 feet), equal to slightly less than half the height. The structure consisted of a 1046-meter (3431-foot) long nonoverflow section and a 426-meter (1396-foot) long overflow weir generally about 3.35 meters (11 feet) below the top of the dam, although part of it was built 0.3 meter (1 foot) lower. The entire mass was built without construction joints despite the irregular profile of the basaltic rock foundation.

Reportedly, no allowance had been made for uplift pressures in the original design, nor was a drainage system provided. To compensate for this, an earth buttress had been added against the downstream face. From all accounts, the earth buttress was not regarded as effective in resisting the unexpected forces.

In 1964, during a severe water shortage, repair of the dam was undertaken by filling the breach with masonry. As a first stage, the opening was closed to an intermediate level to enable the functioning of the dam as a canal diversion works. In the following year reconstruction was resumed and extended to the crest. To provide an additional margin of safety, new features were incorporated. A pervious earth berm with a sand filter was placed against the downstream face of the masonry. Spillway capacity was improved by building a radial-gated structure with 11 spans of 12.1 by 4.3 meters (40 by 14 feet) on the ogee crest. As a further safeguard, a shallow earthfill was constructed at one end as a fuse plug, in effect, to be overtopped or breached if necessary in an extreme flood. The crest of the old nonoverflow masonry dam was raised by adding 1.8 meters (6 feet) of masonry.

Lower Otay Dam

On January 27, 1916, a flood overtopped the Lower Otay Dam, a part of the San Diego, Calif., water supply system. It was dumped-rockfill structure about 40 meters (130 feet) high and 172 meters (565 feet) long. Within a few minutes, the downstream zone of the fill eroded away, and the central steel diaphragm was torn from the top downward. The upstream embankment section opened like a pair of gates. And in 48 minutes the flood wave moved the 16 kilometers (10 miles) down the valley. Thirty people died.

The dam was completed in August 1897 on Otay Creek at the head of a rocky gorge 3.2 kilometers (2 miles) long, about 32 kilometers (20 miles) southeast of San Diego. The foundation at the damsite was a porphyry formation that is shattered and seamy.

The original plans for the dam contemplated the use of masonry, and initial construction proceeded on that basis. The base block was reported to have been placed to a height of 3.7 meters (12 feet), presumably as measured above streambed, where the top length was about 26 meters (85 feet) and the thickness was about 19 meters (63 feet). At this point the decision was made to change to a rockfill dam with a steel core. This was to be accomplished by placing a diaphragm of steel plates in the center of the fill, extending across the canyon and to the top of the embankment. The base block was used as the footing for the steel web, which was placed 1.8 meters (6 feet) from the upstream face of the masonry. Originally, the two faces of the fill were to have slopes of 2 to 1. This was changed to about 1½ to 1, with a crest thickness of approximately 3.7 meters (12 feet).

Rock for the embankment was obtained by blasting in a quarry on the west side of the canyon, about 30 meters (100 feet) downstream from the dam. This material was transported by a cableway and distributed by derricks. Large stones were placed on the downstream side of the steel plate, while smaller stones and earth were placed on the upstream side. The total volume of rock in the dam was about 107 000 cubic meters (140 000 cubic yards).

The bottom plates of the diaphragm were riveted to an angle iron anchored to the top of the base block by 25-millimeter (1-inch) bolts. The plates were 1525 millimeters (5 feet) wide, 5330 millimeters (17.5 feet) long, and the three lowest rows were 8 millimeters (0.33 inch) thick. At higher levels, the thickness was reduced in steps. After the diaphragm had reached a 15.2-meter (50-foot) height, plates 2440 millimeters (8 feet) wide and 6100 millimeters (20 feet) long and 6 millimeters (1/4 inch) thick were used. The steel core was constructed to a height of 39.6 meters (130 feet). All of the plates were hot-riveted with a single row of 16-millimeter (5/8-inch) rivets on 75-millimeter (3-inch) spacing.

On the upstream side, the steel diaphragm was coated with hot asphalt and covered with burlap. A harder grade of asphalt was applied over the burlap. The diaphragm was enclosed in the center of a rubble concrete wall which was 0.6 meter (2 feet) thick except for the bottom 2.4 meters (8 feet) of its height where it was tapered from 0.6 to 3.7 meters (2 to 12 feet) thick at its base. Keeping the plates straight was difficult because of thermal expansion and contraction. As a result, the diaphragm was not consistently in the middle of the wall; however, a minimum concrete cover of 150 millimeters (6 inches) was maintained. The wall was extended into the notched walls of the abutment, anchored with bolts leaded into the rock, and then protected by masonry.

The dumped fill was allowed to establish its own natural slope. It contained a large amount of fines. The embankment was built to a height of 40.5 meters (133 feet). It then settled about 0.3 meter (1 foot), leaving the crest approximately 0.6 meter (2 feet) above the top of the core wall. A log boom was installed at the upstream face to control wave wash.

A concrete spillway was located in a depression on the east bank several hundred feet from the dam. It had a trapezoidal section 11.55 meters (37.9 feet) wide at the bottom and 12.65 meters (41.5 feet) wide at the top, with depth of 2.4 meters (8 feet). Spillway capacity was about 68 cubic meters (2400 cubic feet) per second.

The outlet was a concrete-lined tunnel 350 meters (1150 feet) long, at an elevation about 15.2 meters (50 feet) above the base of the dam. At approximately its midlength, a shaft 31.7 meters (104 feet) high extended to the ground surface, for access to the sluice gate which controlled the outlet. Downstream from the shaft a 1200-millimeter (48-inch) diameter steel pipe was located in the tunnel. This conduit was the only available means of draining water from the reservoir below the spillway level.

Until the fateful flood, the reservoir level had not risen higher than 4.6 meters (15 feet) below the spillway crest. A minor leak had been flowing at a rate of about 0.017 cubic meter per second (25 miner's inches or 0.6 cubic foot per second). This was returned to the reservoir by a pump located at the toe of the dam.

The storm that brought the disaster was without precedent in that area. Mean rainfall at San Diego for a 65-year period had been 243 millimeters (9.57 inches) annually and 47 millimeters (1.84 inches) for the month of January. From January 15 to 20, 1916, the rainfall at the dam was 142 millimeters (5.60 inches). Following this, there was a respite of dry weather for a few days. Then on January 25, 26, and 27, a rainfall of 90 millimeters (3.54 inches) was registered at the dam, and on January 27 the rain gage at San Diego measured 56 millimeters (2.19 inches). The runoff from this heavy downpour destroyed many homes and bridges in the area.

Before the storm the water level in the reservoir was about 8 meters (26 feet) below the spillway crest. The heavy runoff raised the water surface to normal operating maximum, and from January 21 on, the reservoir surface continued to rise in spite of the spillway discharge. On the morning of January 27, the day of the failure, the spillway was discharging about 43 cubic meters (1500 cubic feet) per second. Around noon the operator opened the outlet gate, but the reservoir continued to rise, and the threat of overtopping was imminent. Men were sent to warn residents in the valley downstream. Alarms were also sounded by telephone. These actions gave people time to reach the safety of higher elevations and undoubtedly saved many lives.

At 4:45 p.m., the reservoir was at the top of the embankment. Shortly thereafter, the overtopping occurred, and water poured down and through the downstream zone of the dam, loosening rocks in the fill. Erosion was rapid and the lower face was quickly washed out, leaving the core wall unsupported. At 5:05 p.m., the steel diaphragm was torn open at its middle and the remainder of the dam gave way like a pair of swinging gates. The reservoir emptied in about 2½ hours. A flood wave estimated as high as 6 meters (20 feet) swept through the valley at about 20 kilometers (12.5 miles) per hour.

When the water had gone, only traces of the dam remained. The diaphragm had been ripped away from both abutments. In the first 0.8 kilometer (0.5 mile) downstream of the damsite, rivet heads were found in many recesses in the rock walls of the canyon. The steelplates were strewn along the valley, many pieces being found at Palm City, 16 kilometers (10 miles) downstream. No corrosive deterioration of the plates was apparent. Most of them had been torn apart at the riveted joints by the force of the water.

Machhu II Dam

Failure of the embankment portions of India's 26-meter[1] (85-foot) high masonry and earthfill Machhu II Dam (fig. 4-7) in the western state of Gujarat on Saturday, August 11, 1979, was caused by overtopping during flood.[2] About 700 meters (2300 feet) of earth embankment on the right side of the dam and 1070 meters (3500 feet) on the left were washed away. Loss of human life during the ensuing flood may have been as high as 2,000 or more.

Many of the fatalities (more than 1,300) were in the city of Morvi, about 8 kilometers (5 miles) downstream, an industrial center of 75,000 inhabitants in the Saurashtra district. Approximately 150,000 people were affected by the flooding that submerged Morvi in 1.8 to 6 meters (6 to 20 feet) of water and hit 68 villages along the Machhu River in Morvi and Malia counties. The Machhu River flows northward, ending in an area of marshland inland from the Gulf of Kutch (Cutch).

Estimating the number of victims was difficult. An initial report by the *Hindustan Times* predicted a death count as high as 25,000 based on the populations of Morvi and the nearby villages of Lilapur and Adepar, which suffered the full force of the waters from the swollen river. While such early reports were officially labeled as exaggerated, one Indian Army source suggested that as many as 20,000 people might have been swept down the river into the Gulf of Kutch. Counting of casualties was hampered by disruption of transportation and communications. Troops and other rescue workers dispatched to the disaster area were delayed by washouts of highways and railroads. Morvi's telephone exchange and all other communications had failed the day before the collapse. Power as well as communication lines were obliterated. Rajkot was then the closest communication point, 64 kilometers (40 miles) away. Flights carrying food and relief crews were planned and then canceled due to the continuing monsoon weather. Reportedly, wind velocities had been as high as 45 meters per second (100 miles per hour) during the storm. Helicopters were used to drop food packages to survivors clinging to trees downstream from Morvi.

The cause of the overtopping was exceptionally high flooding due to monsoon rains in Machhu II's 1930-square-kilometer (745-square-mile) catchment area, according to government officials. Rainfall was about 530 millimeters (21 inches) in 21 hours. The ranges of precipitation at 24 gaging stations for 21 hours, August 11-12 (recording time 8 a.m.) were 415 to 555 millimeters (16.34 to 21.85 inches) in the watershed above Machhu I, and 425 to 643 (16.73 to 25.31 inches) between Machhu I and Machhu II [catchments of 736 and 1194 square kilometers (284 and 461 square miles), respectively]. A rainfall of 660 millimeters (26 inches) in 21 hours was recorded at Rajkot, about 64 kilometers (40 miles) from Machhu II.

Maximum inflow into the reservoir was reported to have exceeded 14 160 cubic meters (500 000 cubic feet) per second (possibly as much as 19 820 cubic meters (700 000 cubic feet) per second), two to three times the peak

---

[1] According to the World Register of Dams, published by the International Commission on Large Dams.

[2] "Floods overtop, breach Indian earthfill," *Engineering News-Record,* August 23, 1979.

Figure 4-7.—Machhu II Dam after failure (Courtesy, Irrigation Department, Government of Gujarat India). P-801-D-79352.

design inflow of 5660 cubic meters (200 000 cubic feet) per second. The maximum outflow was about 13 450 cubic meters (475 000 cubic feet) per second just before the collapse.

The Machhu II Dam, designed and built by the Gujarat State government and completed in 1972, consisted of a central masonry spillway in the main river section, with a 2347-meter (7700-foot) long earthfill structure on the left side and a 1524-meter (5000-foot) earthfill on the right. The embankments had 6.1-meter (20-foot) top widths, nominal slopes of 3 to 1 upstream and 2 to 1 downstream, clay core extending through alluvium to rock, "murum" (sandy loam with small gravel fraction) shells, and 0.6 meter (2 feet) of hand-placed riprap on 0.6 meter (2 feet) of rock spalls and sand bedding on the upstream face.

The masonry section consists of a 206-meter (676-foot) long spillway and a 92-meter (301-foot) long nonoverflow section. The ogee crest of the spillway has 18 radial gates, 9144 millimeters (30 feet) long and 6096 millimeters (20 feet) high. Spillway capacity was about 5550 cubic meters (196 000 cubic feet) per second.

The full reservoir level was 57.3 meters (188 feet) at the top of gates while the maximum specified flood level was 57.6 meters (189 feet). The reservoir had a gross storage capacity of 101 000 000 cubic meters (81 900 acre-feet). Due to the high value placed on stored water in this arid area, the reservoir level at the time of the emergency was being held at 56.7 meters (186 feet), in accordance with the operations manual.

Trouble started on August 10 around 9 p.m., when the 30.5-meter (100-foot) high Machhu I Dam, about 48 kilometers (30 miles) upstream from Machhu II Dam, started spilling. By noon, August 11, the water level at Machhu I had risen 2.4 meters (8 feet) over the designed high flood level. At Machhu II, by 1:30 p.m., the water had risen to 60.5 meters (198.5 feet), 2.9 meters (9.5 feet) above the maximum designed high flood level and 0.46 meters (1.5 feet) above the top of the dam. With overtopping, the embankment on both sides was washed out.

The deputy engineer in charge was stationed at Morvi. On August 10, he left for Machhu I Dam at 7 p.m. He got to Wankane at 8 p.m. Rain was starting at that time. He called Machhu II Dam and asked that all gates be raised from 150-millimeter (6-inch) opening to 1829-millimeter (6-foot) opening to hold the reservoir at the 56.7-meter (186-foot) level. He then tried to go to Machhu I but changed his mind when he saw the river rising, and he headed back toward Machhu II. He stopped at Wankane and called Morvi to tell them to sound the siren that would warn the people that all gates at Machhu II might be opened fully. This was normal practice. The siren was blown at midnight on August 10. The deputy engineer reached Machhu II at 11 p.m. on August 10. He found that 13 gates were open 3658 millimeters (12 feet). The water surface had already risen to 57.6 meters (189 feet). The crew tried manually to raise the gates further. They called for more laborers. Eventually 15 gates were fully opened, and the other 3 were opened partially. Most of the gate raising was accomplished by auxiliary power (diesel generators) since the electrical system in the area failed 2 days before the collapse. Motors on the last two gates burned out and the manual efforts to get those gates fully open failed.

At 8:30 a.m. on August 11, city officials from Morvi arrived at the dam and conferred with the project personnel. The deputy engineer returned to Morvi with them to warn the people to get to high ground, leaving his seven operators at the dam. He headed back to the damsite at noon, but was cut off by 1.5 meters (5 feet) of water over the access road to the dam. He waited for 2 hours, and then headed toward Wankane, but the bridge on that road had washed out. He was marooned. He finally got to Wankane at 8 p.m. on August 11 and called his supervisor at Rajkot to report on the flood. But at that time he was still not aware that Machhu II had failed. The collapse had happened at 3 p.m. The flood peak reached Morvi within an hour. The rate of rise in water level at Morvi was said to be 0.3 meter (1 foot) per minute.

The dam withstood overtopping for about 2 hours. The seven operators at the spillway tried to escape by running across the embankment toward the right abutment. However, they saw the danger of failure since water was already running over the top, so they retreated and took refuge in the operating house atop the masonry structure. They were isolated there for 48 hours, and survived, while the flood overtopped the entire dam, including the masonry.

The communications center at Rajkot was not informed of the failure until 8 a.m. on August 12, when a messenger, who finally was able to make his way to Wankane, called in.

Data pertinent to the disaster are summarized below:

	Machhu I	Machhu II
Design flood	2747 m³/s (97 000 ft³/s)	5663 m³/s (200 000 ft³/s)
Observed flood inflow	11 327 m³/s (400 000 ft³/s)	More than 14 158 m³/s (500 000 ft³/s) (Possibly 19 822 m³/s (700 000 ft³/s))
Outflow	9203 m³/s (325 000 ft³/s)	Approximately 13 450 m³/s (475 000 ft³/s) just before collapse
Design flood depth over spillway crest	2.7 meters (9 feet)	2.4 meters (8 feet)
Observed actual flood depth over spillway crest	4.9 meters (16 feet)	4.6 meters (15 feet)
Gates	Ungated	18 - 9144 mm (30 ft) long by 6096 mm (20 ft) high
Gates open	— — — — —	15 - fully open 1 - 4877 mm (16 ft) open 1 - 1829 mm (6 ft) open 1 - 1219 mm (4 ft) open
Overtopping depth	1.2 meters (4 feet)	0.6 meter (2 feet)

	Machhu I	Machhu II
Extent of failure	Remained intact (all-masonry structure)	1067 meters (3500 feet) of left embankment failed 701 meters (2300 feet) of right embankment failed
Average annual rainfall (50 years)	508 mm (20 in)	508 mm (20 in)
Maximum annual rainfall (50 years)	1067 mm (42 in)	1067 mm (42 in)
Minimum annual rainfall (50 years)	102 mm (4 in)	102 mm (4 in)
Service provided by reservoir	8094 hectares (20 000 acres) of irrigation	7284 hectares (18 000 acres) of irrigation Water supply to Morvi and other towns.

Machhu II Dam reportedly cost approximately $4 million (30 million rupees), compared with $2 million (17 million rupees) for Machhu I Dam. Initial estimates of the cost of remedial work were $19 million for reconstruction of Machhu II to pass 19 822 cubic meters (700 000 cubic feet) of water per second, another $19 million for flood protection, plus $5 million for a corresponding increase in discharge capacity at Machhu I Dam. At Machhu II, the initial tentative plan was to provide 24 additional gates 12 497 millimeters (41 feet) long by 8230 millimeters (27 feet) high, on the right side to augment the capacity of the existing 18 gates 9144 millimeters (30 feet) long by 6096 millimeters (20 feet) high.

Morvi was known as "the Paris of Saurashtra," for its broad avenues and green parks. The destruction left nearly all of it under several feet of mud and debris. As the floodwaters passed within a few days, thousands of shattered buildings were left, some with mud up to the second floor. Approximately $15 million in crops were damaged or destroyed in the disaster; 12,000 houses were destroyed and about 7,000 others partially damaged.

Malpasset Dam

Malpasset, a 61-meter (200-foot) high concrete arch dam (fig. 4-8) on the Riviera in the Cannes District near Fréjus in southern France, failed on December 2, 1959. The flood wave left behind total destruction in its 11-kilometer (7-mile) course to the Mediterranean and caused 421 deaths.

At the time of the structure's completion in 1954, it was reported to be the thinnest arch dam of its height, with a maximum thickness of about 6.7 meters (22 feet). The dam had a 105-meter (344-foot) axis radius and a central angle of 121 degrees between two rock abutments on Le Reyan River. On the right bank, the arch abutted a high rocky mass. On the left, a wingwall was necessitated by the site topography.

Construction was started in the spring of 1952. The outlet valve at the base was closed on April 20, 1954, to begin filling the reservoir, which had a

Figure 4-8.—Malpasset Dam, right abutment, showing remnant (Courtesy of L. B. James). P-801-D-79353.

normal capacity of 22 000 000 cubic meters (17 800 acre-feet). The dam led a relatively uneventful life until the fall of 1959, when the watershed was drenched by heavy rains. By mid-November, the water level in the reservoir was at elevation 95.20 meters (312 feet) above sea level, 5.20 meters (17 feet) below the normal maximum water surface. At that time, the operator discovered seepage at the right abutment about 20 meters (65 feet) downstream from the dam.

Beginning on November 28, there was another intensive rainfall. The next day, the reservoir level had risen to elevation 95.75 meters (314 feet). Runoff from heavy rains on the night of the 29th brought a further increase to elevation 97 meters (318 feet) at 6 p.m. on November 30. The seepage had increased.

By noon on December 2, the reservoir level was at elevation 100 meters (328 feet). At 6 p.m. the outlet valve was opened and the reservoir water surface was at elevation 100.12 meters (328.5 feet). An hour and a half later a lowering of 0.03 meter (0.1 foot) from this maximum was noted. The care-taker left the dam at 8:45 p.m. and went to his home on the hillside, about 1.6 kilometers (1 mile) downstream from the reservoir. At about 9:10 p.m., he heard a loud cracking noise, and at about the same instant, a violent blast blew open doors and windows. There was a bright flash observed by the caretaker, and the lights went out. The powerline of the Malpasset Dam was reported to have broken at 9:13 p.m.

After the failure, the right side of the dam and the base of the central part remained. The left side was gone except for the wingwall and part of the concrete abutment, or thrust block. The latter had separated from the wingwall and had slid downward. The surviving elements of the dam were cracked and the joints were displaced. A large volume of left abutment rock downstream from the dam had disappeared. The exposed formation was ex-tensively cracked.

An awesome manifestation of the dam's movement during failure was the large gap opened between the upstream face and the right abutment rock (fig. 4-9). The arch apparently rotated about its right end. Investigators were able to enter the gap to a depth of about 8 meters (25 feet).

No witnesses to the collapse were found. One report, which is perhaps of questionable accuracy, said that the guard fled and left the telephone intact. The police at Fréjus were said to have received a warning on an ap-proaching flood but did not have a way to sound a timely warning.

Most investigators soon focused on the weak rock of the left abutment as a primary contributor to the failure. But various other possibilities had to be checked, even though some could be dismissed quickly. There appeared to be no validity in the hypotheses of earthquake, explosion, meteorite, or abnormal vibration of the outlet valve. The seepage seen on the left bank after the collapse was also pointed to as suspect. However, presumably reliable reports say that the caretaker passed along that side just before the disaster without seeing any alarming leakage. This convinced some analysts that the seepage under the left end of the arch did not precede, but was instead a consequence of, the foundation movement associated with the dam's collapse.

At the direction of the Inquiry Commission appointed by the French Government, the postfailure positions of the remaining dam elements were

Figure 4-9.—Malpasset Dam, right abutment, showing gap caused by pivoting (Courtesy of L. B. James). P-801-D-79354.

measured. This survey showed that there had been a rotation of the dam around a pivotal point on the right abutment. On the left side, the concrete abutment was displaced more than 2 meters (6.6 feet) in a direction approximately tangent to the arch. Uplift pressure in the rock abutment was regarded as a possible causative factor. In view of the fine fissures in the formation, which would not accept grout but still transmitted water, this theory warranted careful analysis.

The Commission concluded: "The construction work was very good, particularly as regards the quality of the concrete and the bondage of the concrete with the foundation rock ***. To sum up, the detailed examination of the conditions prevailing during the construction do not reveal any element which could explain the catastrophe."

The designers' computed stresses were found to have been within acceptable limits. Investigators judged the anchorage of the concrete abutment itself to have been adequate. The break did not occur at the contact between the concrete and the rock but within the foundation. Rock tests were therefore conducted at the site and in the laboratory. The Inquiry Commission decided that an extreme overloading caused the abutment and the underlying rock to move relatively to a plane of shear. The arch could not deform as much as required to transmit the thrust to the deformable rock. It was estimated that the slow redistribution of stresses in the arch, which was the first phase of the rupture, may have taken several days or even weeks. The second phase — the final collapse — took just a a few seconds.

Professor Karl Terzaghi commented on the Malpasset failure in February 1962 as follows:

"The left abutment of this dam appears to have failed by sliding along a continuous seam of weak material covering a large area. A conventional site exploration, including careful examination of the rock outcrops and the recovery of cores from 2-inch boreholes by a competent driller, would show — and very likely has shown — that the rock contained numerous joints, some of which are open or filled with clay.

"From these data an experienced and conservative engineer-geologist could have drawn the conclusion that the site is a potentially dangerous one, but he could not have made any positive statement concerning the location of the surface of least resistance in the rock and the magnitude of the resistance against sliding along such a surface ***. All foundation failures that have occurred, in spite of competent subsurface exploration and strict adherence to the specifications during construction, have one feature in common. The seat of the failure was located in thin weak layers, or in 'weak spots' with very limited dimensions.

"None of the methods of exploration, including those used by mining and petroleum engineers, provides adequate information concerning such minor geological details."

Witnesses at court testified that there were two faults on the left abutment forming a dihedral angle which was displaced by water pressure. There was also testimony that measurements of dam deflections were made annually in the period 1955 to 1959. A displacement at the dam base of 17 millimeters (0.67 inch) was reported. Yet the chief designer had judged the tolerable deflection at that point to be only 10 millimeters (0.39 inch). The court learned that a firm of specialists had conducted the displacement survey, but the design office had not been informed of the excessive deflection until after the failure.

The fatal plane of weakness in the rock of the left abutment was about 30 meters (100 feet) from the dam and had not been regarded as a threat to the safety of the arch. An unrecognized characteristic of the gneiss rock at the damsite, however, was that it has a permeability from 100 to 1,000 times higher in tension than in compression. As a result, the stresses in the abutment in response to the reservoir loading on the dam caused a rapid infiltration and development of water pressures on planes normal to the weak foundation plane. This forced a wedge of rock to "blow out" and triggered the collapse of the dam.

Mill River Dam

Failure of the Mill River Dam near Williamsburg, Hampshire County, Mass., on May 16, 1874, demonstrated the consequences of inadequate professional supervision. No engineers were employed in the design or construction of this structure, which was completed by the Williamsburg Reservoir Company in 1865.

At the time of failure, the lake level was approximately 1.2 meters (4 feet) below the dam crest. The collapse occurred between 7 and 8 a.m. on May 16th. In about 20 minutes, probably three-quarters of the stored water surged from the reservoir, corresponding to a discharge of roughly 1700 cubic meters (60 000 cubic feet) per second. The sudden spilling of reservoir water into the steep and narrow valley caused the deaths of 143 people.

The damsite was on a branch of the Mill River, a small stream which joins the Connecticut River at Northampton. The dam was a 13-meter (43-foot) high earthfill structure with a masonry core wall. Total length of the embankment was about 183 meters (600 feet). It had a crest width of 4.9 meters (16 feet) and slopes of 1.5 to 1. The core wall was 0.6 meter (2 feet) thick at the top and 1.75 meters (5.75 feet) at the base. It was founded on soil in a trench 0.9 meter (3 feet) deep. A spillway 10 meters (33 feet) wide was built in one abutment, and a 400-millimeter (16-inch) outlet pipe was laid through the dam at the lowest point for service to the mills downstream.

The foundation at the site was reported to be a very compact hardpan, overlain by about 0.6 meter (2 feet) of coarse gravel and a few inches of soil. The material for the dam, a gravel containing a little loam, was excavated from the hill immediately above the dam. Considering the nature of the foundation and the permeability of the fill, the integrity of the structure depended heavily upon the masonry core wall and its close contact with the hardpan. The specifications called for trenches 0.9 meter (3 feet) deep, or of sufficient depth to ensure a firm bottom which would preclude settlement of the masonry. Apparently, the important need to prevent underseepage was recognized but given insufficient attention in the design. The constructors probably assumed that the embankment could be made tight enough to keep seepage from reaching the base of the wall. The specifications required that the fill be placed in thin, watered layers, and tamped or beaten with a maul for a distance of 1.5 meters (5 feet) from each side of the wall. The remainder of the embankment was constructed in 1.5-meter (5-foot) layers using this permeable material, on a specified outer slope of 1.5 to 1, which was clearly inadequate by accepted standards.

In addition to the weaknesses in design, inspection of the construction was reportedly limited or nonexistent. The remains of the dam showed glaring defects in workmanship. In some places the base of the wall did not even extend to the hardpan. In violation of the specifications, the site had not been properly stripped of gravel and soil prior to placement of the embankment. At the inquest, evidence was presented that the masonry had been placed dry and grouted in 1.5-meter (5-foot) stages. The grout was of poor quality and had not fully penetrated the voids in the stonework.

From all this evidence, the cause of failure was clear. Seepage had gradually carried away fill under and adjoining the core wall, creating dangerous cavities. Leakage under the masonry at a point about 30 meters (100 feet) from the outlet pipe had triggered a slide in the embankment downstream from the wall. In sequence, the masonry, deprived of its support on that side, collapsed under the pressure of the upstream fill. The breach quickly expanded as the embankment eroded and other sections of the rock wall toppled.

The caretaker of the dam was first alerted when he discovered the slide on the downstream slope. The outlet gate was opened immediately to release the water. Meanwhile, another slide occurred, and within half an hour the entire dam appeared to be moving. In retrospect there was little doubt that the embankment had suffered from internal erosion for several years before its collapse.

In summary, a member of a committee of inquiry stated: "The result is this: the company paid towards the educating of that contractor at least $15,000, to which is now to be added at least $1,000,000 for damages caused by the failure. Men were employed who were ignorant of the work to be done, and there was nothing like an inspection, although money and life depended upon it. I do not believe, however much we are an evolved species, that we are derived from beavers; a man cannot make a dam by instinct or intuition."

Möhne Dam

On May 17, 1943, bombers attacked the Möhne Dam, a 40.3-meter (132-foot) high concrete gravity structure constructed in 1913. This raid near Möhne, Germany, was coordinated with attacks on the Eder and Sorpe Dams in an attempt to flood the industrial complex in the Ruhr Valley.

The bombing occurred at a time when the Möhne Reservoir was filled to its capacity of 134 000 000 cubic meters (109 000 acre-feet). Through a breach 77 meters (253 feet) long and 22 meters (72 feet) deep, 116 000 000 cubic meters (94 000 acre-feet) of water flowed within 12 hours. The initial flow rate was estimated to have been 8778 cubic meters (310 000 cubic feet) per second. The narrow Möhne Valley was swept by a wave 10 meters (33 feet) high which caused widespread destruction and killed about 1,200 people. Bridges were washed away as far as 50 kilometers (31 miles) downstream from the dam. The powerplants at the dam were obliterated. Where the Ruhr River joins the Rhine about 150 kilometers (93 miles) downstream, the flood wave — reportedly still 4 meters (13 feet) high — passed by 25 hours after the air raid.

Since the Möhne Dam is a key facility in the water system for the densely populated Ruhr Valley, the breaking of the dam was a critical blow. However, this was compounded with the consequent flooding out of most of the other waterworks along the Ruhr River down into the Essen area.

The Möhne Dam breach was repaired later in 1943. Water was released on an interim basis through two outlet conduits. Then in 1950 the reconstruction of the powerplants was begun.

Nanaksagar Dam

The 5-year-old Nanaksagar Dam on the Deoha River 257 kilometers (160 miles) east of New Delhi, India, failed on September 8, 1967, and released a flood that caused destruction in 32 villages and the deaths of about 100 people. The city of Shahjahanpur was threatened, and extensive low-lying areas had to be evacuated.

The breaking of the dam came in the midst of a monsoon which brought the worst flooding in 20 years to northern India. Estimates of the toll of human lives during the storm exceeded 500. The Taj Mahal in Agra was reported to be nearly encircled by the waters of the flooded Jamuna River and its tributaries.

Bursting of the Nanaksagar Dam poured such a flood into the valley that the Garra River, 16 kilometers (10 miles) from Shahjahanpur, rose 0.91 meter (3 feet) during the afternoon. The death toll would have been greater except for the efforts of hundreds of runners sent out to sound the alarm.

The dam, a homogeneous earthfill structure, was constructed on a pervious foundation. There was not an effective cutoff to inhibit seepage under the dam, but a partial impervious earth blanket was placed upstream of the embankment. The 2 to 1 upstream slope of the dam was protected in some reaches with hand-placed boulders ranging from 152 to 355 millimeters (6 to 14 inches) in size, while the remainder of the face was covered by a wire reinforced layer of bricks placed over a sand-and-gravel filter bedding. This extended up the slope to a 0.3 meter (1-foot) high parapet on the crest.

Heavy seepage was discovered on the downstream slope of the 16-meter (52-foot) high earthfill structure on August 27, 1967. By September 5, water was gushing from the embankment. Workmen of the government of Uttar Pradesh, which built the dam, were able to stem the flow temporarily by placing inverted filters. But 2 days later a new muddy leak placed the 3.2-kilometer (2-mile) long embankment in jeopardy. A crack had appeared on the downstream slope at about noon on September 7. During the following night, conditions at the 210 000 000 cubic meter (170 000-acre-foot) reservoir worsened. Then at about 1 a.m. on September 8, an intensely discharging boil appeared 16.8 meters (55 feet) downstream from the toe. This enlarged quickly and was soon beyond control. The embankment started to settle. A large quantity of water began to flow through the filter berm. As settlement continued, the dam was overtopped at 1:30 a.m. and a breach opened in the embankment at a point 1.5 kilometers (0.9 miles) west of the spillway, where the dam was 10.4 meters (34 feet) high. The break expanded rapidly to a width of about 150 meters (500 feet).

The reservoir level evidently was not lowered during the prefailure emergency. Operations personnel reportedly considered taking this precaution when seepage was first detected, but they were reluctant to take the responsibility for the consequent downstream flows.

The part of the embankment where foundation erosion began had been constructed over an old channel of the river. Recognizing the potential weakness at that point, the builders had provided a downstream filter blanket, as well as 50-millimeter (2-inch) diameter relief wells near the downstream toe on a spacing of 15.2 to 30.4 meters (50 to 100 feet). These precautions were not enough to control seepage through the permeable strata of the riverbed.

India's Ministry of Irrigation and Power concluded that the break was due to a foundation failure in the area of the old streambed. The body of the earthfill itself was generally judged to have been in satisfactory condition. The Ministry reasoned that otherwise the entire dam would have collapsed.

There was some question as to the condition of the earthfill prior to the fatal storm for it had been in jeopardy before. Intense wind and rain during the preceding October, in an unusually wet monsoon, had caused impairment of the embankment. During the night of October 4, 1966, high winds caused severe wave wash on the upstream slope of the embankment. At that time the reservoir surface was only 1.68 meters (5.5 feet) below the normal maximum level. Both the riprap and the brickwork revetment were damaged and the earthfill was eroded to depths of 1.5 meters (5 feet) or more in several places. The boulder protection was almost completely washed away, and the brick paving and the sand-and-gravel bedding suffered extensive erosion and bulging. Federal officials inspected the structure following the storm and recommended a remedial program. The corrective measures accomplished in the subsequent months were clearly inadequate to eliminate the basic design defects.

Orós Dam

Just before midnight on March 25, 1960, an earthfill dam under construction at Orós in the State of Ceará in northeastern Brazil was overtopped and breached by a flood resulting from rainfall of more than 635 milllimeters (25 inches) in less than a week.

The embankment, which is just upstream of the Klus gorge, was at a vulnerable stage when the storm arrived. Between March 21 and March 25, heavy rains increased the flow of the Jaguaribe River to approximately 2265 cubic meters (80 000 cubic feet) per second, of which about one-fifth could be discharged through the diversion tunnel. Construction crews could not raise the dam fast enough to prevent overtopping. More than 765 000 cubic meters (1 000 000 cubic yards) of material were washed out in a few hours, leaving a breach about 201 meters (660 feet) wide. The peak flow through the gap was about 9630 cubic meters (340 000 cubic feet) per second. Floodwaters cut a path of destruction through the Jaguaribe Valley before arriving at the Atlantic 338 kilometers (210 miles) away.

The Orós Dam is nearly semicircular in plan with an axis radius of 150 meters (492 feet). It is 54 meters (177 feet) high and 620 meters (2034 feet) long. The crest width is 10 meters (33 feet). The core of the embankment is a compacted clay zone with slopes of 1 to 1 and a base width of about 100 meters (328 feet). There is an intermediate zone of compacted sand on each side of the core. The dam was finished with rockfill sections having outer slopes varying from 2.5 to 3.0 to 1 upstream and 2.0 to 2.5 to 1 downstream. Total volume of fill is approximately 3 000 000 cubic meters (3 900 000 cubic yards). Reservoir capacity is 4 billion cubic meters (3 240 000 acre-feet).

Rainfall on the watershed is generally limited to the period from January to May. The construction program for the dam was scheduled to minimize the risk of flood damage during these months. Estimated total construction time was 14 months, which by necessity would span one flood season. Completion was planned for November 1959.

In October 1958, construction was begun. When the peak runoff of early 1959 came, the work was still at a low level and no significant damage was done. However, from that point on, the job began to lag behind schedule. By the time the next rainy season arrived, there was an accumulated delay of about 4 months. In early March 1960, the top of the incomplete fill was approximately 30 meters (98 feet) above the foundation, or at elevation 185 meters (607 feet) above sea level. Then the rainstorms hit the watershed.

On March 12, 1960, the reservoir water surface was 165 meters (541 feet) above sea level, at the invert or bottom of the 6-meter (19.7-foot) tunnel for the future powerplant. With the tunnel functioning as a diversion conduit of limited capacity, the water level continued to rise. Between March 12 and 21, the water surface rose to an elevation of 180 meters (590.6 feet) above sea level, corresponding to an increase in storage from 3 000 000 to 180 000 000 cubic meters (2432 to 145 900 acre-feet). In the period March 19 to 25, there was a total of 648 millimeters (25.5 inches) of rain on the drainage area. On March 23, the reservoir inflow from the Jaguaribe River attained an estimated maximum of about 2265 cubic meters (80 000 cubic feet) per second. In contrast, the capacity of the diversion tunnel probably was not much more than 396 cubic meters (14 000 cubic feet) per second.

With the water surface approaching the top of the unfinished embankment at a rate of about 3 meters (10 feet) per day, two alternative courses of action were considered on March 22: (1) to sacrifice the dam by cutting a spillway in the crest or (2) to attempt to save the dam by the rapid construction of a narrow superimposed earthfill. The latter option was chosen, in the hope that the crest elevation could be kept ahead of the rising water until the storm abated.

From March 22 on, construction crews labored around-the-clock to raise the uncompacted emergency fill. The best they could do was to raise it about 1 meter (3 feet) per day, because the intense rainfall limited the use of heavy equipment. By March 25 the situation was hopeless. At noon on that day, the rising water was within about 1 meter (3 feet) of the top of the emergency dike, which had been built to elevation 190.50 meters (625 feet). Since the beginning of the battle on March 22, storage had increased from 200 000 000 cubic meters (162 000 acre-feet) to 650 000 000 cubic meters (527 000 acre-feet). On the afternoon of March 25, with overtopping accepted as unavoidable, a channel was bulldozed in the fill at the right end with the expectation that erosion would be slower there than at the center. That evening, metal sheets were dropped onto the embankment from aircraft to provide resistance against overwash. About 11:30 p.m. on March 25 the emergency embankment was overtopped, and by 3 a.m. the next morning the storage elevation was 190.85 meters (626.2 feet). With water spilling over the entire length of the crest, erosion accelerated. Later in the day of March 26, discharge attained an estimated peak of 9600 cubic meters (339 000 cubic feet) per second. By noon on March 27, about 90 percent of the reservoir had been emptied; and a 200-meter (660-foot) long central section of the dam was gone.

Evacuation of about 100,000 people from the Jaguaribe Valley was started on March 22. After the overtopping of the dam on March 25, alerts were transmitted by radio to all the endangered areas. The flood wave hit with full force the settlements occupying the narrow valley just downstream from the dam. Castanheiro, a little village at the confluence of the Jaguaribe and Salgado Rivers, was reportedly washed away. The larger town of Jaguaribe, about 75 kilometers (47 miles) downstream from Orós, was struck by the wave approximately 12 hours after the failure started. Six hundred out of a total of about 1,200 dwellings were ruined or severely damaged. Nine hours later the torrent invaded Jaguaribara, 35 kilometers (22 miles) farther downstream, where more than 700 homes were damaged or completely destroyed. Estimates were that about 50,000 people were left homeless in the flooded valley. Reports on the number of dead vary widely.

After repairs, the construction of Orós Dam was resumed. The reservoir went into operation in February 1961.

Panshet Dam

On July 12, 1961, the nearly completed Panshet Dam in the vicinity of Poona in India was overtopped and washed away. The released waters poured into the Khadakwasla Reservoir a few miles downstream, causing it to overflow and then break. Continuing down the Mutha River, the flood hit the city of Poona and wrought widespread destruction. Approximately 5,000 homes were either damaged or demolished.

The 51-meter (168-foot) high Panshet Dam, under construction since 1957, was a zoned earthfill structure with an impervious central core. Total volume of embankment was 2 700 000 cubic meters (3 500 000 cubic yards). A side spillway having a capacity of 487 cubic meters (17 200 cubic feet) per second was constructed. There were plans for future addition of crest gates to enlarge the reservoir. An outlet was located in a trench on the left abutment. The upper part of this outlet conduit was a semicircular masonry arch which was supported on concrete sidewalls. Outlet releases were to be controlled by two sluice gates in the base of the tower at the upstream end.

On June 18, 1961, the monsoon came early to the Western Ghats, with heavy rain falling on the Panshet watershed. Some of the work on the dam was not yet complete when the storm arrived. The intense rainfall seriously slowed construction. Riprap could not be placed at the temporary spillway, and the excavated channel of the permanent one was still irregular from blasting. Outlet gates and hoists were not yet fully installed. The bridge providing access to the outlet tower was only partly completed and therefore unusable. Concrete had not been placed on the invert, or floor, of the outlet conduit. Excavation at the discharge end of the conduit had not progressed deep enough to provide for a safe release of the water.

Between June 18 and July 12, there was a rainfall of 1778 millimeters (70 inches). The storm runoff caused such a rapid rise in the reservoir level that the new embankment had only minimum time to adjust to the new loading. The water rose initially at a daily rate of about 9 meters (30 feet), and later it came up 24 meters (78.5 feet) in 12 days. Wave wash was apparent on the upstream side of the dam, being most noticeable at the temporary spillway.

Due to the rough condition of the outlet conduit, the flow through it was unsteady, with pressure surges. A leakage, estimated between 0.14 and 0.28 cubic meter (5 and 10 cubic feet) per second, developed through the downstream rock toe above the conduit. This flow did not appear to be increasing at the time. In the first few days of July 1961, settlement was discovered on the upstream face of the embankment above the outlet conduit. There were cracks along the edges of the settled sections perpendicular to the axis of the dam. The area of subsidence was approximately 9 meters (30 feet) long and 6 meters (20 feet) wide. At about 5:30 p.m. on July 6, two more cracks appeared. These were roughly 30 meters (100 feet) apart and alined almost directly over the sides of the outlet trench.

The rate and magnitude of settlement were alarming, amounting to as much as 1.4 meters (4.5 feet) in one 2½-hour period. By 5 a.m. on July 11, the crest was only about 0.6 meter (2 feet) above the reservoir water surface. Earth-filled bags were dumped in the depression but to no avail, because the rate of subsidence was too great, and the inevitable overtopping occurred at about 6:30 a.m. on July 12, 1961.

Official investigation of the failure concluded that it was neither attributable to insufficient spillway capacity nor to any foundation defect. The disaster was determined to be caused primarily by the incomplete condition of the facilities and the unsuitable design of the outlet works for the emergency operation to which they were subjected. The outlet gates had been designed by the governmental organization and fabricated by departmental workshops. Design deficiencies were compounded with manufacturing defects.

The gates had been placed in the guides and suspended on temporary chains pending delivery of essential parts. They were set to discharge at a partial opening of 0.6 meters (2 feet) which was a tragic error. Operating in the intermediate position, the gates vibrated violently, and there were air entrainments which created a sharply pulsating flow in the conduit. The large pressure fluctuations caused progressive disintegration of the masonry arch of the conduit, which led to collapse of the embankment overlying the outlet. With this support removed, the continued subsidence and the overtopping of the dam were just a matter of time.

Puentes Dam

The original Puentes Dam was a rubble-masonry structure built during the years 1785 to 1791 on the Río Guadalentín in Murcia Province in Spain. It failed on April 30, 1802.

The dam was 50 meters (164 feet) high and 282 meters (925 feet) long. It was a massive barrier, nearly 11 meters (36 feet) thick at the crest and 44.2 meters (145 feet) at the base. In plan, it was composed of three straight sections, 53 meters (174 feet), 124 meters (407 feet), and 105 meters (345 feet) long forming a dam with a polygonal shape, being convex upstream. The upstream face of this gravity dam was vertical. The structure had a mortared rubble-masonry core, and its faces were made of cut stones.

The original plan was to found the dam entirely on rock. However, early during construction a deep deposit of earth and gravel was discovered in the foundation at the river. Hundreds of wood piles, 6 meters (20 feet) long, were driven into the alluvium and were braced with horizontal timbers at their tops. The masonry of the dam base extended into this grillage to a depth of about 2 meters (7 feet). To avert erosion of the alluvial channel by discharges from the sluiceway and outlet, the network of piling and bracing was extended 40 meters (131 feet) downstream from the dam. This was covered with about 2 meters (7 feet) of masonry which was overlain by planks.

During the period 1791 to 1802, a maximum water depth of 25 meters (82 feet) had accumulated in the reservoir. During these 11 years, the dam evidently did not show signs of distress. Then early in 1802, the impoundment began to increase and by the end of April had reached a depth of 47 meters (154 feet). At this point the dam failed.

Hydrostatic pressure had blown a hole all the way through the alluvium under the dam, which was so large that the stored water was quickly drained. Six hundred eight people were drowned in the flood that disgorged from the reservoir.

A witness at the scene gave this account:

"About half-past two on the afternoon of the 30th of April, 1802, it was noticed that on the down-stream side of the dam, towards the apron, water of an exceedingly red colour was issuing in great quantities in bubbles, extending in the shape of a palm-tree. Immediately someone was sent to inform Don Antonio Robles, the director of the works. About three o'clock there was an explosion in the discharge-wells that were built in the dam from top to bottom, and at the same time the water escaping at the down-stream side increased in volume. In a short time a second explosion was heard, and, enveloped by an enormous mass of water, the piles, beams and other pieces of wood which formed the pile-work of the foundation and of the apron were forced upwards.

"Immediately afterwards a new explosion occurred, and the two big gates that closed the scouring-gallery, and also the intermediate pier, fell in; at the same instant a mountain of water escaped in the form of an arc; it looked frightful and had a red colour, caused either by the mud with which it was charged, or by the reflection of the sun. The volume of water which escaped was so considerable that the reservoir was emptied in the space of one hour. The water reached Lorca ahead of the messenger sent to tell the director of the first signs of the disaster; as the flood caught up with him, he was obliged to escape up a nearby hill.

"The dam presents since its rupture the appearance of a bridge, whose abutments are the work still standing on the hillsides, and whose opening is about 17 metres broad by 33 metres high.

"At the moment of the accident the effective depth of the water was 33.4 metres. Its surface was 46.8 metres above the base of the dam, the lower 13.4 metres being taken up by the deposited material."

Some years later, most of Puentes Dam was removed. In 1884, a new concrete gravity dam was completed on the site. It is 69 meters (226 feet) high and 291 meters (954 feet) long.

St. Francis Dam

The 62.5-meter (205-foot) high St. Francis Dam (fig. 4-10), a part of the water supply system of Los Angeles, Calif., collapsed (fig. 4-11) on the night of March 12, 1928 without warning. Floodwaters killed 450 people. The concrete dam, an arched concrete-gravity type, was built in San Francisquito Canyon, roughly 72 kilometers (45 miles) north of Los Angeles. The reservoir, which began filling in 1926 and reached practically full capacity at EL 559.2 meters (El. 1834.75 feet) on March 5, 1928, had a volume of 46 900 000 cubic meters (38 000 acre-feet) and was essentially full at the time of the failure.

One of the caretakers reportedly was observed on the crest of the dam at 11 p.m., just an hour before the collapse, so presumably at that time no alarming condition had been discovered. The caretakers disappeared in the flood. There was no living witness to the dam's last minutes.

At 11:47 p.m., only 10 minutes preceding the disaster, the operator at the powerplant above the reservoir had called the lower plant, but nothing unusual was reported by the staff on duty there. The break came just before midnight. At 11:57:30 p.m. that night in the city of Los Angeles, there was a momentary flickering of lights. At 11:58 p.m., a break occurred in the power transmission line of the Southern California Edison Company in the canyon downstream from the dam. Farther south at the Company's Saugus Substation, an oil circuit breaker reportedly exploded, and a lineman was sent up San Francisquito Canyon to find out what had happened.

On a hill above the powerhouse and downstream from the dam, the home of one of the workmen for the Bureau of Power and Light was shaken. The residents waited stunned for a moment as the windows rattled. The rumbling became more ominous, and the entire house was vibrating strongly. Then the lights went off.

Down in the canyon, another employee was roused from his sleep by a thunderous sound. He hurried outside, and then a tremendous flood wave bore down upon the place. The roof of a building washed toward him, and he clambered onto it. After a short and turbulent ride, he was thrown onto the canyon slope and was able to climb out of reach of the waters. Searching hopelessly for his family, he wandered until dawn, when he found a woman and her young son, the only other survivors from the company settlement at the powerplant.

There is no doubt that the collapse came all at once. In about 70 minutes, practically the entire 46 900 000-cubic-meter (38 000-acre-foot) storage in the reservoir had been spilled. The flood wave attained an estimated maximum depth of 38 meters (125 feet) in the first 1.6 kilometer (1 mile) below the dam. At the lower powerplant, about 0.8 kilometer (0.5 mile) farther downstream, an even greater depth was reported. The heavy concrete

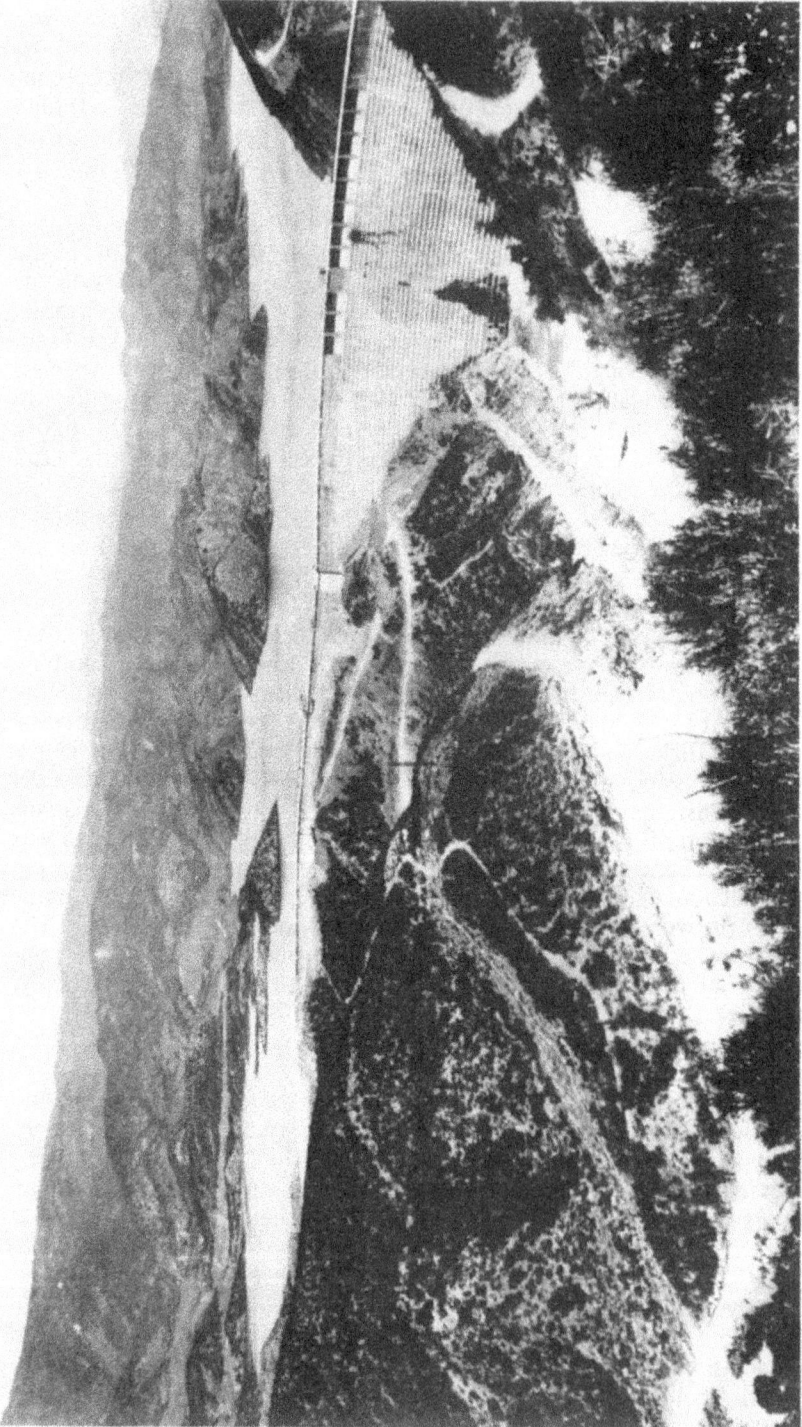

Figure 4-10.—St. Francis Dam before failure (Courtesy, Calif. Dept. of Water Resources). P-801-D-79355.

Figure 4-11.—St. Francis Dam after failure (Courtesy, Calif. Dept. of Water Resources). P-801-D-79356.

powerhouse was torn away down to the generator floor; also, the homes occupied by the workmen and their families were washed away. The torrent rushed down San Francisquito Creek 14 kilometers (9 miles) from the dam and then down the Santa Clara River 70 kilometers (43.5 miles) to the ocean.

Peak discharge just below the dam probably was greater than 14 160 cubic meters (500 000 cubic feet) per second. The enormity of this sudden deluge so engulfed the dark and narrow canyon that very few of its inhabitants escaped with their lives, even though they were close to the safety of the bordering hills. At a construction camp of the Southern California Edison Company about 26 kilometers (16 miles) below the dam, more than 80 out of about 140 people died. The flood buried large expanses of the valley in mud and debris. Great damage was done to highways, bridges, and the railroad along the Santa Clara River. Fortunately, no trains were traveling on the railroad track that was inundated, and not many automobiles were on the highways that late at night. Three hours after the failure, the wave struck the town of Santa Paula, 61 kilometers (38 miles) downstream from the damsite; the town suffered badly.

Great blocks of the concrete mass of the dam, weighing many thousands of tons, were washed thousands of feet down the canyon. Practically all the fragments from the westerly side of the structure were transported downstream (fig. 4-12). Many of the large pieces from the easterly side moved just a short distance and came to rest against the base of the solitary 30-meter (100-foot) long section still standing. However, other large blocks from that end were moved as far as those from the westerly end, being identifiable from inclusions of schist. One large section from the west side was lying bottom up, and the rock embedded in it showed that it had been founded on the fault contact between the schist and conglomerate formations at the damsite.

The single erect monolith (fig. 4-13) survived essentially unmoved from its original position, despite the violent shock of having both ends torn away. That this section had been subjected to tilting or twisting was evidenced by a measured displacement at the top of 140 millimeters (5.5 inches) in the downstream direction and 150 millimeters (6 inches) toward the east abutment. On each side of this block was a large gap approximately 90 meters (300 feet) wide. The breach on the right or west side was stripped clean by the flood. The gap on the left or east side, in contrast, contained large blocks of concrete. The difference in the condition of the two gaps suggested to some observers that the failure may have started on the right side.

A survey of the remaining portion of the west wingwall showed that it had risen as much as 91 millimeters (0.3 foot). Downslope from the wall, in the lower part of the west abutment, the conglomerate was washed away to a depth of as much as 9 meters (30 feet), and on the east side of the standing monolith the schist was cut to an estimated maximum depth of 12 meters (40 feet).

A huge slide was active on the east abutment for at least 2 weeks following the disaster. This was a manifestation of even greater movement within the mountain. Cracks were discovered 61 meters (200 feet) farther up the slope, indicating the total magnitude of dislocation of the rock mass.

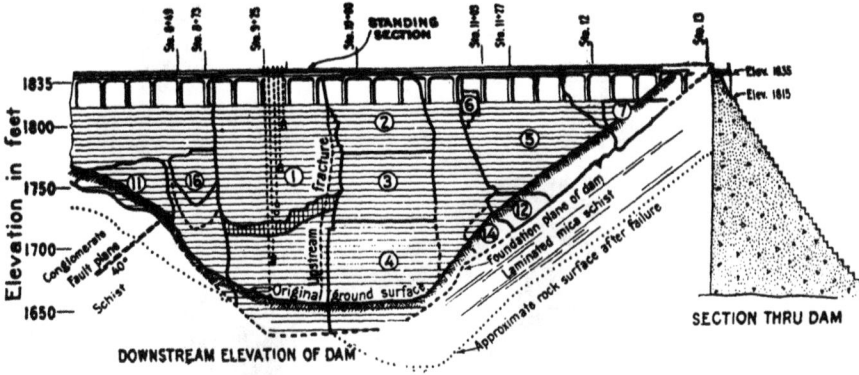

DIAGRAM SHOWING EROSION
AND
ORIGINAL POSITION OF IDENTIFIED FRAGMENTS

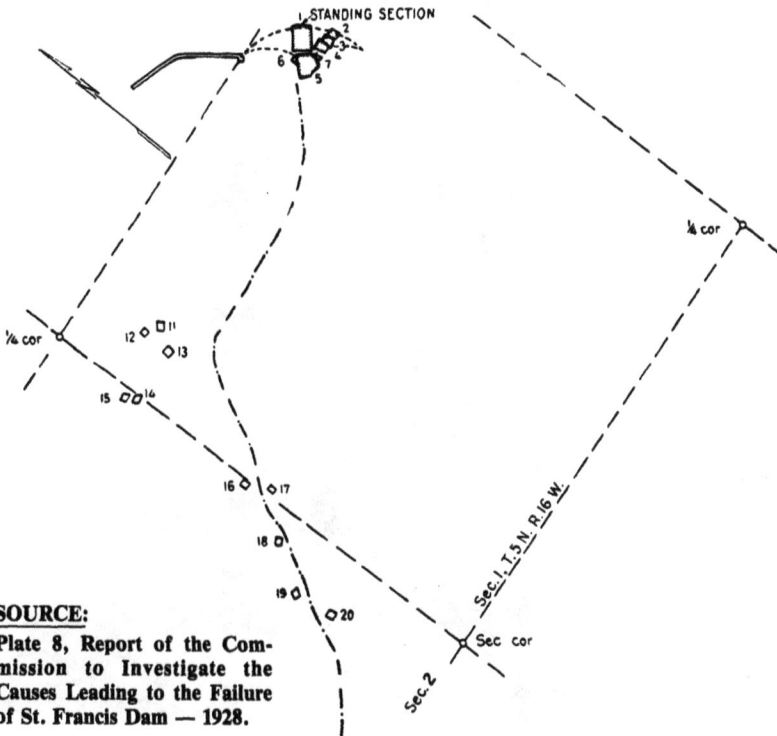

SOURCE:
Plate 8, Report of the Commission to Investigate the Causes Leading to the Failure of St. Francis Dam — 1928.

ST. FRANCIS DAM
DISTRIBUTION OF FRAGMENTS
Survey by H. Wildy
SCALE OF FEET

Figure 4-12.—St. Francis Dam, distribution of fragments. P-801-D-79357.

Figure 4-13.—Surviving monolith of St. Francis Dam (Courtesy, Calif. Dept. of Water Resources). P-801-D-79358.

However, there was evidence suggesting that the sliding started above the dam rather than under it. Rescue parties arriving at the site early in the morning after the failure found debris at an elevation on the slopes 1.2 meters (4 feet) higher than the maximum shoreline. Such an exceptional wave wash could have been attributable to movement of the slide mass into a full reservoir.

Some observers conjectured that a massive landslide on the left abutment may have caused the failure. The moving rock mass could have forced out that end of the dam, leaving part of it downstream from the monolith that stayed in place. There were such concrete remnants. If the left side failed first, somewhere down the canyon there should have been pieces of the dam from that end lying under fragments from the right side. There was such evidence. One concrete block from the east end of the dam was found about 1067 meters (3500 feet) downstream. This was about as far as any of the large fragments were carried by the flood.

Indication of a high velocity of flow toward a breach on the left side early in the failure was the breaking of the 300-millimeter (12-inch) stilling well pipe about 7.6 meters (25 feet) below the dam crest and its bending toward the left abutment. While this evidence, along with the pattern of movement of the concrete remnants, is not conclusive, it provides a reasonable theory as to the sequence of failure.

There was some speculation that the foundation failure may have started near or at the fault contact between the conglomerate and schist at the west side of the channel as a result of seepage through this part of the foundation. The supposition was that such a flow would have softened the conglomerate under the dam. Assuming either a blowout underneath or settlement of the dam at that point, collapse of the remainder of the structure could have ensued quickly.

At the time of its destruction, the St. Francis Dam was less than 2 years old. Construction, which began in April 1924, was completed on May 4, 1926. The 62.5-meter (205-foot) high concrete gravity dam (fig. 4-14) was arched on a radius of 152.4 meters (500 feet) to the upstream face at the crest. At the right abutment, the barrier was extended by a low wall along a narrow ridge, ending at a point about 150 meters (500 feet) from the main mass of the dam. The crest thickness of the dam was 4.88 meters (16 feet), and at the maximum section, the base thickness was 53.4 meters (175 feet). The length of the main mass was about 213 meters (700 feet) along the curved crest.

Eleven spillway openings, each 6.10 meters (20 feet) long and 0.46 meter (1.5 feet) high, were built in two series near the center of the dam. There were five 750-millimeter (30-inch) diameter outlet pipes at vertical intervals of about 11 meters (36 feet) which were controlled by slide gates on the upstream face.

The dam had no inspection gallery, and the foundation was not pressure grouted. Uplift pressure relief under the dam was provided only at the river channel. About 9 meters (30 feet) in from the upstream face of the dam, three holes were drilled in line parallel to the dam axis at a spacing of approximately 6 meters (20 feet); and then at about 14 meters (45 feet) downstream from the upstream face another row was drilled, with seven holes at about the same spacing as in the first row. These holes were

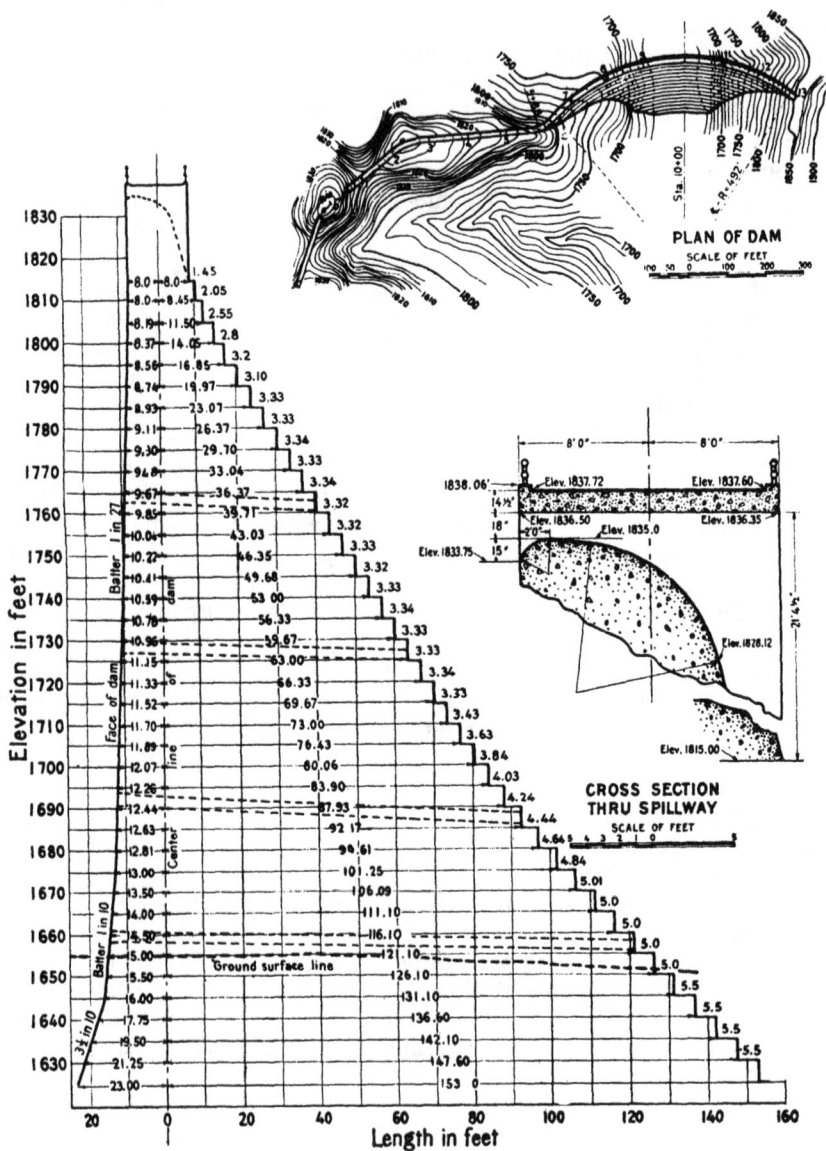

PLAN OF DAM

SCALE OF FEET

CROSS SECTION
THRU SPILLWAY

SCALE OF FEET

PLAN AND PROFILE
OF
ST. FRANCIS DAM

SCALE OF FEET

CITY OF LOS ANGELES
DEPARTMENT OF PUBLIC SERVICE
BUREAU OF
WATER WORKS AND SUPPLY
ENGINEERING DEPARTMENT

SOURCE:
Plate 4, Report of the Commission to Investigate the Causes Leading to the Failure of St. Francis Dam — 1928.

Figure 4-14.—St. Francis Dam, plan and profile. P-801-D-79359.

variously reported as from 4.6 to 9.2 meters (15 to 30 feet) in depth. A small section of 100-millimeter (4-inch) pipe was installed in the top of each hole and connected to a central drainpipe which discharged at the downstream toe of the dam. The drainage from this network was reported to have been small and had been conveyed to the caretaker's home, where it provided domestic service. Most of this system was located under the monolith which remained standing.

Excavation of the foundation during construction had not required any blasting. High-pressure hoses had been used to sluice off the rock. Picks and steel bars usually had sufficed to complete the foundation preparation. Power equipment was needed only in the bottom of the canyon and on the wingwall ridge.

Two kinds of rock were predominant in the foundation at the site. The canyon floor and the left abutment are composed of a relatively uniform mica schist known locally as "graywacke" shale. On the right side of the canyon is a red conglomerate. The contact between the two formations is a fault which, at the damsite, has a strike approximately parallel with the stream and outcrops on the right abutment about 15 meters (50 feet) above the channel. The dam was placed across this fault, the existence of which had been well known. It had not undergone any movement during the life of the dam. There evidently was no basis for blaming the disaster on a fault shift, even though the foundation at this point was exceptionally weak. Along the contact was a band of serpentine which was so softened by water after the failure that it could be dug out by hand. The fault contains a clay gouge as thick as 1.2 meters (4 feet). When dry, it is rather hard, but upon wetting it becomes soft and plastic.

The schist on the left abutment splits easily into thin sheets and weathers readily. In many places it has been sheared, roughly parallel with the planes of lamination. In these shear zones, the rock has been reduced to flaky material that is quite fragile. The steep slope of the left abutment can be attributed to the laminated structure of the schist. This rock can resist large loads perpendicular to its planes of lamination, but is weak against loading parallel with those planes. Under such loads the schist tends to slip as would a stack of chips. The flood cut away a large volume of this rock on the left side of the canyon. Extensive sliding on that slope was perhaps due to combined saturation and undercutting.

On the right abutment above the fault, the conglomerate is laced with fractures, some of which contain gouge or gypsum. Many of the pebbles in the conglomerate have been sheared. The whole mass has been so crushed that it has lost much of its strength. In the dry condition, fragments of considerable size can be broken out and trimmed with a hammer. When immersed in water, however, absorption proceeds rapidly, air bubbles appear, crumbling begins, the water becomes turbid with suspended clay, and usually in less than an hour a specimen the size of a baseball has disintegrated almost completely.

Soon after the failure, the surface of the conglomerate at the damsite showed marked softening due to saturation. After a few days of drying, this same rock exhibited a smooth surface when broken, and some of the samples rang under the hammer. Of two specimens selected from the firmest part of the eroded foundation, one fractured while being readied for testing and the other was found to have a compressive strength of only 3606 kilopascals (523 pounds per square inch). This was no more than one-fourth the strength of the concrete in the dam.

With such a geological setting, the collapse of the St. Francis Dam was inevitable unless water somehow could have been prevented from entering the foundation. This however was impossible. The pressure grouting, drain wells and galleries, and deep cutoff walls used at other dams to control seepage probably could not have been fully effective at this site, although they might have postponed the day of failure.

In view of the many deficiencies of the site, the survival of the structure for 2 years is remarkable indeed. Storage of water began on March 1, 1926. From the beginning of 1928, when the water surface elevation was 555.0 meters (1821 feet) above sea level, storage was increased until March 5, when the reservoir was essentially filled to capacity, at elevation 559.2 meters (1834.75 feet) or 0.076 meter (0.25 foot) below the spillway crest. It was maintained at that level until the failure at a few seconds past 11:57 p.m. on March 12, 1928.

Photographic evidence and the reports of witnesses confirmed that there was not much seepage through the dam. Some cracking appeared in both the gravity mass and the wingwall, and minor seeps were noted in certain of these places. But none of this flow through the structure itself was viewed as alarming. However, more significant seepage had developed through the dam's foundation. As the reservoir level rose, the magnitude of this discharge increased considerably. It had reportedly reached a maximum of nearly 57 liters (2 cubic feet) per second on the afternoon before the disaster.

Seepage along the contact between the dam and its foundation, especially on the west abutment, was noted as soon as impoundment began. Two or three weeks before the catastrophe, flow under the wingwall increased, and some drains were installed. In the 24 hours before the collapse, the seepage in that area accelerated.

On the morning of March 12, 1928, the Chief Engineer and his assistant went to investigate the leakage at the dam which was reported by the dam tender. The two engineers and the caretaker inspected the dam and its abutments. They found that the seepage was clear, which indicated that foundation material was not being eroded from beneath the dam. There was nothing in the performance of the structure that they judged to be hazardous.

In the documented evidence of the failure, the first indication of trouble was shown by the automatic water level recorder located on top of the surviving erect monolith. It recorded an apparent accelerated falling of the water level starting at about 11:20 p.m. and totaling about 90 millimeters (0.3 foot) just before midnight, when the collapse was depicted on the graph by a sudden drop. The chart-indicated fall in the lake level during the final 40 minutes represented approximately 228 000 cubic meters (185 acre-feet) of storage. Five minutes before the break, the recorder indicated a flow of as much as 425 cubic meters (15 000 cubic feet) per second. But if the discharge had been that high, the canyon would have been flooded from wall to wall. Workmen at the powerplant and settlement just below the dam presumably would have been alarmed by the flood and would have evacuated their families to high ground. The fact that witnesses passed through the canyon below the dam during the half hour before the break also suggests that this was a misleading reading.

There was speculation that some of the apparent drop on the reservoir chart might be attributable to a rising dam rather than a falling water level. A slight rotation of the structure about its toe might have caused it. There were other possibilities. The gage pencil was actuated by a float in a 300-millimeter (12-inch) pipe mounted on the upstream face of the dam block that survived. Rocking or tilting of this monolith as adjoining sections began to fail may have caused abnormal tension in the chart paper. Only a slight distortion could have invalidated the recording.

The disaster was subjected to several inquiries. All attributed the cause of the failure to defective foundations.

A committee selected by the District Attorney of Los Angeles County on about March 15, 1928, concluded:

"The defective foundation material referred to *** as conglomerate became softened by absorption and percolation of water from the reservoir and was by hydrostatic pressure pushed out from under the dam structure, permitting a current of water of high velocity to pass under this sector of the dam. This current, by eroding the soft foundation material quickly extended the opening under this portion of the structure to such an extent that a part of the Westerly section of the dam collapsed through lack of support ***.

"Following failure in the Westerly sector, the escaping water swirled across the downstream toe of the dam to the Easterly wall of the canyon below the Easterly abutment, causing a slide which broke the already weakened bond between the Easterly sector and the side wall of the canyon, allowing the Easterly sector of the dam to collapse ***.

"A portion of the Easterly sector of the dam fell and now lies upstream from the original face of this sector, indicating that this portion failed after the water level had dropped enough to materially relieve the pressure from the upstream face."

Consultants retained by the Santa Clara Water Conservancy District filed an engineering and geological report on April 19, 1928, stating:

"The old slide against which the dam rested at the east, its composition of mica schist bedded at an angle of, or approaching 45° with the horizontal, practically paralleling the general slope of the canyon wall, and shattered by previous movement, subject to lubrication with water from the reservoir as well as with infiltering rainwater, offered only insecure support to the dam, and this was rendered more precarious by the adoption of a design which did not include adequate foundation drainage.

"At the west end of the dam, the material on which it rested is more or less pervious and softens when wet. It remains doubtful whether any precautions (such as cut-off walls and weep-holes) could have made this a satisfactory foundation ***.

"The appearance, as early as January 1928, of cracks in the dam other than temperature cracks, indicated a movement of the dam. Apparently the significance of these cracks was overlooked ***."

Engineer S. B. Morris, in a letter to the American Society of Civil Engineers in April 1930, reported that:

"*** a sample of water leaching from the ravine immediately down stream from the westerly abutment of the dam, was taken on March 30, 1928, and another sample was taken on the same date from the Los Angeles Aqueduct. The chemical analyses *** (showed) a great increase in dissolved calcium, magnesium, sulfates, chlorides, and silicates. Owing to the presence of numerous seams of gypsum in the conglomerate, it is quite natural that the greatest increase in dissolved solids should be that of the sulfate radical which was 29.4 times as great in the water leaching from the abutment as in the Aqueduct water. While, of course, it is now difficult to form any idea as to the actual quantity of dissolved solids in the larger leakage from the westerly abutment prior to failure of the St. Francis Dam, it is still interesting to note the possible effect of such leakage and the dissolved solids contained therein.

"The total evaporated solids increased from 294.60 to 2319.00 or 2024.4 parts per million. At this rate, a leakage of 1 cu. ft. per sec. would carry in solution 10930 lb. per day, or approximately 2½ cu. yd. per 24 hours, which is about 900 cu. yd. per year. By dissolving such a quantity of cementing material the strength, resistance, and imperviousness of many times this volume may be seriously affected.

"These tests indicate the importance of mineral analysis of seepage water from dam foundations, particularly where such foundations are of rocks containing solvent minerals."

Consulting Engineer E. L. Grunsky, in a letter published by the American Society of Civil Engineers in May 1930, stated that:

"*** there was plenty of evidence on the ground *** that portions of the dam were in movement prior to the failure, and that this movement was accelerating just before the failure occurred ***.

"*** the foundation material under the easterly end was a laminated mica schist, with a dip approximating the slope of the east abutment, but *** in addition, it was the face of an old slide which had originally moved down some hundreds of feet and finally had come to rest on the canyon floor ***. This slide was in more or less stable equilibrium until the dam was built against it, and water percolating under pressure into the fissures and laminations of the more or less broken schist, lubricated the sliding planes, whereupon movement began. This movement along the cleavage planes caused a large block of the mountain side to act as a wedge against the sloping base of the east portion of the dam, thus causing uplift. Early in January 1928, more than two months before the failure *** two diagonal cracks (were noticed) running from the top of the dam, at angles of approximately 45°, down to the foundation at the sides. That these cracks were known and were of considerable magnitude is shown by the fact that in a large fragment of the ruins from the east end of the dam, one of these cracks could be seen, and the writer took oakum packing from the crack at the downstream face. The breadth of this packing indicated that the crack was 1/2-inch or more in width where it ran out at the abutment surface.

"The diagonal cracking of the east portion of the dam is evidence that the hillside was in motion under the action of the water possibly from rains and from the reservoir, months before failure, and that this part of the dam was slowly being lifted from its foundation ***.

"The second long diagonal crack (approximately 45 °) in the west or right bank portion of the dam indicated that there had also been a lifting of that part. Since the failure, re-surveys have disclosed that swelling ground raised the long parapet-like western extension of the dam about 0.3 feet. The swelling material is the same conglomerate as that found under the west portion of the main dam.

"The two diagonal cracks evidenced the fact that the dam was being subjected to uplift forces and that portions of its foundation were in motion months prior to the failure ***.

"It is the writer's opinion that the forces which were acting on the dam at the time of its failure *** finally broke the contact between the dam and its foundation, resulting in a rapid increase of the hydrostatic uplift pressure. There was thereupon a sudden tilting of the dam downstream. Such tilting would account for the apparent drop in elevation of the water surface as shown by the automatic gauge on top of the dam just prior to complete failure (attributed in most of the reports to a drop in water surface). This tilting continued with the toe as a fulcrum until a large slab about 15 feet in thickness was spalled off from the down-stream face, and in all probability complete failure of the two ends of the dam occurred at or about the same time. The consequent relief of pressure due to the release of water from the reservoir then permitted the central portion of the dam to settle back on its foundation. That this tilting did occur is shown by the spalled off lower down-stream face of the central portion which remained upright; by the fractured condition of the down-stream or heel portion of the exposed cross-section of this central block; and by a crack extending along the base thereof at bed-rock and along the up-stream face for a considerable distance. The crack had opened up sufficiently to trap and crush a ladder with 4-inch sides when the monolith settled back into place ***."

The jury drawn by the Coroner of Los Angeles County asserted that:

"A sound policy of public safety and business and engineering judgment demands that the construction and operation of a great dam should never be left to the sole judgment of one man, no matter how eminent, without *** checking by independent experts ***."

In 1929, the State of California enacted legislation with this intent.

San Ildefonso Dam

The San Ildefonso Dam, approximately 435 kilometers (270 miles) southeast of La Paz in the interior mountains of Bolivia, collapsed in mid-March, 1626. One of the largest dams in the vicinity of Potosí, it was reported to be 8 meters (26 feet) high and about 500 meters (1640 feet) long. Storage capacity was approximately 430 000 cubic meters (350 acre-feet). The reservoir was said to have been practically full when the disaster struck. All this water was spilled in about 2 hours.

Downstream about 3.2 kilometers (2 miles), the town of Potosí was swept by the full force of the flood. Accounts of the number of human lives lost in this remote mining district ran as high as 4,000. There apparently was no way to verify such estimates, but the maximum figures may have been exaggerated. However, another indicator of the severity of losses was the extensive property damage. Roughly 95 percent of the many mills along the stream were reported to have been destroyed or badly damaged.

A dam was rebuilt at the San Ildefonso site, and reservoir operation was resumed. However, the town of Potosí had suffered irreparable damage.

Sempor Dam

On December 1, 1967, the partially complete Sempor Dam in an irrigation project on the Djatinegara River in Central Java washed out during monsoon rains. The water pouring from the broken rockfill structure flooded three villages, taking the lives of about 200 people.

The damsite is near the villages of Gombong and Magelang, approximately 40 kilometers (25 miles) from the south coast of Central Java and about 16 kilometers (10 miles) west of the city of Jogjakarta. The flood severed the railroad connecting that city with the capital at Djakarta.

The design engineers called for the rockfill dam to be 54 meters (176 feet) high and 228 meters (748 feet) long. Work on the project had been suspended about 2 years previously because of insufficient funding. With additional financing provided in 1967, the construction had been resumed; but the work of the various small contractors had been delayed often by slow deliveries of cement and by the heavy rainfall.

South Fork (Johnstown) Dam

On May 31, 1889, a major catastrophe occurred in the Allegheny Mountains of Pennsylvania, when the South Fork Dam (fig. 4-15), about 14 kilometers (9 miles) above Johnstown, suddenly broke and loosed a flood down the narrow valley of the Conemaugh River. A large part of the city of Johnstown was ruined, and 2,209 people died.

At the time of the failure, Johnstown was already inundated to depths up to 3 meters (10 feet). Because of this, accounts of the disaster toll attributable to the collapse of the dam were difficult to reconcile. Damage at the reservoir site was easier to assess. The breach in the embankment was 128 meters (420 feet) long. About 69 000 cubic meters (90 000 cubic yards) of earth and stone had been washed out along with as much as 19 000 000 cubic meters (5 billion gallons) of water that had been stored in the lake.

The dam was an earthfill structure 21.9 meters (72 feet) high and 6 meters (20 feet) wide at the crest. Slope of the downstream face was 1½ to 1, while the upstream slope was placed at 2 to 1 and was protected with a light riprap.

The South Fork Reservoir was built by the State of Pennsylvania to supply water to the navigable canal from Johnstown to Pittsburgh, part of the railway and water transport system from Philadelphia to Pittsburgh. A design of the dam had been developed in 1839 by a young State engineer named William E. Morris, who selected the site. He proposed an embankment 259 meters (850 feet) long and 19 meters (62 feet) high.

The specifications written by Mr. Morris were thorough and left little question of his intentions, as may be seen in the following excerpts:

"The Dam will be constructed as represented in the plan, with a slope of 2 to 1 on the upper side, and 1½ to 1 on the lower side, it will be raised 10 feet above the water line, and be 10 feet on top. The lower angle will be composed entirely of stone, of such nature as to resist the decomposing action of air, frost and water.

"In the outer portion of the stone, for 4 feet thick at top and 20 feet thick at bottom, no stone must be used which does not contain at least 4 cubic feet. The remaining part of the stone may be of any size. Next to the stone will be a body of slate rock or coarse gravel 3 feet thick on top, and 30 feet thick at bottom.

"The remainder of the bank or dam, being that between the water and the slate, shall be composed of the best water-tight, solid and most imperishable material that can be procured, within 1/4 mile of the dam. No light, spongy, alluvial, or vegetable matter will be used in its construction. Neither will any coarse gravel or stones larger than 4 inches square be permitted to form any part of it. The whole material of the dam, viz., stone, slate and earth, shall be brought and deposited in the proper place in carts and wagons, and no portion of the dam shall be made by transporting the material in barrows, by schutes, or upon a railway. If it shall be deemed necessary by the engineer, a puddle course of the best fine river gravel, 20 feet in width, shall be carried from bottom of puddle ditch to 4 feet above water line, which said puddle course shall be kept 1 foot higher than the other portions of the embankment, and at all times to be well wet and carted upon, and next the walls, if necessary, well pounded, with a 4-inch rammer. The whole bank shall be made in layers 2 feet thick, be started at the same time, and carried up together, without troughs or hollows, and as nearly level as practicable throughout its whole extent. No part of it shall be made in freezing weather. If, during the progress of the work, any part of the embankment, by long exposure or too frequent passage upon it by carts or wagons, shall become so compact upon the surface as to be incapable of uniting completely with the material above to be deposited upon it, such surface shall be well plowed, and, if thought necessary by the engineer, puddle ditches cut, at the expense of the contractor.

"In the embankment *** a wall of rubble masonry, made of well-shaped quarried stone, laid in a full bed of cement, with the faces undressed, and the beds and joints close and free from spalls, shall be carried up in the puddle course before-mentioned (or if it be omitted) in the earth embankment. This wall will be started 3 feet below the surface of the rock, if it should be found in excavating puddle ditches, and made completely to fill a trench excavated in the rock for the purpose. It will be 6 feet thick at the bottom, 25 feet high, and 2 feet thick on top; made with buttresses upon each side at intervals of 20 feet, and the difference of thickness between bottom and top, disposed of in offsets of 6 inches in width. No stone shall be used in the wall, of a less size than 2 feet long by 1 foot wide, and 6 inches thick, larger stones to have similar proportions ***.

"A slope of wall of dry masonry will be built upon the upper slope of the dam, 15 inches in thickness, backed in by a layer of slate rock or coarse gravel, 6 inches thick. No stone shall be used in its construction which do not reach through the wall, nor any that are of a less size than 4 inches thick by 8 inches wide. This wall shall be neatly laid, the beds of the stone at right

DAM AND SLUICE
for
RESERVOIRS.

BY

Wm E. Morris, Civil Engineer.
1839

SCALE 16 FT TO AN INCH

— Reference —

a. — Represents valve rods.
b. — " wall to support pipes.
c. — Protection wall.
d. — Stop cocks.
e. — Pipes.
f. — Grate.

PLATE XLVIII
TRANS. AM. SOC. CIV. ENGRS
VOL. XXIV. NO 477
REPORT ON
SOUTH FORK DAM.

"Copy of a Plan on file in the Department of Internal Affairs of Pennsylvania."

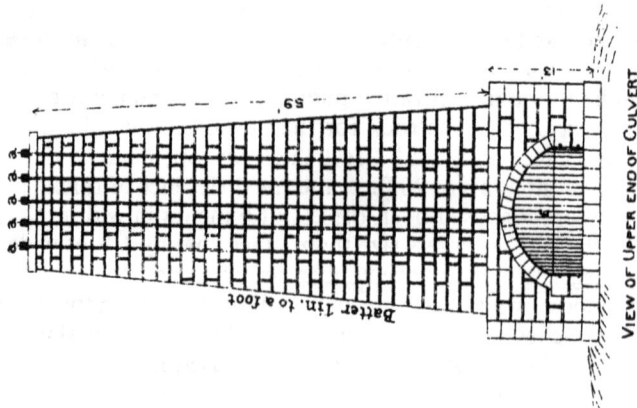

NOTE:
The original design called for a masonry tower, however a wooden tower was substituted.

SECTION THROUGH CENTRE

VIEW OF UPPER END OF CULVERT

Batter 1 in. to a foot

Figure 4-15. — South Fork Dam, cross sections (1 of 2). P-801-D-79360.

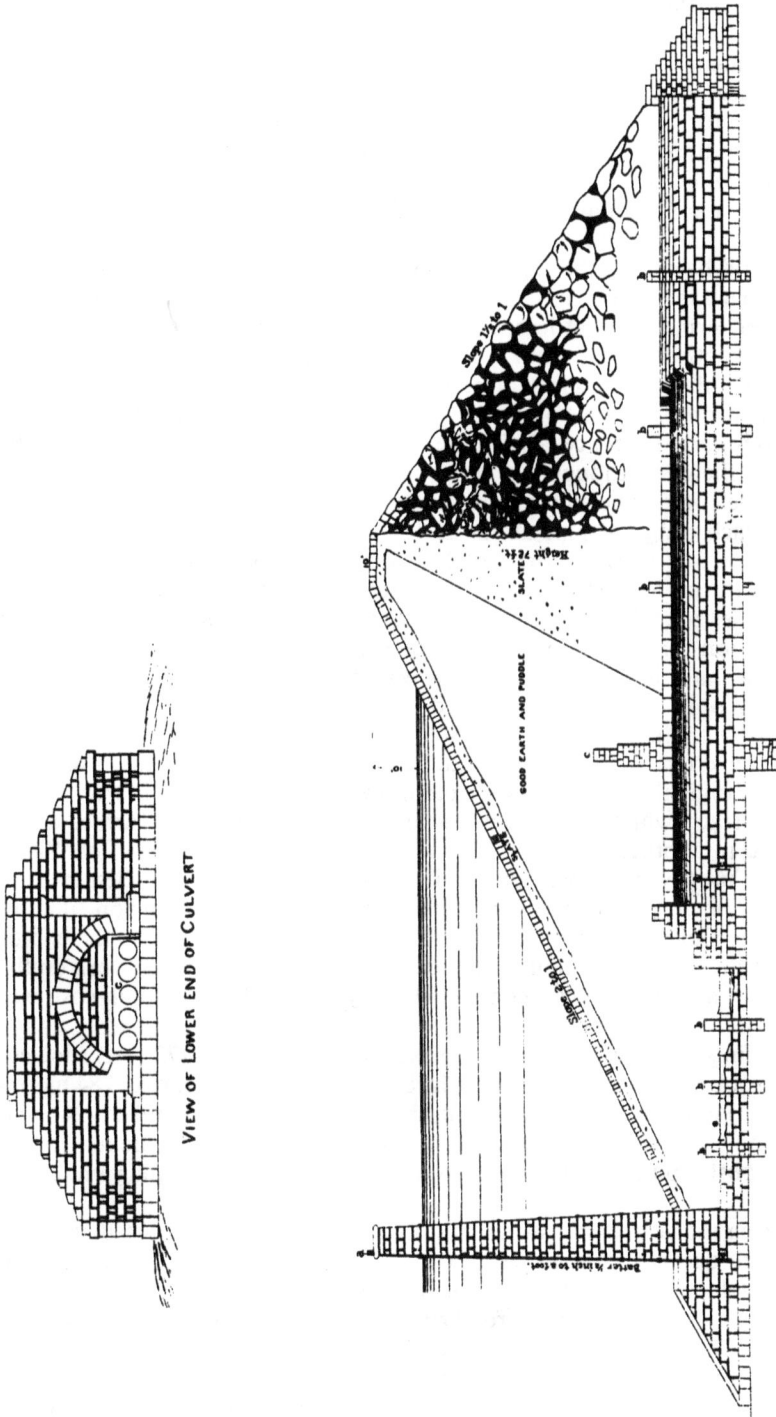

Figure 4-15.—South Fork Dam, cross sections (2 of 2). P-801-D-79361.

angles with the face of the bank, the joints close and free from spalls. A paving of 18 inches depth laid in a similar manner, will cover the top of the embankment ***.

"A waste or waterway will be excavated in the hill at one or both ends of the dam, for the discharge of surplus water in the time of floods, the aggregate width of channels will not be less than 150 feet. The earth covering the rock will first be stripped off, the channel will then be excavated in the rock, leaving for an abutment or guard bank, between the ends of the embankment and the inner slope of the channel, a mass of solid rock, of such width and height, as the engineer may think sufficient. The entrance to the waste or wastes will be as close above the dam as its safety will permit, and its lower termination at least 50 feet beyond the outer slope of the dam."

Construction of the dam was started in 1840. However, the State of Pennsylvania was so short of funds in 1841 that the Legislature severely limited appropriations for public works. On November 30, 1841, William Morris reported: "Since last fall the contractors have steadily pushed on the work at the dam, though, from the smallness of the appropriation, with a moderate force. The sluice walls are raised sufficiently high to receive the pipes, each range of pipe about 80 feet long, has been laid and tested by a head of 300 feet." Work on the dam was suspended soon thereafter.

In 1846, more money was appropriated, and an order was given to expedite the job. Morris was asked to draw new plans and specifications in preparation for a bid letting on March 3, 1846. The only significant change which he made in his original scheme was to substitute a frame tower for the masonry one. The State, burdened by the expense of repairing damages caused by the spring flood of 1846, postponed the work again; and the money reverted to the general fund. In 1850, a new appropriation was made and the original contractors were permitted to proceed with the work under the contract which had been suspended. However, still further delay resulted in another reversion of funds.

More money was voted on April 15, 1851; and work was resumed on May 1. The project then was expedited as much as possible; and on June 10, 1852, the sluice gates were closed to begin storage of water. By the end of August the water was about 12 meters (40 feet) deep. The final work on the reservoir apparently was completed early in 1853. The total cost was approximately $167,000.

The finished work included a spillway about 22 meters (72 feet) wide cut in the rock of the east abutment. The embankment was just over 284 meters (930 feet) long. At about midlength, there were five 610-millimeter (24-inch) cast-iron pipes set in stone masonry and controlled at a wooden tower. These outlet pipes discharged into a masonry conduit running under the dam.

During the year 1854, two breaks occurred which were reported as minor and were quickly repaired. However, details of the character and position of the defects apparently were not readily available.

The new Pennsylvania Railroad gave the State's canal system strong competition. Water transport traffic declined. In 1857, the railroad bought the State's transportation facilities in the region, including the Main Line, the Portage Railroad, and the South Fork Reservoir. Having no need for the dam, the railroad evidently did not do much to maintain it.

In the following years, leaks developed where the pipes entered the masonry conduit; then in 1862 there was a break at this point. In the late spring of that year, the mountains of Pennsylvania had been hit by intense thunderstorms. The streams in the South Fork watershed flooded; and on June 10 a sudden surge of water washed out the upper end of the masonry conduit, the cast-iron pipes, and part of the embankment. A short length of the lower end of the outlet remained. The washout was blamed on a foundation defect near the masonry conduit. The released water did not do much damage since the storage level had been low and, just before the failure, a caretaker had opened the valves.

In 1875, the Railroad sold the dam and reservoir. A sportsmen's club in Pittsburgh finally came into possession of the property. Then in late 1879, repair was begun by closing off the remainder of the conduit and backfilling with mud, rock, and brush. On Christmas Day, 1879, a sudden runoff erased most of this work. Repair was interrupted until the next summer.

When work was resumed in 1880, the remnants of the five lines of 610-millimeter (24-inch) sluice pipes were removed. The remains of the outlet masonry were left. Timber sheet piling was placed across the lower part of the breach. The original plan of constructing the downstream zone of the embankment of rock was followed, and large stones were dumped into the breach. Since there were no outlet works, the runoff percolated through this rockfill at first; and the water in the reservoir rose as the placement of the upstream zone progressed. The washing of the fill through the dumped rock was prevented by covering its face with brush and hay. The material for the impervious embankment was clay and shale, which was deposited against the upstream side of the rockfill in layers extending so as to restore the full thickness of the original dam. Hauling by teams on the freshly dumped material, wet from the rising water, gave this zone some compaction. The fill was not finished that season, and it suffered some flood damage in the following winter. During the completion stage in 1881, both of the outer slopes were covered with heavy riprap.

The original specifications called for a crest 3 meters (10 feet) wide and 3 meters (10 feet) above the normal water level. The crest of the undisturbed end sections of the dam after the failure in 1889 averaged about 0.6 meter (2 feet) lower than this, which is in accord with reports that, when the breach was repaired in 1880 to 1881, the top of the dam was lowered to build a 6-meter (20-foot) roadway over it. This was wide enough for two carriages to pass comfortably. However, the lowering had significantly reduced the ultimate spillway freeboard from 3 meters (10 feet) to perhaps no more than 2 meters (7 feet). As rebuilt, the fill also sagged 0.3 to 0.6 meters (1 to 2 feet) in the middle at the location of the old breach. Further reducing the discharge capacity was a screen of iron bars, each about 12 millimeters (1/2 inch) in diameter, installed between bridge columns in the spillway to prevent fish loss down the creek. This was susceptible to clogging with debris.

In the period May 30 to June 1, 1889, an intense rainstorm hit Pennsylvania, extending over an area of about 38 850 square kilometers (15 000 square miles), focusing on the Allegheny Mountains, where there was as much as 250 millimeters (10 inches) of precipitation in 36 hours. The watershed above the South Fork Reservoir was drenched. Soon the threat to the dam was apparent. The president of the South Fork Fishing and Hunting Club inspected it on the morning of May 31, 1889, and expressed some doubt that it would survive the day. He put his men to work in excavating an emergency spillway at the western end. Digging in the rock, they made disappointing headway. A crew was sent to remove the trash which was clogging the fish screens in the main spillway. By 11 a.m., the reservoir water was at the crest of the dam and was beginning to erode the loose material that had been hurriedly thrown up by plow and shovels. Near the downstream toe of the dam, several serious leaks were observed.

In the early afternoon, the emergency spillway was discharging a stream of water about 7.6 meters (25 feet) wide. Flow in the main spillway was 1.8 meters (6 feet) deep or more. Since the dam was already starting to wash away, nothing more could be done. Because the outlet pipes had been removed in the reconstruction of 1880, the operators no longer had any means of controlling the level of the lake.

The hopelessness of the final hours was voiced by the resident engineer, John G. Parke, Jr., in a statement on August 22, 1889:

"The water in the lake rose until it was passing over the breast, notwithstanding that the lake had then the two outlets (the waste-weir and the one cut by the laborers). The breast was slightly lowered in the center and the water washed away our temporary embankment thrown up by the plow and shovels, and the water was passing over in many places in a distance of 300 feet about the center of the breast; the men stuck to their task and worked until the water was passing over in nearly one sheet, and then they became frightened and got off the breast. I saw what would be the consequence when the water passed over the breast and rode to South Fork Village and warned the people in the low lands there, and had word telegraphed to Johnstown that the dam was in danger. The people in South Fork heeded the warning and moved out of their houses. When I left South Fork to return it was just twelve o'clock noon, and the water had been flowing over the dam for at least a half hour. I rode back up to the lake 2½ miles through the valley and found the men had torn up a portion of the flooring of the waste-weir bridge and were endeavoring to remove the V-shaped floating drift guard that projected into the lake. It was a light affair and was built to float on the surface of the lake and catch twigs, leaves, etc., and prevent their clogging up the iron screens ***. I crossed the breast at this time and found the water was cutting the outer face of the dam, but not as badly as I feared it would, its greatest effect was on some portions of the roadway which crossed the breast where the roadway had been widened on the lower side by the addition of a shale earth or disintegrated shale, upon which the action of the water was instantaneous, but the heavy riprapping on the outer face of the dam protected this wash and the water cut little gullies between each of the large stones for riprap. I did not stay on the dam when it was in that condition, but went on to the end of the dam and found that over its entire top it was serried by little streams where the water had broken through our little embankment and was running over the dam. I went on to the new waste-weir we had cut and found it carrying off a great volume of water and at a great velocity. I with difficulty waded it and found that it was up to my knees or 20 inches deep. I felt confident that nothing more could be done to save the dam unless we were to cut a wasteway through the dam proper at one end and allow it to cut away in but one direction, and that towards the center of the dam, but this I would not dare to do, for it meant the positive destruction of the dam, and the water at the time was almost at a stand, owing, without doubt, to the large increase of outlet by the overflow on the breast, and I hoped that it would not rise, but yet expected it to rise for it had been raining most all of the morning, and consequently we had more water to expect. I hurried to the club house to get my dinner and to note the height of the water in the lake, and found that it was a little over a stake, that from my level notes of a sewer I was constructing I knew was 7.4 feet above the normal lake level. I returned to the dam and found the water on the breast had washed away several large stones on the outer face, and had cut a hole about 10 feet wide on the outer face and about 4 feet deep, the water running into this hole cut away the breast in the form of a step both horizontally and vertically, and this action went on widening and deepening this hole until it was worn so near to the body of the water in the lake that the pressure of the water broke through, and then the water rushed through this trough, and cut its way rapidly into the dam at each side and the bottom; and this continued until the lake was drained. I do not know the actual time it consumed in passing through the breach, but it was fully 45 minutes. It did not take long from the time that the water broke into this trough until there was a perfect torrent of water rushing through the breast, carrying everything before it, trees growing on the outer face of the dam were carried away like straws. The water rushed out so rapidly that there was a depression of at least 10 feet in the surface of the water flowing out, on a line with the inner face of the breast and sloping back to the level of the lake about 150 feet from the breast ***. There is one thing I want to impress on every one's mind, and that is, that the dam did not break, but was washed by the water passing over it from 11:30 o'clock A.M. until nearly 3 P.M. until the dam was made so thin at one point, that it could not withstand the pressure of the water behind it, and the water once rushing through this trough nothing could withstand it.''

An investigative committee appointed by the American Society of Civil Engineers examined the remains at the site and reported that the dam was well built, but that the disaster was due to deviation from the original plan of construction. The design of 1839 was for a watertight embankment of stone and earth with five lines of sluice pipes and a spillway cut in solid rock at one or both ends of the dam with an aggregate width of not less than 46 meters (150 feet). The spillway was, however, not constructed as specified but instead had an effective width of less than half the amount. The committee therefore concluded that the disaster was caused by deficiency of the outlet and spillway. Estimated maximum inflow to the reservoir during the flood was 283 cubic meters (10 000 cubic feet) per second. The committee calculated that if the spillway had been built according to specifications, and if the original outlet pipes had been available for discharge at their full capacity, there would have been no overtopping.

Teton Dam

The Bureau of Reclamation's Teton Dam failed on June 5, 1976, during initial impoundment when the reservoir was nearly full (figs. 4-16, 4-17, and

191

Muddy flow at about El. 5045, 20-30 c.f.s., 8:30 a.m. P-801-D-79362 (top).

Increased flow, dozers sent to fill hole at El. 5200, about 10:45 a.m. P-801-D-79363 (bottom).

Figure 4-16.—Teton Dam failure sequence, June 5, 1976.

Dozers lost in hole, about 11:20 a.m. P-801-D-79364 (top).

Dam crest breaching, 11:55 a.m. P-801-D-79365 (bottom).

Figure 4-16.—Teton Dam failure sequence, June 5, 1976—Continued.

Early afternoon. P-801-D-79366 (top).

Late afternoon. P-801-D-79367 (bottom).

Figure 4-16.—Teton Dam failure sequence, June 5, 1976—Continued.

Figure 4-17.—Teton Dam, downstream view after failure (P549-100-215A). P-801-D-79368.

4-18). Deaths of from 11 to 14 persons and property damage estimated at about 400 million dollars have been attributed to the failure.

The dam (fig. 4-19) was situated on the Teton River, 5 kilometers (3 miles) northeast of Newdale, Idaho. It was a central-core zoned earthfill, with a height of 93 meters (305 feet) above the riverbed and 123 meters (405 feet) above the lowest point in the foundation. Provisions for seepage control included a key trench in the abutments and a cutoff trench to foundation rock in the river bottom. A grout curtain extended below these trenches. The embankment was topped out November 26, 1975. Filling of the reservoir began October 3, 1975, and continued until the failure on June 5, 1976.

Immediately following the dam's failure, the U.S. Secretary of the Interior and the Governor of Idaho joined in ordering an investigation and established for this purpose the Independent Panel to Review Cause of Teton Dam Failure.[3] The Panel conducted a comprehensive study of the failure and concluded "(1) that the dam failed by internal erosion (piping) of the core of the dam deep in the right foundation key trench, with the eroded soil particles finding exits through channels in and along the interface of the dam with the highly pervious abutment rock and talus, to points at the right groin of the dam, (2) that the exit avenues were destroyed and removed by the outrush of reservoir water, (3) that openings existed through inadequately sealed rock joints, and may have developed through cracks in the core zone in the key trench, (4) that, once started, piping progressed rapidly through the main body of the dam and quickly led to complete failure, (5) that the design of the dam did not adequately take into account the foundation conditions and the characteristics of the soil used for filling the key trench, and (6) that construction activities conformed to the actual design in all significant aspects except scheduling." The Independent Panel published its report "Failure of Teton Dam" in December 1976.

A second investigating team was the Interior Review Group (IRG) established by the Department of the Interior and composed of representatives of six Federal agencies.[4] This group, supported by the forces of the Bureau of Reclamation, continued its studies throughout the period 1976 to 1979, extending the field excavations and testing to the left abutment. The Independent Panel had focused on the right abutment, where the failure occurred. However, the Panel also inspected the site in 1978 during final stages of the field investigation, and provided the views of its members on disclosures found on the left abutment.

The initial left abutment remnant investigations were conducted between April and October of 1977 and involved removal of material between

---

[3] Wallace L. Chadwick (Chairman), Arthur Casagrande, Howard A. Coombs, Munson W. Dowd, E. Montford Fucik, R. Keith Higginson, Thomas M. Leps, Ralph B. Peck, and H. Bolton Seed. Robert B. Jansen was Executive Director.

[4] The members of the IRG were Dennis N. Sachs (succeeded later by F. William Eikenberry) Chairman, Harold G. Arthur, Neil F. Bogner, Floyd P. Lacy, Jr., Robert Schuster, and Homer B. Willis.

Figure 4-18.—Teton Dam, upstream view after failure (P549-100-218A). P-801-D-79369.

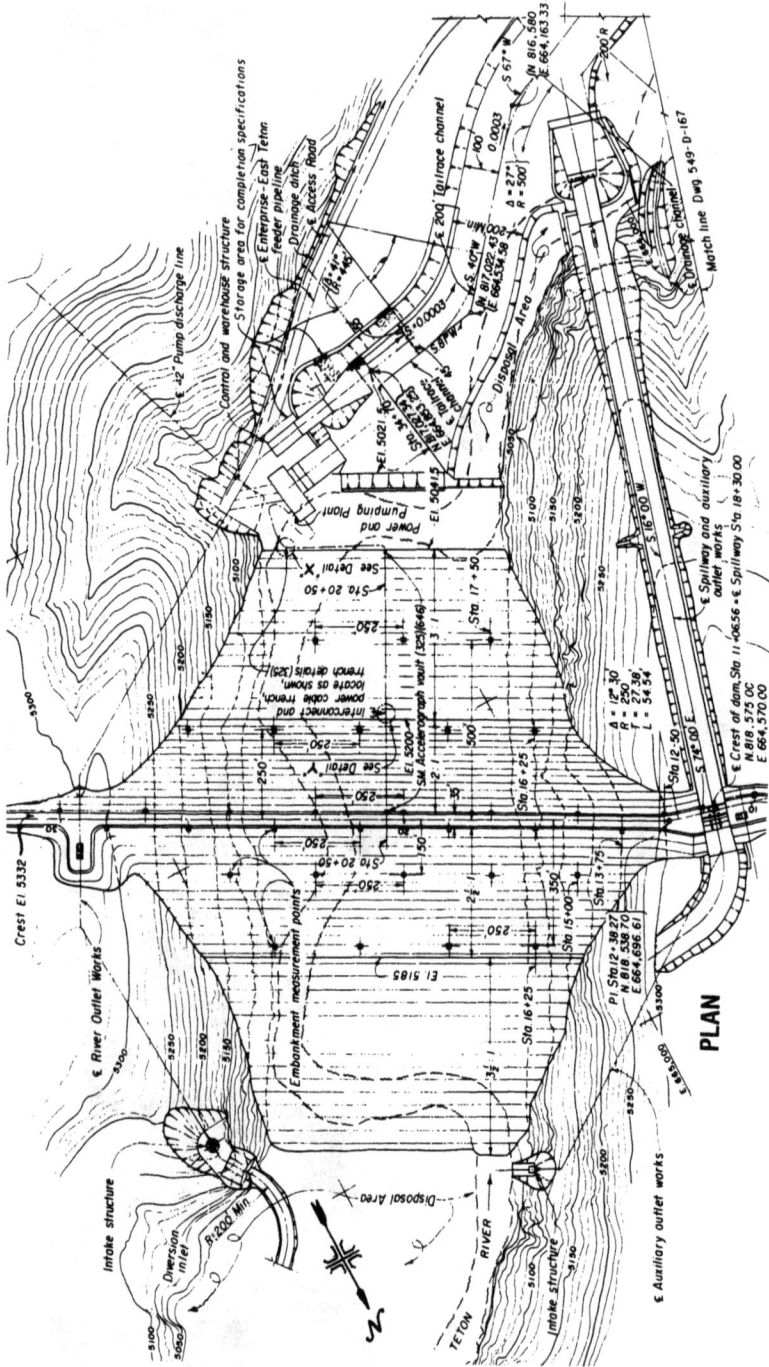

PLAN

Figure 4-19.—Teton Dam, plan and cross sections (1 of 2). P-801-D-79370.

198

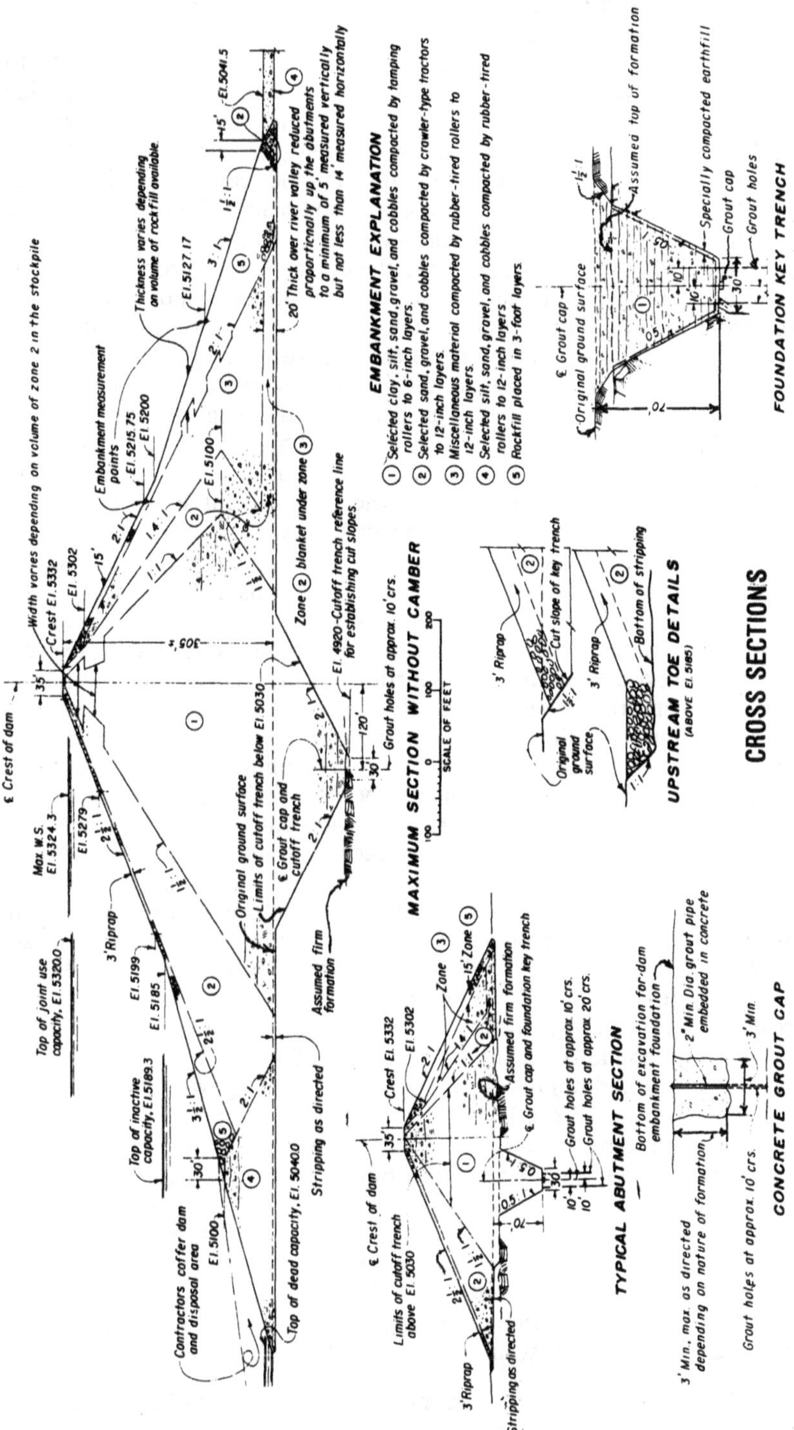

Figure 4-19.—Teton Dam, plan and cross sections (2 of 2). P-801-D-79371.

199

STAS[5] 6 + 90.67 and 8 + 92.45 (Stas. 22 + 66 and 29 + 28) of the left key trench. A "wet seam" was discovered in October of 1977 and investigations continued until January 1979. The seam, as discovered, was a 75- to 200-millimeter (3- to 8-inch) thick lens exposed in the face of the left side core excavation. A freshly exposed face of the seam would begin to "sweat" and then to transmit water while the adjacent soil remained comparatively dry. This seam extended 122 meters (400 feet) in the 167.6-meter (550-foot) thick core with wet areas also being discovered in the downstream zone 2 and zone 3 materials. The single exposed seam was discovered only 1.2 meters (4 feet) from the bottom of the proposed excavation and at a time when it was thought that all possible data had been extracted from the left abutment area. Three 1.8 meter (6 foot) and one 33.5 meter (110 foot) exploratory adits were excavated and 30 borings and numerous other testing, both field and laboratory, were performed to obtain information on the extent and properties of the seams. These generally were found to be of somewhat lower density and higher permeability, and a few degrees cooler, than the surrounding material in the embankment. In June of 1978, the former members of the Independent Panel visited the damsite and received a briefing on the wet seams.

Nothing was found during the left abutment excavation that would physically confirm or refute the initiating mechanisms hypothesized for the failure on the right abutment.

Loose blocks of the welded tuff, up to approximately 1.5 meters (5 feet) on a side, were exposed in the upper portions of the trench, and continuous joints open to several inches existed with the loessial material packed into them. The intense jointing in the upper portions of the key trench was similar to that on the right abutment. A general comparison of the geologic mapping of the left and right abutments reveals a slightly greater number of joints crossing the bottom of the left key trench but with the openness and joint appearances being very similar on each abutment.

The volcanic rocks at the Teton damsite are highly permeable and moderately to intensely jointed. Water was therefore free to move with almost equal ease in most directions, except locally where the joints had been effectively grouted. Thus, during reservoir filling, water was able to move rapidly to the foundation of the dam. Open joints existed in the upstream and downstream faces of the right abutment key trench, providing potential conduits for ingress or egress of water.

The wind-deposited nonplastic to slightly plastic clayey silts used for the core and key trench fill are highly erodible. The Independent Panel considered that the use of this material adjacent to the heavily jointed rock of the abutment was a major factor contributing to the failure.

The records show that great effort was devoted to constructing a grout curtain of high quality, and the Independent Panel considered that the resulting curtain was not inferior to many that have been considered acceptable on other projects. Nevertheless, the Independent Panel's onsite tests and other field investigations showed that the rock immediately under the grout cap, at least in the vicinity of STAS 3 + 96.24 to 4 + 57.20

---

[5]Abbreviation STA indicates stations in meters (1 station equals 100 meters) and Sta. indicates stations in feet.

(Stas. 13 + 00 to 15 + 00), was not adequately sealed, and that additional unsealed openings may have existed at depth in the same locality. The leakage beneath the grout cap was capable of initiating piping in the key trench fill, leading to the formation of an erosion tunnel across the base of the fill. The Panel considered that too much was expected of the grout curtain, and that the design should have provided measures to render the in-evitable leakage harmless.

The geometry of the key trenches, with their steep sides, would have been influential in causing transverse arching that reduced the stresses in the fill near the base of the trenches and favored the development of cracks that could open channels through the erodible fill. Arching in the longitudinal direction, due to irregularities in the base of the key trenches, and arching adjacent to minor irregularities and overhangs, probably added to the reduction of stress.

Stress calculations by the finite element method indicated that, at the base of the key trench near STAS 4 + 26.72 and 4 + 57.20 (Stas 14 + 00 and 15 + 00), the arching was great enough that the water pressure could have exceeded the sum of lateral stresses in the impervious fill and the tensile strength of the fill material. Thus, cracking by hydraulic fracturing was a theoretical possibility.

A close examination of the interior of the auxiliary outlet tunnel showed no distress of any kind such as would be expected had the right abutment, through which the tunnel passes, been subjected to significant settlement or other structural change. Geodetic resurveys showed only minor surface movements as a result of reservoir filling and emptying. Accordingly, dif-ferential movements of the foundation are not considered to have contributed to the failure.

The Independent Panel's investigations were directed particularly to determining the most probable manner in which piping erosion started. The Panel believed that two mechanisms were suspect. Either could have worked alone or both could have worked together. One is the flow of water against the highly erodible and unprotected key trench filling, through joints in the unsealed rock immediately beneath the grout cap near STA 4 + 26.72 (Sta. 14 + 00) and the consequent development of an erosion tunnel across the base of the key trench fill. The other is cracking caused by differential strains or hydraulic fracturing of the core material filling the key trench. This cracking would also result in channels through the key trench fill which would permit rapid internal erosion. In either case, leakage occurring through the key trench ultimately initiated further erosion along the downstream contact of the core and the abutment rock. Since the core material was both easily erodible and strong, any erosion channels in the core, along the contact with the rock, readily developed into large tunnels or pipes before becoming visible along the downstream parts of the dam.

The fundamental cause of failure may be regarded as a combination of geological factors and design decisions that, taken together, permitted the failure to develop. The principal geologic factors were (1) the numerous open joints in the abutment rocks, and (2) the scarcity of more suitable materials for the impervious zone of the dam than the highly erodible and brittle windblown soils. The design decisions included among others (1) complete dependence for seepage control on a combination of deep key trenches filled with windblown soils and a grout curtain; (2) selection of a geometrical configuration for the key trench that encouraged arching, cracking, and hydraulic fracturing in the brittle and erodible backfill; (3) reliance on special compaction of the impervious materials as the only protection against piping and erosion of the material along and into the open joints, except some of the widest joints on the face of the abutments downstream of the key trench where concrete dental treatment was used; and (4) inadequate provisions for collection and safe discharge of seepage or leakage which inevitably would occur through the foundation rock and cutoff systems.

Teton Dam was located in a steep-walled canyon incised by the Teton River into the Rexburg Bench, a volcanic plateau draining into the Snake River Plain. The exposed rocks are almost entirely of volcanic origin, but these are covered on the high lands flanking the canyon by a layer of aeolian sediments up to 15 meters (50 feet) thick.

The walls of Teton Canyon at the damsite consist of late-Tertiary rhyolite welded tuff which has undergone various degrees of welding. Alluvium has been deposited in the channel of the canyon to a depth of about 30 meters (100 feet).

During excavation of the dam foundation, large openings were uncovered in left and right abutment key trenches. Near the right end of the dam, two large fissures were exposed near STAS 1 + 08.20 and 1 + 32.28 (Stas. 3 + 55 and 4 + 34). Both fissures trend generally east-west and cross the axis.

The fissure near STA 1 + 32.28 (Sta. 4 + 34) was entered and explored by a Bureau of Reclamation employee for a distance of about 30 meters (100 feet) both downstream and upstream of the dam axis and an estimated 30 meters below the key trench invert. He described the cavity downstream as fairly consistently about 1.2 meters (4 feet) wide having a floor strewn with angular blocks of rock measuring up to 1.2 or 1.5 meters (4 or 5 feet) on a side. Upstream from the key trench, the roof and floor of the cavity were reportedly lined with stalactites and stalagmites up to 10 millimeters (3/8 inch) in diameter. About 30 meters upstream from the key trench wall the fissure pinched and turned so that the end could not be seen. It was reported that in winter vapor could be seen emerging from the downstream segment and that this segment was warm and could be entered in winter without a coat. Conversely, upstream of the key trench the air was reported to be cold. The end of the downstream segment was blocked by a large rock "the size of a pickup truck." A room or passage could be seen beyond, but the opening into it was too small to enter. In one place the cavity walls were described as covered with a red coating "which rubbed off on our clothes."

The interior of the fissure at STA 1 + 08.20 (Sta. 3 + 55) was not examined since it was too narrow to permit entry. High-slump concrete was poured into both fissures during project construction. The extent to which the uppermost parts of the cavities may have been sealed by this procedure is uncertain. However, the concrete-rock contact was drilled at three points during postfailure exploration, and in each instance the rock cores obtained displayed a tight bond between grout and rock.

From preconstruction geologic evaluation of the damsite, it was apparent that much of the foundation bedrock to depths of at least up to 30 meters was highly pervious, and that curtain grouting would be difficult, extensive, and costly. To obtain a quantitative assessment of the problem, a pilot grouting program was carried out on the left abutment in 1969. This program showed that it would be extremely costly to attempt to curtain-grout the upper 21 meters (70 feet) of foundation bedrock for the dam above elevation 1554.5 meters (5100 feet). Accordingly, the decision was made to excavate the relatively ungroutable rock to a depth of about 21 meters on both abutments from elevation 1554.5 meters upward to the ends of the dam, and to begin the grout curtain under a concrete grout cap in the center of the excavation. The trenches, for economy, were designed to be deep, narrow, and steep-sided.

In addition to the major adjustment to site conditions of locally substituting a key trench filled with impervious zone 1 for a grout curtain through highly jointed, pervious bedrock, the designers concluded that extensive curtain grouting beneath key and cutoff trenches would be required. To indicate the scope, the bid items included provision of 6230 cubic meters (55 000 barrels) of cement, and 1300 cubic meters (1700 cubic yards) of sand, together with 7360 cubic meters (260 000 cubic feet) of pressure grouting. Actual quantities of cement injected were over twice the bid quantities.

The drawings and specifications called for three rows of deep grout holes along most of the axis of the key and cutoff trenches, with wide latitude retained by the Bureau to direct and modify specific details. The center row of grout holes, intended to form the impermeable curtain, was provided with a concrete grout cap, nominally 1 meter (3 feet) wide by 1 meter (3 feet) deep in a drilled and blasted notch in rock. In the key trenches the specified grouting sequence was: First, the downstream row of holes on 6-meter (20-foot) centers; second, the upstream row of holes on 6-meter (20-foot) centers; and third, closure along the center row of holes working through the grout cap. These center holes were spaced on 3-meter (10-foot) centers, with split spacing where the primary holes did not indicate a tight curtain. It is important to recognize that, as this procedure was actually carried out, neither the upstream nor the downstream rows constituted grout "curtains" as the term is conventionally understood. Actually, full closure along the two outer rows was neither attempted nor attained.

The specifications required "blanket grouting" in the key trench and cutoff trench, as directed by the Bureau. It entailed drilling and grouting both uniformly and randomly spaced and angled holes to shallow depths of 6.1 to 10.7 meters (20 to 35 feet) to intercept and plug open joints. The scope of the blanket grouting done was limited, and the areas so treated were almost exclusively in the bottom of key and cutoff trenches, and at only a few local spots. A major exception was at the spillway crest structure where a close pattern of 24.3 meter (80-foot) deep "blanket" grout holes was placed under the entire structure.

A review of the drawings and specifications failed to show that it was expected to do slurry-concrete treatment of open bedrock joints at the core-to-bedrock contact. The Bureau's Teton Project Office records indicate, however, that a total of about 1400 cubic meters (1830 cubic yards) was placed at the instigation of that office. This was accomplished principally by pouring slurry into open joints and the more obviously open cracks. This procedure was discontinued above about elevation 1588 meters (5210 feet).

It was particularly evident that, in the failure area, significantly large open joints existed at the top of the downstream face of the key trench at axis STA 4 + 26.72 (Sta. 14 + 00) and near the downstream toe of zone 1 opposite STA 4 + 57.20 (Sta. 15 + 00), where slurry takes totaled 76 cubic meters (100 cubic yards) or more.

These large takes under gravity placement conditions identified the rock as being extremely pervious, indicating that grave incompatibility existed between the highly erodible zone 1 fill and its underlying intensely jointed foundation.

The Drawings and Specifications, as interpreted, permitted slope wash materials, overburden, and talus to be left in place under the dam, in all areas outside of the zone 1 fill. Under the zone 1 material, stripping to bedrock was required. Stripping of overburden was subject to the Construction Engineer's judgment as to the amount necessary to uncover reasonably strong materials.

It seems conclusive, from examination of as-excavated cross sections taken at 3-meter (10-foot) intervals along the axis between STAS 4 + 41.96 and 4 + 72.44 (Stas. 14 + 50 and 15 + 50), as well as construction photos and from oral inquiry, that a substantial depth of talus was left under the dam downstream from the zone 1 fill. The project construction staff confirmed that substantial volumes of talus were left in place under the outer zones of the dam.

A large volume of bouldery talus along the canyon wall under the downstream zones of the dam in the vicinity from STA 4 + 57.20 to 5 + 18.16 (Sta. 15 + 00 to 17 + 00) could have provided an exit conduit for the initial relatively restricted leakage across or under the key trench at about STA 4 + 26.72 (Sta. 14 + 00), and increasing leakage flows could have had explainable exits at the groin of the dam at elevation 1585 meters (5200 feet) and 1537.7 meters (5045 feet) where the large flows of muddy water were seen on June 5.

Low-level reservoir discharge was to be accomplished through the river outlet works in the left abutment and by the auxiliary outlet works in the right abutment. The design capacities of these facilities at a maximum water surface elevation of 1622.84 meters (5324.3 feet) were 96.3 and 24.1 cubic meters (3400 and 850 cubic feet) per second, respectively.

Even though the approved construction schedule required construction to be completed by March 31, 1976, only the auxiliary outlet works were in operation through June 5, 1976. This resulted in virtually no control of the reservoir filling rate during the late spring of 1976.

The records of Teton River hydrology were well known to the Bureau of Reclamation. The design criteria recognized that it would be necessary to have the river outlet works in operation after May 1, 1976, to control the rate of filling so as not to exceed a 0.3-meter (1-foot) per day increase when the reservoir surface elevation was above 1585 meters (5200 feet). This design fill rate was relaxed to 0.6 meter (2 feet) per day on March 23, 1976, but the new rate was exceeded on 3 days in April and during the entire period from May 11 to June 5.

Any satisfactory explanation of the failure must be in accordance with the known chronology and eyewitness accounts. The facts are summarized as follows:

Before June 3, no springs or other signs of increased seepage were noticed downstream of the dam. On June 3, clear-water springs appeared at distances of about 400 and 460 meters (1300 and 1500 feet) downstream, issuing from joints in the rock of the right bank.

During the night of June 4, water evidently flowed down the right groin from about elevation 1585 (5200) inasmuch as a shallow damp channel was noticed early on the morning of June 5. Shortly after 7 a.m. when the first observations (see figs. 4-20 and 4-21) were made on June 5, muddy water was flowing at about 0.6 to 0.8 cubic meter (20 to 30 cubic feet) per second from talus on the right abutment at about elevation 1538 meters (5045 feet), and a small trickle of turbid water was flowing from the right abutment at elevation 1585 meters (5200 feet). Both flows were at the junction of the embankment and the abutment, referred to as the groin, and both increased noticeably in the following 3 hours.

At about 10:30 a.m., a large leak of about 0.4 cubic meter (15 cubic feet) per second appeared on the face of the embankment, possibly associated with a "loud burst" heard at that time, at elevation 1585 meters (5200 feet) about 4.5 meters (15 feet) from the abutment and adjacent to the smaller leak previously observed at the same elevation. The new leak increased and appeared to emerge from a "tunnel" about 1.8 meters (6 feet) in diameter, roughly perpendicular to the dam axis approximately opposite dam axis STA $4+64.82$ (Sta. $15+25$) and extending at least 11 meters (35 feet) into the embankment. The tunnel became an erosion gully developing headward up the embankment and curving toward the abutment.

At about 11 a.m., a vortex appeared in the reservoir at about STA $4+26.72$ (Sta. $14+00$), above the upstream slope of the embankment. At 11:30 a.m., a small sinkhole appeared temporarily, ahead of the gully developing on the downstream slope, near the top of the dam. Shortly thereafter, at 11:57 a.m., the top of the dam collapsed.

The open nature of the joints on the upstream face of the right abutment key trench was confirmed by the Independent Panel's investigation after removal of the key trench fill. At the failure section, part of the abutment rock had been removed by the action of the escaping floodwaters. There was no reason to believe that the removed rock was less open-jointed than that which remained. Hence, there could be no doubt that reservoir water had ready access to the entire upstream face of the key trench, including the portion adjacent to the failure section.

Beneath the level of the base of the key trench, the rock was also jointed and pervious, as judged by water tests in exploratory drill holes and by grout takes in the curtain. Since the curtain was confined to the key trench, there is no doubt that the rock at depth, upstream of the curtain, was pervious, although possibly less so than that in the upper 21 meters (70 feet). Inspection of the face of rock remaining along the right abutment after the erosion by the escaping floodwaters disclosed many open joints below key trench level, some partially filled with grout, both upstream and downstream of the grout curtain.

Ponding tests conducted by the Independent Panel demonstrated that water could flow readily beneath the grout cap at several locations near the failure zone. Water tests in drill holes on the center line of the grout curtain

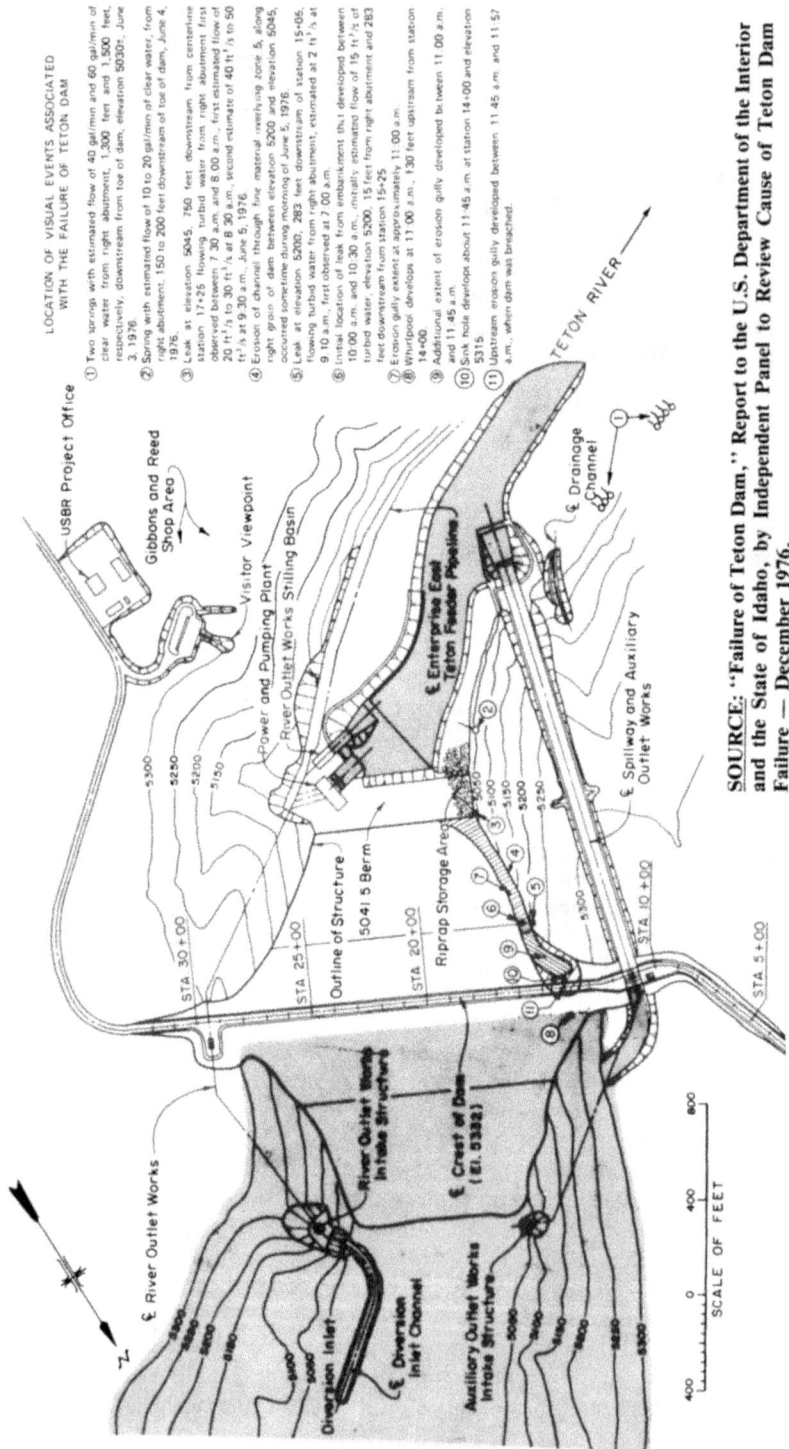

LOCATION OF VISUAL EVENTS ASSOCIATED
WITH THE FAILURE OF TETON DAM

① Two springs with estimated flow of 40 gal/min and 60 gal/min of clear water from right abutment, 1,300 feet and 1,500 feet, respectively, downstream from toe of dam, elevation 5030±, June 3, 1976.

② Spring with estimated flow of 10 to 20 gal/min of clear water, from right abutment, 150 to 200 feet downstream of toe of dam, June 4, 1976.

③ Leak at elevation 5045, 750 feet downstream from centerline station 17+25 flowing turbid water from right abutment first observed between 7:30 a.m. and 8:00 a.m. first estimated flow of 20 ft³/s to 30 ft³/s at 8:30 a.m. second estimate of 40 ft³/s to 50 ft³/s at 9:30 a.m. June 5, 1976.

④ Erosion of channel through fine material overlying zone 5, along right groin of dam between elevation 5200 and elevation 5045, occurred sometime during morning of June 5, 1976.

⑤ Leak at elevation 5200, 283 feet downstream of station 15+05, flowing turbid water from right abutment, estimated at 2 ft³/s at 9:10 a.m. first observed at 7:00 a.m.

⑥ Initial location of leak from embankment that developed between 10:00 a.m. and 10:30 a.m. initially estimated flow of 15 ft³/s of turbid water, elevation 5200, 15 feet from right abutment and 283 feet downstream from station 15+25.

⑦ Erosion gully extent at approximately 11:00 a.m.

⑧ Whirlpool develops at 11:00 a.m. 130 feet upstream from station 14+00.

⑨ Additional extent of erosion gully developed between 11:00 a.m. and 11:45 a.m.

⑩ Sink hole develops about 11:45 a.m. at station 14+00 and elevation 5315.

⑪ Upstream erosion gully developed between 11:45 a.m. and 11:57 a.m., when dam was breached.

Figure 4-20.—Teton Dam, prefailure leakage — location of visual events. P-801-D-79372.

SOURCE: "Failure of Teton Dam," Report to the U.S. Department of the Interior and the State of Idaho, by Independent Panel to Review Cause of Teton Dam Failure — December 1976.

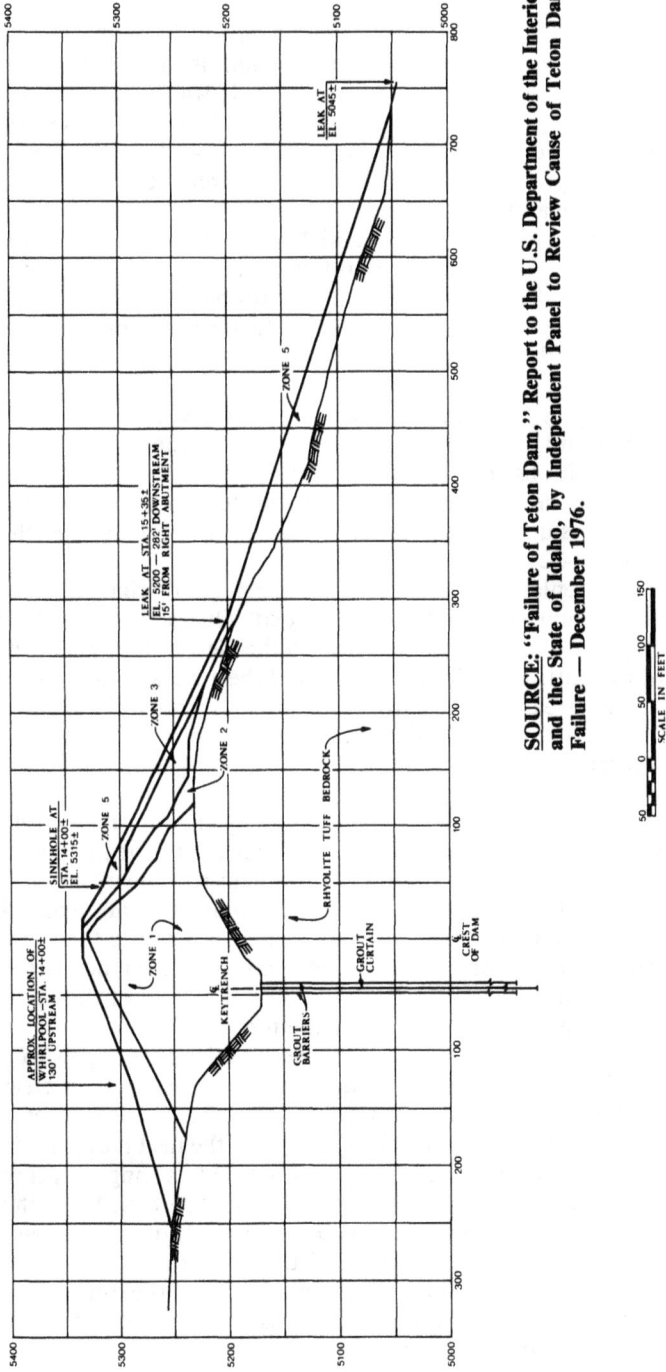

Figure 4-21.—Teton Dam, section along approximate path of failure. P-801-D-79373.

SOURCE: "Failure of Teton Dam," Report to the U.S. Department of the Interior and the State of Idaho, by Independent Panel to Review Cause of Teton Dam Failure — December 1976.

near the failure section also demonstrated the existence of passages through which water emerged downstream.

The key trench fill was investigated extensively as the remnant on the right abutment was excavated. The material, as indicated in the specifications, consisted of windblown clayey silts. It was compacted generally on the dry-side of optimum, contained occasional lenses or layers more plastic than the rest, and was placed against the rock walls of the key trench with no rock treatment or transition. Loose zones were noted beneath occasional overhangs or against open joints.

Between about STAS 4 + 08.4 and 4 + 57.2 (Stas. 13 + 40 and 15 + 00) on the right abutment, particularly unfavorable conditions existed; i.e.: (1) a geometry of the key trench especially favorable to arching and cracking of the embankment core material; (2) significant water passages through the rock just beneath the grout cap and possibly through the grout curtain at greater depth; (3) a concentration of throughgoing joints beneath and alongside the key trench; and (4) an erodible fill within the key trench and in contact with the jointed rock downstream from the key trench. As a result of these conditions, one or more erosion tunnels probably formed across the bottom of the key trench which permitted water to flow readily from the open joints upstream to those downstream of the key trench and grout curtain (fig. 4-22).

As erosion enlarged the tunnel or tunnels, the discharge, being of an increasing amount and containing eroded silty soils, could escape only through passages of appreciable size. In the view of the Independent Panel, some of the outflow undoubtedly entered the generally interconnected joint system downstream of the cutoff and spread through the rock mass, but a large part passed nearly horizontally, through or around the rock between the downstream face of the key trench and the right abutment wall. Part of the flow emerged from the rock against the zone 1 fill on the right abutment, turned downstream, and flowed along the interface until it reached zone 2 or the pervious surficial soils and talus left beneath zone 2. Another part of the flow passed through the abutment rock near the surface, where weathering and relaxation left more open joints than at greater depth, and then emerged into zone 2 or the talus beneath it. Once the pervious zones were reached and as long as the outflow did not exceed their capacity, water flowed through the pervious materials near the groin of the right abutment and through the riprap stockpile at the toe.

During the night of June 4, however, the leakage began to exceed the capacity of the pervious materials, whereupon it emerged at elevation 1585 meters (5200 feet) and flowed briefly down the surface, where dampness and slight erosion were noted along the groin the next morning. Early in the morning, as flow continued to increase, muddy springs appeared at both elevation 1537.7 meters (5045 feet) and 1585.0 meters. Soon the spring at elevation 1585 meters was seen to be the mouth of an erosion tunnel extending along the rock at the base of the earthfill close to the groin. Progressive erosion led to continued increase in size of the tunnel until finally at about 10:30 a.m., the water pressure was great enough to break suddenly and violently through the zone 2 fill and erupt on the face of the dam. Thereafter, the tunnel became a gully, working headward first up the groin and then along the initial passage through the key trench. The gully ex-

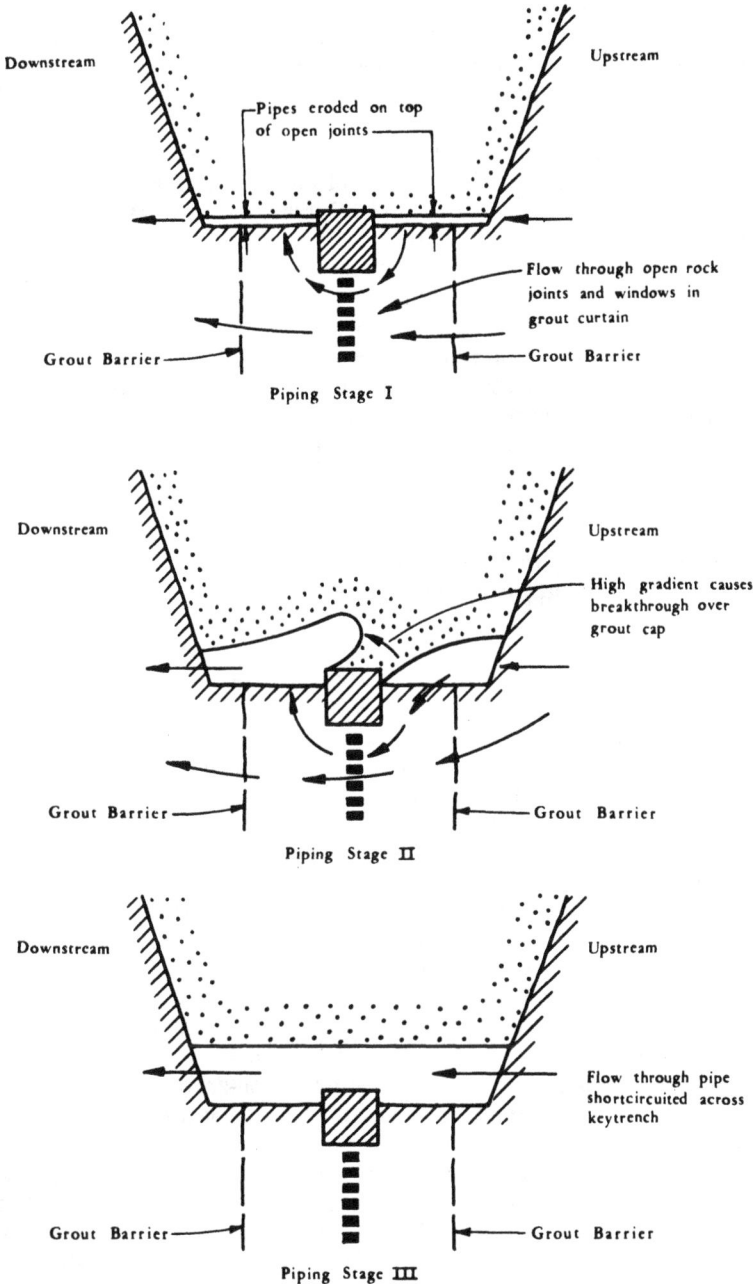

SOURCE: "Failure of Teton Dam," Report to the U.S. Department of the Interior and the State of Idaho, by Independent Panel to Review Cause of Teton Dam Failure — December 1976.

EROSION DIAGRAMS

Figure 4-22.—Teton Dam, key trench erosion diagrams. P-801-D-79374.

tended upstream by successive collapses of the roof of the tunnel, including the sinkhole that appeared briefly at elevation 5315, toward the vortex over the upstream end, culminating in collapse of the roadway at the top of the dam.

The Interior Review Group, in addition to paralleling and extending the technical investigations by the Independent Panel, examined organizational and procedural factors that related to the Bureau's work on the project. The findings in its 1977 Report included:

"The design incorporated a feature, key trenches in the abutments, that significantly departed from past Bureau of Reclamation practices."

"The geometry of the abutment key trenches was conducive to developing stress patterns that could have allowed cracking of the impervious core.

"The narrow width of the sealed foundation rock at the bottom of the key trench, combined with the high permeability of the rock foundation on either side of the key trench, produced steep hydraulic gradients across the trench.

"Zone 1 compaction at the contact with foundation rock surfaces was generally equivalent to that within the body of the fill. Joints and fractures in the key trench walls were numerous, with openings up to 6 inches wide. There was little evidence of surface grouting in joints exposed in the key trench walls."

"Surface grouting stopped at El. 5205. Neither the designers nor the liaison engineer were aware of the decision to stop surface grouting until after the failure of Teton Dam. According to field personnel, the field geologists also played no part in the decision to stop surface grouting."

"The rock surface was not adequately sealed under the impervious core upstream and downstream from the key trench."

"The open fractures in the abutment foundation rock allowed direct access by reservoir water to the impervious core on the upstream side of the key trench. Any water flowing through the impervious core could exit into open fractures on the downstream side of the key trench."

"The laboratory tests, *** confirmed the highly erodible nature of the zone 1 material and its brittle characteristics when compacted dry of optimum moisture content."

"The design failed to provide a defense against flow through embankment cracks or against erosion of the impervious core at rock surfaces."

"Defensive measures were within the state-of-the-art of dam design at the time Teton Dam was designed, and should have been used."

"Design notes, developed early in the design process, identify and report a variety of potential design problems and possible design alternatives. However, there are no records, documents, or reports that show: (1) the logical resolution of each of the identified design problems, (2) why a particular design alternative was considered satisfactory and selected in preference to others, and (3) why an identified design problem was subsequently judged important or not important and omitted from, or included for, further consideration."

The following conclusions from the IRG's January 1980 "Failure of Teton Dam, Final Report" were presented to supplement those in the 1977 IRG report.

"In the April 1977 report Failure of Teton Dam, A Report of Findings, the IRG concluded that the dam ' *** failed as a result of inadequate protection of zone 1 impervious core material from internal erosion.' None of the findings resulting from this continued investigation have changed this primary conclusion. A safe dam could have been constructed if the designers had provided a defensive design with proper embankment filtration and drainage, and appropriate foundation surface treatment."

"Weaknesses existed in construction practices. Key trench rock-slope surfaces were not provided with adequate treatment, the upper ends of a number of drill holes in the grout curtain were not grouted, and inspection procedures did not always provide adequate control of zone 1 fill placement. While construction supervision and control was as good or better than normally provided on major dams, there is evidence of a period at the beginning of the 1975 construction season when it was not adequate.

"The left abutment embankment remnant excavation revealed areas where moisture penetration into zone 1 was significantly greater than expected for the short life of the reservoir. The density-permeability relationships of zone 1 show generally larger permeabilities and faster reservoir water penetration into zone 1 than would normally have been assumed from past experience or from test data available prior to this investigation.

"No piping had occurred within the material removed by the left abutment excavation.

"The open jointed condition of rock in the key trench walls shows that piping of zone 1 fill into foundation rock could proceed undetected for a significant period of time.

"Discovered conditions within zone 1 fill support the following physical modes by which piping could have been initiated:
    a. Seepage of reservoir water along either the fill/rock contact surface or near the top of the grout curtain in the right key trench, or
    b. Seepage of reservoir water through a low density, high permeability lens located within or adjacent to the right key trench.

"No cracks associated with reservoir-induced hydraulic fracturing or differential settlement within the embankment were discovered. However, evaluation of hydraulic-fracture test data indicates the possible occurrence of embankment stresses low enough for hydraulic fracturing to occur.

"Overhangs and ledges of rock on slopes within the key trench were not, according to practices used by other Federal dam building agencies, prepared adequately to receive compacted fill.

"The grouting program did not establish the impractical objective of providing an impervious barrier in rock. Localized areas were found where seepage could pass through the grout curtain. Grout containing calcium chloride in excess of 2 percent of the weight of the cement used in the mix will usually not be adequate for the control of seepage.

"The wet seams are the result of thin, low-density, high-permeability layers placed in the embankment. Their low density is a function of fill placement procedures and/or unusual material properties possibly related to varying percentages of amorphous minerals or calcium carbonate equivalent in the soil's clay size particles.

"Fill with low dry densities was placed in zone 1 and accepted as suitable because procedures for evaluation of density from construction control tests were based solely on D ratio.

"The Bureau practice of using the word "tight" to define those joints open less than 1/2 inch or 0.1 foot can lead to false interpretation of the strength, permeability, and deformation modulus of a rock mass."

In its summary, the Independent Panel concluded that the physical conditions were fully satisfied for water flowing under high pressure to attack the lower part of the key trench fill along open joints, some of which were found to transmit water freely through the grout curtain, particularly through the upper part near the grout cap. The attack was fully capable of quickly developing an erosion tunnel across the key trench. Arching at local irregularities, loose zones of fill at re-entrants, and local cracking may have contributed to the success of the attack and determined the precise location. Conditions were favorable for escape of the water and eroded solids into the joints of the rock downstream, for discharging the water against and along the interface of the right abutment of the dam and the embankment, and for development of the erosion features that ultimately breached the dam.

In the Independent Panel's words: "The precise combination of geologic details, geometry of key trench, variation in compaction, or stress conditions in fill and porewater that caused the first breach of the key trench fill is of course unknown and, moreover, is not relevant. The failure was caused not because some unforeseeable fatal combination existed, but because (1) the many combinations of unfavorable circumstances inherent in the situation were not visualized, and because (2) adequate defenses against these circumstances were not included in the design."

The tentative conclusions of Independent Panel member Thomas M. Leps following his examination of the left abutment disclosures are noteworthy:

"I. 'Wet seams' in an 'impervious', layered embankment constructed of compacted silt are a normal and inevitable consequence of the necessarily layered construction, of varying weather conditions, and of accepted latitudes in the practices of fill moisture and quality control.

"II. The 'wet seams' at Teton Dam do not appear to be sufficiently different from fill layers above and below them, or to be so extensive or so importantly located, as to constitute significantly weak horizons, structurally.

"III. The possibility, and even the probability, of existence of a weather-caused 'wet seam' in Zone 1 fill contacting the right abutment keyway at about axis station $14+20$, and over which 1975 fill construction subsequently progressed, seems on balance to have been more realistic than not, but no direct proof of such presence seems to have been detected in available records.

"IV. Whether or not a throughgoing, horizontal, 'wet seam' was present at axis station $14+20$ is judged to be not significant. The only basically significant factor in the failure continues to be the absence of any thorough provision downstream from the Zone 1 fill for preventing the progressive removal of that fill through open joints in the downstream bedrock, under the action of any type of leakage, whether through hydraulic fractures or through possible channels above or below the grout cap.

"V. An hypothetical 'wet seam' in the Zone 1 fill near axis station $14+20$ (at the right abutment) could be conceded to have created local conditions which might have accelerated failure of the dam, but in and of itself could not have been the basic cause of failure. In our judgment, the failure occurred because complete plugging of open bedrock joints, and engineered filtering and drainage of the downstream surface of Zone 1 at the bedrock interface, had not been provided."

Some of the other views expressed on the occasion of the Independent Panel members' visit at the Teton damsite in 1978 were:

Panel member Ralph B. Peck: "Since the borrow materials for all embankment dams are products of nature and vary from point to point; since efficient modern construction requires rapid placement and treatment of large quantities of these materials; and since weather conditions are variable and often unpredictable, no zone of any earth dam can be expected to be perfectly uniform or homogeneous. Very few dams have been dissected in such a way as to permit the comprehensive investigation of variability and properties undertaken in the left remnant of Zone 1 of Teton Dam. The few dissected dams that have come to my attention have been appreciably more variable than Teton. I consider that the control of Teton Dam was above average, that the embankment of the dam was probably more uniform than that of many successful dams, and that the occasional wet seams within the embankment should not be considered to be unusual defects but, rather, to be representative of nonuniformities likely to be found in virtually all earth dams. In my judgment, the presence of variations of this sort should be assumed by designers, and the design and specifications should take them into account."

Panel member H. Bolton Seed: " *** it is not difficult to visualize that low density non-plastic seams which were placed in a dry condition ultimately had their water contents raised either by percolation of water from above or by squeezing of water from adjacent wet plastic zones, to the point where they became saturated and retained the water which now shows up if the zone is exposed.

"I suspect that many dams given the detailed scrutiny to which Teton has been exposed would show deviations from intended placement conditions similar to those found at Teton. Certainly this is the case in many hydraulic fill dams and it is likely to be so in compacted earth fill dams."

Panel Chairman Wallace L. Chadwick: " *** the frequency with which wet spots were observed during drilling and excavation of the left remnant suggest strongly that many embankments may include such spots or seams, and that safety analyses should allow for some lack of design homogeneity in each constructed embankment. It seems probable that even with the best of construction techniques, an embankment of the homogeneity envisioned in design is not fully attainable. No design should be considered complete until the embankment has been built and tested through at least one filling cycle."

Tigra Dam

At 4 p.m. on August 4, 1917, a new dam near the village of Tigra on the River Sank in Madhya Pradesh, India, was breached during a flood. Apparently few records are available on the extent of damage and the number of casualties downstream.

Heavy runoff caused the straight gravity dam to be overtopped from abutment to abutment by a flow of about 850 cubic meters (30 000 cubic feet) per second. Under this unprecedented water pressure, the structure slid out on its base.

This 26-meter (86-foot) high barrier, built in the period 1913 to 1917, was composed of a rubble-masonry interior placed in lime mortar and faced with coursed rubble masonry, cement-jointed on the upstream side. It was 1341 meters (4400 feet) long, including 305 meters (1000 feet) of uncontrolled spillway. The elevations of parapet, top of dam, and spillway crest were 226.9 meters (744.5 feet), 226.2 meters (742.0 feet), and 225.6 meters (740.0 feet), respectively.

The geologic formation at the site is a stratified sandstone. Excavation for the dam was generally limited to about a 0.6-meter (2-foot) depth, except that weaker rock zones reportedly were dug deeper and filled with concrete. The lack of a continuous cutoff of significant dimensions left the structure vulnerable to uplift pressures and sliding on the essentially horizontally bedded rock.

While the dam was overtopped in the monsoon season of each year during construction, it was not taxed fully until the final year. However, even before the reservoir was filled to its capacity, serious cracking appeared in the masonry.

On August 4, 1917, with a reservoir water surface elevation of 227.0 meters (744.68 feet) and with spilling beginning over the parapet, a long section of the dam slid out. Much of the spillway portion breached. The nature of the movement is evidenced by two large monoliths which still remain at the site, standing erect but shifted downstream. These blocks are of such sound construction that defective workmanship has been discounted as a causative factor. There appears to be general agreement that failure was attributable to tension associated with hydrostatic uplift and accompanied by scouring from overflow — all of which set the stage for sliding of the dam mass on its stratified foundation.

In 1929, a new dam was constructed at the site just upstream from the remains of the old one. It was designed for the same reservoir capacity, but this time gates were provided for spillway control. The cross section of the dam is nearly identical to the original, with vertical upstream face and 2 to 3 downstream slope, but additionally a concrete cutoff with a maximum depth of about 4.6 meters (15 feet) was extended into the foundation at the heel. Another new feature was a clay blanket approximately 37 meters (120 feet) wide placed just upstream of the dam. Safeguarded by these improvements, the replacement structure has been operated for half a century without adverse event.

Vaiont Dam

One of the most damaging reservoir disasters of all time occurred on October 9, 1963, at the Vaiont Dam (fig. 4-23) near Belluno in Veneto Province in Italy, when 2,600 human lives were lost. During the night a tremendous rockslide fell into the reservoir. The impact of the great mass moving with terrifying speed raised gigantic waves which overtopped the structure. Tremors caused by the slide triggered seismological instruments throughout a vast area of western and central Europe. The dam itself sustained no major damage even though it was hit by a total force of about

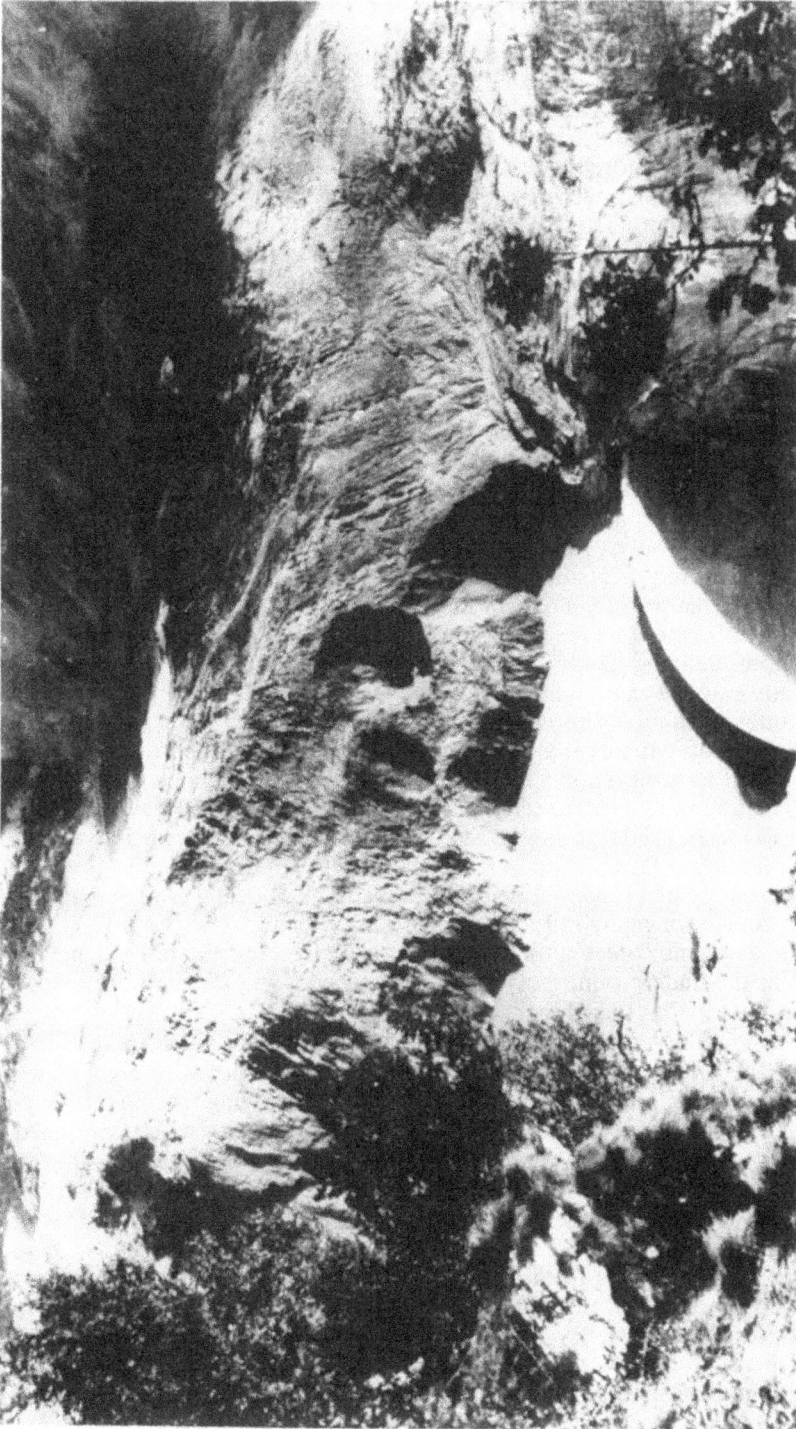

Figure 4-23.—Vaiont Dam after failure, showing massive slide upstream of dam. P-801-D-79375.

36 000 meganewtons (4 000 000 tons) from the slide and overtopping pressures.

The dam, with a height of 265 meters (869 feet) was said to be the world's highest thin arch and the second highest dam of any kind. It was completed in the fall of 1960. The arch is 3.4 meters (11.2 feet) thick at the top and 22.7 meters (74.5 feet) thick in the bottom of the canyon. It has an overflow spillway which, before the disaster, had a two-lane concrete bridge over the crest. There was an underground powerplant in the left abutment. Reservoir capacity was about 150 000 000 cubic meters (122 000 acre-feet).

The volume of the slide, which had been the left wall of the canyon, exceeded 240 000 000 cubic meters (314 000 000 cubic yards). This material filled the reservoir for a distance of 1.8 kilometers (1.1 miles) and up to heights of 150 meters (490 feet) above reservoir level. The quick movement of the rock mass, with a velocity as great as 30 meters (100 feet) per second, created an air draft that blew water and rocks up the right canyon wall to a height of 240 meters (780 feet) above the normal reservoir surface. Water displaced by the slide material was thrown up the right canyon wall to the village of Casso, where it washed through buildings. It spilled over the dam to a height of about 100 meters (330 feet) above the crest, which was at elevation 722.5 meters (2370.4 feet). The spillway bridge was torn away. Figures 4-24 and 4-25 show Vaiont Dam after the overtopping.

Those at the site at the time of the slide included 20 technical personnel in the control center on the left abutment and about 40 occupants of the office and hotel building high on the right abutment. But evidently nobody who actually saw the mountain come down was counted among the survivors. After demolishing the hotel at elevation 780 meters (2559 feet), much of the displaced water apparently surged back across to the left abutment and rose there to at least elevation 820 meters (2690 feet). Giant waves converged at the dam and went over the crest in a massive spill. The flood wave was more than 70 meters (230 feet) high where the Vaiont River enters the Piave River 1.6 kilometer (1 mile) downstream. After obliterating the town of Longarone at that junction, the flood wave left practically total devastation in its course for many miles down the Piave valley.

The awesome forces acting in the disaster taxed imagination. Structural steel in the underground powerplant was twisted and sheared. Steel doors were ripped from their hinges.

A resident of Casso, on the right canyon wall over 260 meters (850 feet) above the lake, reported that at about 10:15 p.m. on that rainy night he was awakened in his second-story room by the roar of moving rocks. He was not alarmed since surficial sliding occurred frequently. The sound continued. Then at about 10:40 p.m., an air blast hit the building, breaking the windows. Soon the roof was lifted and water and rocks came into the room. He had scrambled to the door when the roof fell onto his bed. The wind abruptly subsided.

Surviving witnesses from Longarone said that a flood wave came down the canyon at 10:43 p.m., and a strong wind broke windows. There were strong earth tremors. By 10:55 p.m. the flood had passed, and the valley was silent.

A combination of adverse geologic conditions led to the slide on October 9, 1963. Strong forces acted on the rock mass, which was 1.8 kilometers (1.1

Figure 4-24.—Vaiont Dam after failure, looking upstream at slide mass. P-801-D-79376.

Figure 4-25.—Vaiont Dam showing damage to top of dam and spillway. P-801-D-79377.

miles) long and 1.6 kilometers (1.0 mile) wide. The mass was a prehistoric slide which had come down from Mount Toc, which lies south of the reservoir. The instability of the slope of this mountain had been recognized by some of the project planners as a serious disadvantage of the reservoir location.

The site is on a deep sedimentary formation composed primarily of limestone with interbedding of claystone and marl. This rock has low shear strength, and the limestone has been subjected to solution by ground water. This had led to extensive cavities and tubes and enlarged joints and bedding planes. Sinkholes developed in the land south of the canyon rim and functioned as inlets for the runoff which replenished the ground-water basin. The labyrinth of solution conduits reduced the integrity of the formation and facilitated a gradual increase in internal water pressures. This in turn diminished friction between rock interfaces. Creation of the reservoir superimposed additional water forces on the subterranean complex. A slow creep was in progress for several years before the disaster.

On November 4, 1960, when the reservoir water depth was about 130 meters (426 feet) a slide of roughly 690 000 cubic meters (900 000 cubic yards) came down the left slope of the canyon near the dam. Cracking ensued above the scarp. This gradually spread and ultimately delimited the mass which brought disaster. The mountain slope was in motion on a tremendous scale.

After the 1960 slide, the lake level generally was not allowed to exceed elevation 680 meters (2231 feet). A network of survey monuments was established on the slope extending 4 kilometers (2.5 miles) upstream. Rapid creep was measured during the 1960 to 1961 period. From time to time, movement of as much as 300 millimeters (12 inches) per week would precede small tremors within the mass, centered at depths ranging up to about 500 meters (1600 feet). The total volume affected by creep was estimated as 200 000 000 cubic meters (260 000 000 cubic yards).

The movement then tended to decrease. Until the fall of 1963, the slope moved slowly. The average rate of creep since the 1960 slide was estimated roughly as 10 millimeters (0.4 inch) per week. In April 1963, evidently the reservoir operating limit was raised to about 20 meters (66 feet) higher than the 680-meter (2231-foot) level, which had been the criterion since the 1960 movement. In September, the creep accelerated. Intense rainfall in August and September caused runoff which recharged the water storage in the rock mass. This increased its weight and the internal water pressures on the planes of weakness.

Beginning on about September 18, 1963, many geodetic monuments were creeping at a rate of approximately 10 millimeters (0.4 inch) per day. The general assumption of observers apparently was that these were separate movements that did not integrate into the sliding of a single, huge mountain slope.

Heavy rainfall resumed on about September 28 and continued past the time of failure on October 9. More runoff entered the unstable mass. At the end of September the reservoir level reached a maximum of 710 meters (2329.4 feet), but prior to the disaster it was lowered again to 700 meters (2296.6 feet). Around the first of October, animals grazing on the north slope of Mt. Toc abandoned the area, presumably sensing the hazard. The mayor of Casso had cautioned citizens to leave the restless slopes because of the possibility of a wave in the reservoir.

On October 8, those responsible for monitoring the geodetic grid recognized that a tremendous mass was in motion, embracing an area five times as large as they had assumed to be affected. On that day the operators started lowering the water from elevation 700 meters (2296.6 feet). Two outlet tunnels in the left abutment were making a combined release of about 142 cubic meters (5000 cubic feet) per second, but heavy reservoir inflow diminished the effect. Total storage at the moment of the slide was approximately 120 000 000 cubic meters (97 000 acre-feet).

On October 9 at 10:39 p.m., the rock mass came down over a length of nearly 2 kilometers (1.2 miles). Its speed was so great that the timespan of all movement was estimated at just 5 minutes.

Valparaíso Dam

On August 11, 1888, a reservoir at Valparaíso, Chile, burst and poured its contents into the city. The estimated death toll included more than 100 people.

This water storage facility was located in a ravine among high hills, about 1.6 kilometer (1 mile) above the city of Valparaíso. It was formed by the construction of an earthfill structure 17 meters (56 feet) high with a base thickness of 40 meters (131 feet) and a crest width of 15 meters (49 feet).

Failure came after an extended period of heavy precipitation. Wet weather had persisted since the beginning of the calendar year, and rain had fallen continuously since the first of August. At 8:30 a.m. on August 11, the embankment collapsed, releasing 64 350 cubic meters (17 000 000 gallons) of water into the populated area. Buildings were demolished under the mass of mud and debris carried by the flood.

Some reports indicate that precarious conditions at the reservoir before the disaster were known to certain responsible officials. Despite this, the facility evidently could not be saved by taking emergency measures.

Van Norman Dams
(San Fernando)

Nearly catastrophic performances of the Upper and Lower Van Norman (San Fernando) Dams (fig. 4-26) during an earthquake near Los Angeles, Calif., on February 9, 1971, were significant in that they led to a major breakthrough in techniques for design and analysis of embankments.

The dams were about 14 kilometers (8.5 miles) southwest of the epicenter. With a Richter Magnitude of 6.4 and approximately 15 seconds of strong vibration, the quake had a maximum acceleration of about 50 percent of gravity (0.5g) in the dam foundations.

The two embankment structures were the key elements of the Van Norman complex, a focal point for distribution of water to the city of Los Angeles. Normal flow through the complex in the latter half of 1970 was 18.4 cubic meters (650 cubic feet) per second, which supplied about 80 percent of the city's water needs.

Figure 4-26.—Lower Van Norman (San Fernando) Dam after earthquake of February 9, 1971 (Courtesy, Calif. Dept. of Water Resources). P-801-D-79378.

The 1971 earthquake caused massive sliding of the lower dam. Practically all the upstream slope and the top 9 meters (30 feet) of the fill slumped and spread into the reservoir. The dam, 43 meters (142 feet) high and 634 meters (2080 feet) long, created a reservoir with a capacity of 25 300 000 cubic meters (20 500 acre-feet). When the quake struck, the water surface was approximately 10.7 meters (35 feet) below the top of the embankment, and storage was at about half of maximum. The slide left roughly 1.5 meters (5 feet) of freeboard between the water and the ragged top of the surviving fill, which had wide and deep longitudinal cracks and a nearly vertical scarp.

The 25-meter (82-foot) high upper dam, just above its companion, suffered severe distortion in the downstream part of the embankment. There was some cracking of the fill, and opening of outlet conduit joints induced serious internal erosion. The dam crest was displaced downstream as much as 1.5 meters (5 feet) and subsided about 0.9 meter (3 feet). Despite this damage, the 2 300 000-cubic-meter (1850-acre-foot) reservoir was able to continue in service following the quake. If this impoundment had failed, the spill undoubtedly would have caused overtopping and collapse of the remaining section of the lower dam.

With the two embankments in such precarious condition, a strong aftershock could have triggered a catastrophe. The fate of 80,000 residents downstream was in the balance. Without delay, the endangered area was evacuated until storage in the lower reservoir could be reduced to safe limits. There is no question that, if conditions had been just fractionally more adverse, this event would have been recorded as one of the worst disasters in history.

The Van Norman Dams were hydraulic fills built in the period 1912 to 1921. The lower dam was completed in 1918, followed by the upper dam in 1921.

The hydraulic fill of the lower dam was topped with a rolled section made of shales from the left abutment. Later successive additions of compacted fill raised the crest, the final one in 1929 to 1930 bringing it to approximately 14 meters (45 feet) above the sluiced embankment. A rolled-earth buttress was built against the downstream slope in 1940.

Until 1966, the lower reservoir was operated throughout its full range. In that year, investigations of the dam and its foundation culminated in setting of a limit 2.9 meters (9.6 feet) under the design maximum. Just before the quake on February 9, 1971, the water level was about 7.6 meters (25 feet) below the spillway or 10.7 meters (35 feet) below the top of the embankment. At this elevation, storage was approximately 13 600 000 cubic meters (11 000 acre-feet), which was slightly more than half the design capacity. Inflow was about 13.4 cubic meters (474 cubic feet) per second, while outflow was about 11.0 cubic meters (390 cubic feet) per second. Maximum discharge from the reservoir during the emergency evacuation, which lasted 4 days, was estimated as 19.8 cubic meters (700 cubic feet) per second.

Movement at the lower dam during the earthquake was indicated on two seismoscopes, one on the east abutment and another on the top of the embankment. Data derived from these instruments show an estimated maximum foundation acceleration of 0.5g. Comparison of the recordings of the two seismoscopes indicates no appreciable amplification between the foundation and the top of the dam.

Strong-motion records were obtained at several other places in the area. One of the most significant was registered at a station on the left abutment of the 113-meter (370-foot) high Pacoima Dam, approximately 8 kilometers (5 miles) south of the epicenter. Its two horizontal components indicated accelerations as high as 1.25g and the vertical component a maximum of 0.72g. These extreme peaks may have been attributable partly to amplification of motion on the narrow ridge where the instrument station was located. However, even after adjusting for this, analysts estimated that base horizontal acceleration at the Pacoima damsite may have reached 0.75g.

Postearthquake investigation of the Lower Van Norman Dam concluded that:

• There was no evidence that foundation displacement triggered the embankment collapse. Pre-existing faults on the right abutment apparently did not move during the earthquake.

• Sliding of the upstream slope was activated by an increase in porewater pressure and consequent liquefaction of the hydraulic fill near the base of the dam.

• The slide occurred after a period of vibration rather than during initial acceleration.

A large part of the embankment evidently liquefied in the later phases of the shaking. This mass was enclosed by stronger material which resisted liquefaction. However, the sudden pressures exerted by the fluid internal zone caused the compacted top and the upstream shell to break away.

Circumstances at the Van Norman complex during the event of February 9, 1971, placed two dams and 80,000 people on the narrow threshold of disaster. A slight change in a delicate balance would have drastically altered the consequences. One or two seconds of continued high-intensity vibration might have loosed a horrible flood onto the densely populated area adjoining the lower dam. On the other hand, if the earthquake had been shorter or just a little less violent, the dam might have been able to undergo appreciable distortion without liquefaction. This would suggest that some hydraulic fills may be capable of resisting seismic forces of moderate intensity and comparatively short duration. Moreover, conventional earth embankments with proper zoning and compaction should perform safely during shocks of greater severity than the one that struck the Van Norman Dams.

Vega de Tera Dam

Failure of the Vega de Tera Dam (fig. 4-27 and 4-28) on January 10, 1959, caused the deaths of 144 people. This 34-meter (112-foot) high slab-and-buttress structure in the district of Zamora in northwest Spain collapsed suddenly during the night. The flood released from its full reservoir destroyed the mountain village of Rivadelago 5 kilometers (3 miles) down the Tera River, a tributary of the Río Duero.

The Vega de Tera Dam, constructed in the period from 1954 to 1957, was a straight barrier with masonry buttresses and concrete slabs. Its builders followed the practice of suspending work during winter. As a result of inadequate preparation of joints on resumption of placement, poor bond was

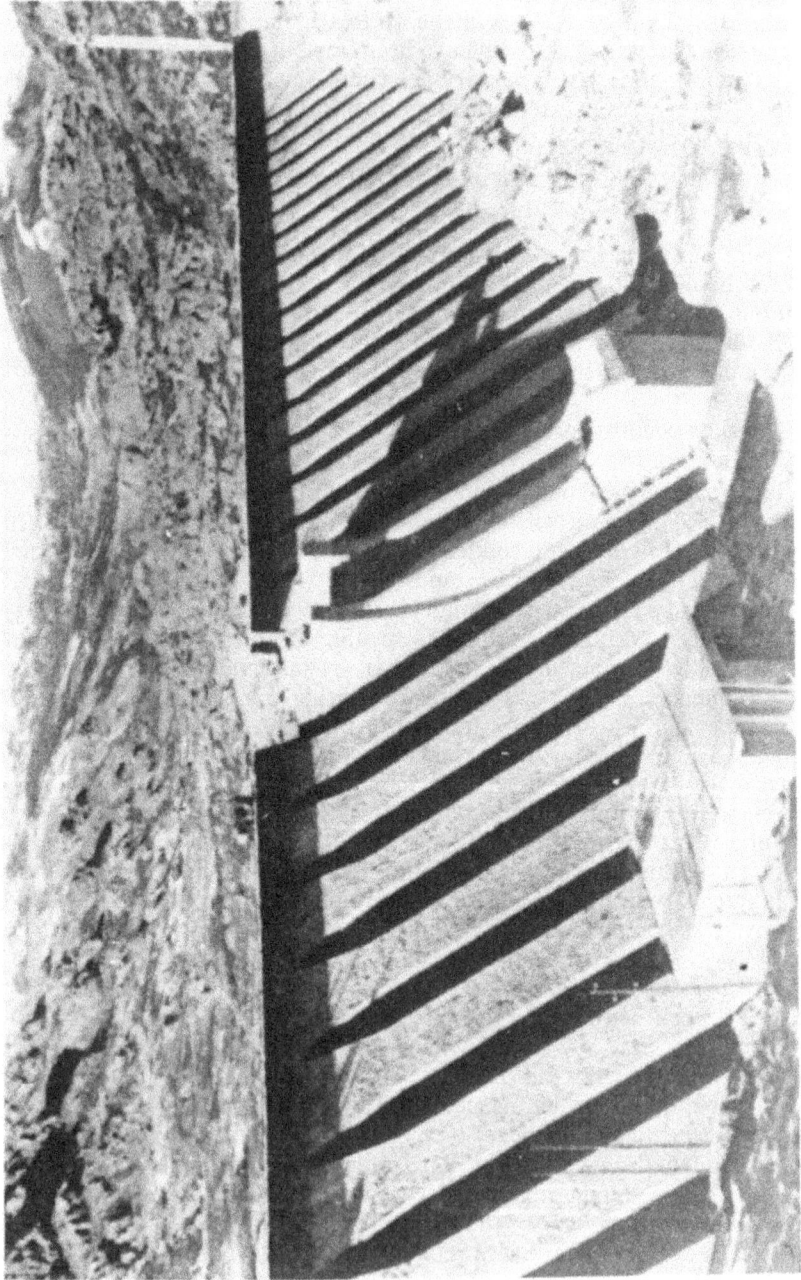

Figure 4-27.—Vega de Tera Dam before failure (Courtesy, Comité Nacional Español, ICOLD). P-801-D-79379.

Figure 4-28.—Vega de Tera Dam after failure (Courtesy, Comité Nacional Español, ICOLD). P-801-D-79380.

established between old and new masonry. The subsequent heavy leakage through the masonry marked the zones of weakness in the dam.

Failure was said to have started in a buttress standing on a sloping foundation near the left abutment at a joint between the masonry and the concrete. This triggered the collapse of 17 buttresses in succession. A 100-meter (330-foot) long section of the structure, including a ski-jump spillway, broke apart and was washed away. The powerplant at the site was also demolished.

Runoff from intense rainfall had just completed the initial filling of the 7 800 000-cubic-meter (6300 acre-foot) reservoir, which had been placed into operation 2 years previously. The dam reportedly was breached at the moment of topping of the crest. Most of the contents of the lake were spilled within a period of 20 minutes. Nearly 8 000 000 cubic meters (2 billion gallons) of water surged down upon the village of Rivadelago, at an elevation 518 meters (1700 feet) below the damsite. The momentum of this flood rushing down the precipitous canyon annihilated about 125 of the town's 150 buildings. Because the deluge struck in the early morning hours when most of the 500 townspeople were still asleep, the list of the dead was long. Only a few were able to climb to higher ground. Others rode out the torrent and survived. The people had been unaware of any danger.

The damsite is in one of the most isolated regions of Spain. Rescue efforts were hampered as the unrelenting rainstorm limited access to the stricken area. Since the catastrophe came in the middle of a severe winter, investigating authorities were not able to make a complete examination at the site of the failure until the spring. An inquiry concluded that the collapse was attributable to differences between the moduli of elasticity of the masonry and the concrete in the structure. Before the official court hearings began, Hidroelectrica Moncabril, S.A., the power company which owned the dam, had settled claims for losses and thereby secured dismissal of legal charges against it. But the public prosecutor still indicted 10 individuals for alleged negligence leading to the failure. Although the court accepted that there were deficiencies in the dam, including those related to the moduli of elasticity, four of the defendants were convicted. The penalty for each was a fine, suspension of civil rights, and 1-year conditional imprisonment. They also lost their jobs.

Walnut Grove Dam

The Walnut Grove Dam on the Hassayampa River, 48 kilometers (30 miles) south of Prescott, Ariz., failed by overtopping on February 22, 1890. About 150 people died in the waters released from the reservoir. The failure was blamed on inadequate capacity of the spillway and poor construction workmanship.

This rockfill dam was constructed to provide water for irrigation and for gold placer mining. The dam had a height of about 34 meters (110 feet) and was 122 meters (400 feet) long. Embankment width varied from 4.6 meters (15 feet) at the crest to 43 meters (140 feet) at the base. The faces were inclined at 0.5 to 1 on the upstream side and 0.6 to 1 on the downstream slope. The base of the structure for a height of 3 meters (10 feet) was composed of rubble masonry set in cement mortar. On top of this base block was a rockfill structure with dry-masonry walls on both faces made of large granite blocks reportedly laid by hand. These walls were 6 meters (20 feet) thick at the base and 1.5 meters (5 feet) at the top. Granite was quarried in the vicinity and was dumped from cars that crossed the site on a timber trestle erected over the axis of the dam. As the height of the rockfill increased, the trestle was raised in stages, the supporting wood frameworks being left in the embankment.

For watertightness, the dam was provided with a plank facing. Cedar logs, inclined at 0.5 to 1, were embedded in the face of the upstream rock wall, spaced approximately 2 meters (6 feet) apart. Horizontal wood beams, 200 by 200 millimeter (8 by 8 inch), were laid across and bolted to the logs, at a spacing of about 1 meter (3 feet). An initial sheathing of 75- by 200-millimeter (3- by 8-inch) planks, extending down the slope, was nailed to the horizontal timbers. This planking was covered with tar paper, which in turn was protected by a second layer of 75- by 200-millimeter (3- by 8-inch) planks, laid horizontally. Joints of the planking were thoroughly calked. The outer surface was coated with pitch and finally with paraffin paint. Despite these rather elaborate precautions, the dam leaked considerably during the early period of impoundment. The leakage eventually diminished.

The outlet tower was constructed of timber and was 1.8 meters (6 feet) square. The gates controlled two 500-millimeter (20-inch) outlet pipes embedded in the masonry base of the dam.

A spillway 1.8 by 7.9 meters (6 by 26 feet) was blasted out of rock on one abutment. In retrospect, this was obviously inadequate. With a drainage area above the damsite of about 1300 square kilometers (500 square miles), a spillway weir 7.9 meters (26 feet) wide did not provide nearly enough discharge capacity. Overtopping was therefore inevitable.

Walter Bouldin Dam

A significant failure, even though it did not result in loss of life, was that of the Walter Bouldin Dam, (fig. 4-29) an earthfill structure near Wetumpka Ala., in February 1975. The dam had a height of 50 meters (164 feet) above the lowest point in the foundation. The collapse of this 8-year-old embankment, operated by the Alabama Power Company, was attributed by some investigators to piping in the downstream shell. Their analyses concentrated on the alleged deficiencies in internal drainage, which relied upon a horizontal drain blanket under the downstream part of the dam, without a sloping drain zone between the core and the shell. This apparent inadequacy was judged to be contributory to poor seepage control and the consequent dangerously high water pressures in the fill. Whether this theory had demonstrable validity or not, the Walter Bouldin Dam case serves to emphasize the thorough attention which must be given to safeguards against internal erosion in an embankment.

In 1976, a FPC (Federal Power Commission) administrative law judge[6] reviewing the failure criticized the performance of both the contractor and the power company inspectors during construction, but added that from the

---

[6] "Dam Failure Inquiry," *Engineering News-Record,* September 2, 1976.

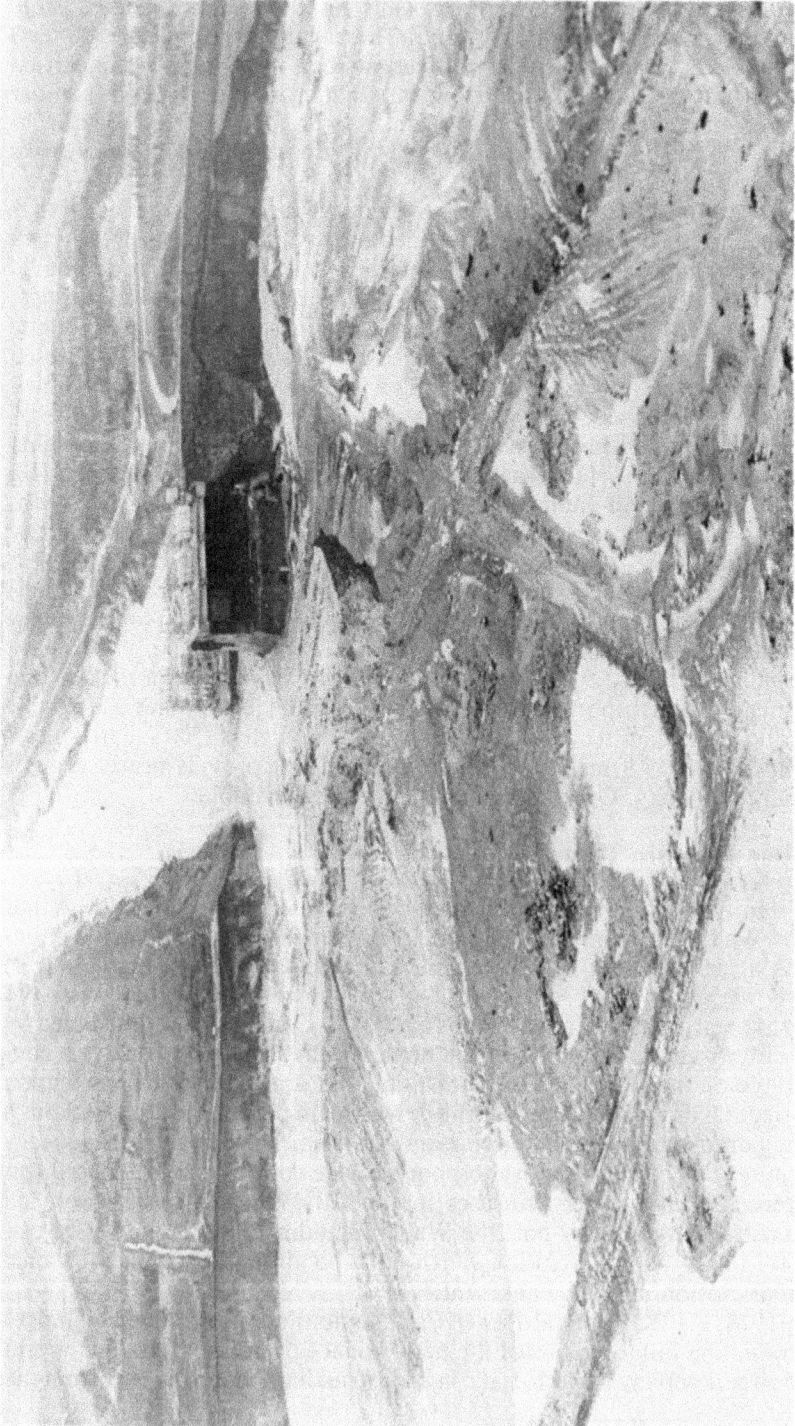

Figure 4-29.—Walter Bouldin Dam after failure (Courtesy, The Alabama Power Company). P-801-D-79381.

evidence he could not point to a single cause of the 91-meter (300-foot) long breach.

In submitting results of his investigation of the collapse of the earthfill, the judge pointed to "serious deficiencies," i.e.: (1) Construction did not comply with design specifications in one or more critical areas of the earth-fill dikes; (2) the Alabama Power Company's inspections did not detect critical deficiencies in construction; (3) FPC reviews were not thorough enough to disclose marginal design criteria; and (4) FPC reviews of construction were not adequate to uncover areas of weakness.

The failure occurred at the 2268-meter (7440-foot) long dam after part of the upstream side of the embankment near the crest slid into the water. Outrushing water destroyed part of the fill and eroded the foundation to 15 meters (50 feet) below the reservoir bottom. Flooding damage was limited, but a 225-mW powerplant was destroyed.

Four factors leading to the dam's failure, according to an FPC regional report, were a weakened foundation, a weakened embankment caused by a 1972 slide in the area of the breach, steep embankment slopes, and poorly compacted fill.

Reconstruction of the dam was accomplished after significant design changes.

Zgorigrad Dam

A 12-meter (39-foot) high earth embankment structure, impounding a sedimentation basin for an ore processing mill in northwestern Bulgaria, failed at about noon on May 1, 1966. The collapse caused a 5-meter (15-foot) flood wave which swept the villages of Zgorigrad and Vratza, reportedly destroying as many as 196 houses and leaving 96 people dead and 25 missing. Some estimates of the toll of human life ran as high as 600.

Reports from the scene of this major disaster mentioned the possibility of sabotage. However, probably more reliable sources attributed the failure of this tailings dam near Zgorigrad to heavy cloudbursts over the capital of Vratza Okrug in the preceding days. The resulting runoff overwhelmed the settling pond and ruptured the embankment. Sudden release of the reservoir water superimposed a fatal wave on the already flooded Leva River.

Part V

PREVENTIVE AND REMEDIAL ENGINEERING

Introduction

What is the common thread, if any, among failures of dams? Some observers would suggest that the blame often lies on unwise economizing on structural dimensions and the limited amount of investigation done on foundations and materials. There is more than a little truth in this. For example, disastrous overtoppings of dams can be prevented by spending money on a large enough spillway. Handling of floods is therefore not one of the most difficult problems to resolve. Other phenomena — particularly geologic hazards, earthquakes, seepage, and difficulties at conduits and structures — have more subtle aspects. A vital key to understanding them and coping with them is surveillance.

The failures and their dreadful toll emphasize the great responsibility that designers and constructors assume when creating a major dam, and how faulty their best efforts can be at times. A careful balance must be struck in reducing the risk to a tolerable minimum without raising the cost to a prohibitive level. Moreover, hazard must be recognized as a variable. The condition of a dam can change, and the consequences of its failure will depend upon developments in the area that might be threatened. An indeterminate degree of risk will always be present.

The history of dam disasters throughout the world reveals that problems often arise from undetected or inaccurately evaluated defects in the foundation. This dictates that engineering must be linked closely with geology in the design, construction, and continuing surveillance of a dam.

Reservoir safety cannot be assured by a uniform code of design practice. The designing of dams generally entails rigorous and sometimes complex studies of forces often based upon assumed material characteristics and structural behavior. Results of these analytical efforts cannot always be precise. The proper margin of safety is assured by application of both mathematical logic and practical judgment.

Of course, the engineering of dams does not end with design. It may become crucial when the construction phase is entered and some of the assumptions about foundations and materials are subjected to comparison with reality. The proudest engineer may grow progressively humble as he follows projects through design and construction into operation and sees how the conditions of dams can change - for some of them are like humans in that they can become weaker with advancing years. Tender care can make the difference between life and death in either case.

Dams may be the victims of various external and internal disorders not unlike the ailments of man, such as high pressure, sluggish drains, and dislocations. The syndrome in some instances may be less susceptible to diagnosis, but the consequences of ineffective remedies can be equally disastrous. To a doctor of medicine, each patient represents one human life. The untreated sickness of a dam can threaten the lives of many people. The care and treatment of water storage facilities therefore involve heavy responsibility.

The potentialities of deterioration of aging dams must be closely watched and analyzed. Many years of safe operation may pass before the attack of water on a faulty foundation becomes apparent. Attention must be given to monitoring the performance of the dam to detect adverse conditions so that remedial action can be taken in time to avert disaster. Regular checkups are essential to the well being of any reservoir.

Bigger and better dams are being built, and they are being placed necessarily on poorer sites. Maximum care is required in evaluating foundation conditions and construction materials to overcome site deficiencies. Complete elimination of defects will not always be possible during construction. Not infrequently, some further work on the foundation will be needed after a period of operation. Openings in the rock may contain erodible or soluble matter and may remain closed only until reservoir pressures are imposed. Consolidation and deformation under structural and water loads also may be detrimental.

These changes can be subtle and difficult to detect. Until better devices can be developed for seeing inside and under dams to diagnose their ills, all prescriptions cannot be infallible. However, there is much that can and must be done to reduce the frequency and consequences of failures.

The work of protection of a dam begins with the first examination of the foundation and continues on the drawing board and through construction and operation. There can never be certainty that all problems have been solved. Also, the nature of the problems will change from time to time. There have been cases of dams which have failed more than once, and sometimes for different reasons.

The causes and processes of dam failure are varied. History discloses some of the most likely causes — overtaxing of spillways by unexpected floods, movement and deterioration of defective foundations, and piping of embankment materials caused by inadequate control of seepage.

Invariably, failures of dams have contributed to advancement of the specialized body of knowledge which is essential to their prevention. The case histories of the misfortunes of dams reveal some remarkable similarities in antecedent conditions and in the process of breakdown. Most troubles have developed over extended periods of time — in some cases months and years. Yet, these conditions went either undiscovered or improperly appraised. Otherwise, corrective measures usually could have been taken.

No one can say how rapidly a dam will fail once the limit of its resistance has been reached. Usually embankments can be expected to fail more slowly than concrete structures. Failure times for fills have varied from a few hours to several days. Concrete dams have been known to collapse almost instantaneously. Assuming that previous adverse trends had gone undetected, the guarding of a dam would have to include round-the-clock inspection to ensure the maximum time available for evacuation of people in time of emergency. At critical sites where urban centers could be threatened, such close observation may be highly desirable. Obviously, though, a monitoring program designed for early diagnosis and prompt therapy must be the cornerstone of any surveillance system. Coupled with the capability to lower the reservoir during a crisis, this should be as beneficial in the long run as a guard who can blow the siren or ring the bell when he finds the water gushing forth.

Quality of inspection is also receiving more attention. The search for superficial signs of distress is only one phase of the examination. The engineering diagnosticians now probe deeply for internal disorders, using instruments such as soniscopes, hydrophones, and television and borehole cameras to examine parts inaccessible to inspectors. These observations are correlated closely with measurements of implanted devices, and the patient's chart is watched carefully for dangerous trends.

In designing dams and in devising surveillance systems, every effort must be made to preclude or minimize the development of conditions which cannot be easily detected. Adverse processes have been concealed by structures placed on or against embankments. For example, dams have been endangered or caused to fail by settlement or piping of fill under spillways or fish ladders.

Monitoring of a dam requires some knowledge of how it adjusts to imposed conditions. For example, loads are transferred within an embankment through displacement. Before initial impoundment of the reservoir, magnitudes of adjustments in the fill zones tend to decline and the dam approaches a state of equilibrium. Then as the reservoir loading is superimposed, a new pattern of accommodation is begun. Compression may develop quickly in the upstream zone as water permeates the fill. This may induce differential settlement progressively as the flow net expands. Such movements may cause pronounced changes in stress and strain in the dam.

Since the integrity of the impervious core is vital to survival of the embankment, its adjustment under load is especially important. In this zone, differential settlement may occur at sharp angles in the abutment. There may also be a tendency for load to be transferred or arched over to the abutment because of large differences in compressive strain between the higher fill over the streambed and the low embankment at the abutment.

A sound principle which is widely accepted specifies that vertical faces should be avoided in foundations or structures against which embankment is to be placed. Such surfaces should be battered at least slightly so that any settlement of the fill will tend to improve the contact between the fill and the underlying surface.

Engineering Geology

The safety of a dam is inseparable from the condition of its foundation. About 40 percent of all dam failures have been caused by inadequate foundations.

Geologic investigation of the damsite and the reservoir area should include identification and evaluation of hazards from:

- Landslides.
- Subsidence.
- Expanding soils.
- Seismicity, including fault offset.
- Soluble foundation rock.
- Foundation caverns and channels.
- Inherent rock stress.
- High primary permeability
- Erodible rock.
- Open fractures.
- Low bearing capacity.
- Weak shearing resistance.

To be acceptable as a foundation for a dam, the rock must be sufficiently strong and bonded to remain intact under forces superimposed by the dam and reservoir, as well as by natural elements. It must also be impervious enough to preclude excessive seepage. To assure these qualities, determinations should be made of crushing strength, mineral composition, cementation, porosity, and resistance to cleavage and slaking.

Texture is usually one of the reliable indicators of rock integrity. Fine-grained rocks such as shales, siltstones, and tuffs are generally not strongly bonded because water does not permeate them readily to deposit cementing agents. Some of these may be merely highly compacted and, although apparently competent in a stable environment, may come apart when exposed to alternate wetting and drying. In a dam foundation, such rock must be covered soon after exposure to minimize deterioration. The resistance of fine-grained sedimentary rocks to rapid seepage also makes them subject to high pore pressures. Coarse-grained sedimentary rocks generally are more strongly bonded, but the interstices of some sandstones and conglomerates may be large enough to permit high rates of percolation.

Cohesion of rock particles varies with the type and quantity of the cementing agent. Silica, calcium carbonate, and iron oxide are relatively strong, insoluble, and durable; but clay and gypsum are not. Rock strength depends upon not only the cementation but also the size, shape, and arrangement of the particles. Compressive strength may range from about 3.4 megapascals (500 pounds per square inch) for some tuff to more than 213 megapascals (31 000 pounds per square inch) for basalt.

Excluding some of the weaker shales and tuffs, most rock has enough intrinsic strength to resist the loads imposed by a dam. But a rock mass may have bedding and foliation planes, joints, shears, and faults. These can be natural channels for seeping water, which may carry away soluble materials and erode openings. The planes may also be deficient in shearing resistance and susceptible to weathering.

Foliation, the tendency to break into thin sheets, is a common characteristic of schists and slates. The cleavage may allow water, air, and other weathering agents to invade the rock mass. The foliation planes are generally conducive to slippage.

Practically all rock formations have joints, which are fractures along which there has not been any slipping. They form the boundaries of individual blocks in the mass. In a dam foundation, joints may be of concern because the condition of the joint fillings is uncertain and the joint has the adverse potential of becoming a conduit for leakage under the dam.

Among the more dangerous elements at a damsite are faults, fractures which have slipped. These are of particular concern because they may have caused physical alteration of the rock to the extent that the load-bearing capacity has been reduced. The fault zone may have been so shattered and crushed that it is unable to support the heavy loads of a reservoir. Its soft filling could be susceptible to squeezing or blowing out. Fault gouge also may hinder the grouting of cracks. Faulting not only alters the condition of the adjoining rock but also displaces foundation blocks so that rocks of contrasting characteristics are side by side. This may bring a hard rock to bear on a soft rock, or a tight rock against one that might leak like a sieve.

The major faults at a damsite or in its environs must be examined to assess the probability of their future movement. Dams constructed on active faults may be stressed severely during such slippage. Disclosure of geologically recent movement at a proposed damsite is usually reason enough for abandonment of the site. Dams have been built at such sites when there was no alternative, but in these cases almost invariably the design was ultraconservative, incorporating features that would allow accommodation of displacement.

Resistance to erosion is an important factor in determining suitability of the rock at a damsite. This may depend more on bedding, foliation, and jointing characteristics than on the inherent strength of the rock. Where the potential planes of breakage are closely spaced, vulnerability to disintegration under water forces may be high. Such weaknesses must be given special attention in areas where outlets and spillways will discharge.

Solubility of the rock underlying the reservoir must also be considered. Limestones and gypsums sometimes present problems when exposed to water under pressure. Limestones may have joints and bedding planes that provide paths of infiltration that facilitate rock solution. However, joint enlargement and cavern development in limestone usually are slow enough to be controllable during the life of a reservoir. The deterioration of gypsum may be rapid enough to create a hazard.

Hales Bar Dam on the Tennessee River, replaced by the Nickajack Dam in the mid-1960s, was founded on cavernous limestone. It suffered serious leakage throughout its life of more than 50 years. Even during the construction period, 1905 to 1913, difficulty was experienced in dewatering the foundations, necessitating compressed air caissons and extensive grouting. During operation, remedial work was done from time to time. Asphaltic grouting effected some limited reduction in flow through the foundation, but other methods did not yield any worthwhile results. The facility was finally removed from service in the interest of safety and economy.

Another case of leakage through limestone was at the Ontelaunee Dam in Pennsylvania. Water began to pass under the core wall of this earthfill soon after the impoundment was raised to capacity. To eliminate the threat to the dam, the reservoir was emptied and thorough grouting of the leaky foundation was accomplished. This corrective work was effective.

Numerous cases could be cited of the problems caused by gypsum in dam foundations. It was present in an abutment at the St. Francis Dam in

California, and its possible contribution to the failure of that structure is still debatable.

At some sites it is practically impossible to discover and to assess all geologic defects prior to construction. Moreover, there is little likelihood that drilling and sampling of the foundation materials will be so selectively accurate as to define the most critical zones completely. Only during construction and operation can there be assurance that the facility and its site have been fully tested.

Not infrequently, problems appear for the first time in the operational phase, despite conscientious efforts to detect them sooner. For example, some reservoirs do not hold water. One of these leaked immediately in the first attempt to impound. The water disappeared into sinkholes as fast as it could be delivered. After sealing of the holes, storage was still unsuccessful. The reservoir was then found to be underlain by a limestone formation with a multitude of solution channels which defied sealing. Experiences such as this point to the need for a broad viewpoint in considering plans for water storage. The necessary perspective embraces the damsite, the reservoir basin, and usually a large part of the environs. To illustrate the possible impact of a reservoir on its environment, the maximum water and silt load at Lake Mead behind Hoover Dam may exceed 36.4 billion metric tons (40 billion tons). Subsidence of the reservoir floor and the immediate vicinity has been reported to be as much as 180 millimeters (7 inches).

Embankment Safeguards

An embankment dam must be an optimum product of the local materials from which it is constructed and must harmonize with its site. If the foundation is not strong enough to support the loads of the structure and water, the inferior materials must be removed or improved. If it is too permeable to serve as an adequate water barrier, it should be sealed by measures such as grouting, cutoffs, or blanketing. A foundation with openings that are difficult to seal may also be treated by drains or relief wells.

In 1967, an American Society of Civil Engineers Committee on Earth and Rockfill Dams recommended guidelines to protect against cracking and consequent piping of embankments:

1. Use of a wide transition zone or of properly graded filter zones of adequate width.
2. Special treatment of foundation and abutment conditions to reduce sharp differential settlement.
3. Arching of the dam horizontally between steep abutment slopes.
4. Adjustment of construction sequence for the different zones or sections.
5. Requiring special placement methods for questionable materials.
6. Thorough compaction of rock shells to avoid inducing tensile stresses in adjacent core material.

The internal distortions that occur in embankments result from compression, shear strain, and/or plastic deformation of the materials in the dam and in its foundation. In an earthfill or rockfill structure with several internal zones having different material characteristics, degrees of compaction, and moisture content, there will be almost inevitably appreciable interzonal adjustments in response to the various imposed forces.

Irregular rock surfaces in the core foundation increase the potential for differential strain and consequent cracking in the core. Therefore, careful attention to foundation treatment under the core and the adjoining transitions is necessary. Overhangs should be eliminated and rock protuberances should be trimmed. To preclude disturbing acceptably sound foundation rock, this excavation should be done preferably without blasting. In conjunction with rock excavation or as an alternative, concrete may be placed under overhangs or at other irregularities to give the foundation an acceptable shape.

Seepage through a rock joint or crack underlying erodible material in the embankment may cause fatal damage. Fine sands, silts, and dispersive clays are susceptible to such erosion. In some cases, an initial contact layer of plastic clay has been placed on the foundation for protection. While the benefits of this may be argued, there should be no question of the value of permanent sealing of foundation openings with grout or concrete to isolate the embankment from potentially damaging underflow. In common practice on many projects, this is accomplished effectively with slush grout, mortar, dental concrete or shotcrete. An additional line of defense is provided by filters and drains in the downstream part of the embankment.

Some aspects of embankment design necessarily depend on assumptions and approximations and therefore call for ample factors of safety. Since these are introduced to compensate for uncertainties, they must not be counted on as reserves of strength to support superimposed loadings. While less liberal safety factors may be used as confidence in data and methods increases, enough conservatism must be retained to cover the remaining unknowns. To cite an obvious but instructive example, the most rapid electronic computer cannot offset imperfections of input data from the field or laboratory. Personnel who tend to be fascinated by sophisticated analytical techniques must pause from time to time to appraise the value of the ingredients.

Earthquake Engineering

In the 20th century there has been a trend toward bigger dams. Especially remarkable is the increasing size of embankment dams. For example, in the year 1910, the highest earth dams had a height of about 30 meters (100 feet). In each decade since, the maximum has been raised approximately 30 meters (100 feet), so that in the 1980's there are embankments of greater than 244-meter (800-foot) height. The risk has increased proportionately and has been compounded at the same time by construction in marginal locations as good damsites have become scarcer. This applies especially to seismically active areas.

Exposure to Earthquakes.—Public confidence in dams is based mainly on the safe performance of thousands of reservoirs under less than the most severe conditions. Only a few dams have been exposed to major earthquakes. In the San Francisco temblor of 1906 no dam was destroyed, although several were in the area of high intensity, including the Upper and Lower Crystal Springs, the Pilarcitos, the San Andreas, and the Temescal Dams. The outlet conduit in an abutment of the San Andreas Dam, adjacent to the San Andreas Fault, was displaced, but the earthfill itself survived without serious damage.

San Andreas and Upper Crystal Springs Dams, both earthfills, are on the San Andreas Fault with axes at approximately right angles to it. The Lower Crystal Springs Dam, a concrete gravity structure, lies parallel to the fault and at the edge of the rift zone. The two earth dams were sheared by the 1906 movement, but the concrete dam survived intact.

There was much evidence of displacement on the fault, including fences, roads, and rows of trees that were offset as much as 4 meters (12 feet). Despite this movement, the leakage caused at the dams was minimal, even though the reservoirs were practically full at the moment of the quake.

The San Andreas and Upper Crystal Springs embankments were built in the 1870's by identical methods. A core trench was excavated to bedrock and backfilled with puddled clay, which was also the material placed in the core of the fill. The outer zones were constructed in layers and compacted with rollers.

San Andreas Dam, which was then 29 meters (95 feet) high above streambed, was displaced about 2.4 meters (8 feet). The structure actually consisted of two embankments separated by higher natural ground through which the fault passes. In a subsequent enlargement, the fill was made continuous over the rise. An outlet tunnel was ruptured by the 1906 shearing on the fault, and the intake tower and other facilities were destroyed or seriously damaged. The embankments suffered longitudinal cracking along their entire length, as well as large transverse cracks in the abutments. The only repair deemed necessary, however, was resurfacing over the fault.

Upper Crystal Springs Dam is a 26-meter (85-foot) high embankment dam. When the earthquake struck, the embankment was not subject to unbalanced water pressure since an open conduit maintained the Upper and Lower Crystal Springs Reservoirs as a common pool. After the earthquake, the dam was found to have been offset about 2.4 meters (8 feet). A brick-lined conduit around the dam was broken for a length of 6 meters (20 feet) at the fault. The dam itself evidently suffered no damage which required reconstruction.

Lower Crystal Springs Dam, constructed in the period 1887 to 1890, is 44 meters (145 feet) high (above streambed) and is curved on a radius of 194 meters (637 feet). It was the first significant concrete dam built in the Western United States. This structure reportedly did not show the slightest sign of distress despite the impact of the 8.3 Richter Magnitude earthquake.

One of the few dams known to have failed under seismic forces was the Sheffield Dam (fig. 5-1) in the Santa Barbara, Calif., earthquake of June 29, 1925, which had a Richter Magnitude of 6.3 and an epicenter about 11 kilometers (7 miles) from the damsite. This 7.6-meter (25-foot) high earthfill was 220 meters (720 feet) long and had slopes of 2.5 to 1. Founded on a silty sand layer about 2 meters (6 feet) thick, the embankment consisted mostly of this same material, except for a clay blanket on the upstream slope. Seepage had saturated the foundation alluvium and the lower part of the fill. In response to the earthquake's estimated maximum acceleration of

Figure 5-1.—Sheffield Dam following failure (Courtesy, Calif. Dept. of Water Resources. P-801-D-79382.

15 percent of gravity (0.15g), and about 15 to 18 seconds of significant vibration, the relatively loose saturated silty sand near the base failed in liquefaction. The consensus of investigators who viewed the broken dam was that the earthquake "had opened vertical fissures from base to top" and had "formed a liquid layer of sand under the dam, on which it floated out, swinging about as if on a hinge." A 91 meter (300 foot) length of the embankment at its center slid downstream about 30 meters (100 feet). As a consequence, approximately 113 000 cubic meters (30 000 000 gallons) of water spilled into the City of Santa Barbara. Apparently no deaths were attributed to the reservoir failure, although the toll for the earthquake included 12 human lives lost. Figure 5-2 shows the damage to Lower Van Norman (San Fernando) Dam caused by the 1971 San Fernando earthquake.

Earthquakes may affect dams in various ways. Seismic forces may be transmitted directly from the foundation to the structure. Overtopping water waves may be generated by landslides or oscillation of the reservoir or sudden movement of the dam foundation.

As demonstrated at the Sheffield Dam, foundations and embankments under certain circumstances may suddenly weaken when subjected to prolonged vibration. Liquefaction of fine-grained cohesionless soils under such conditions can place a dam in jeopardy. Since the Anchorage, Alaska and the Niigata, Japan earthquakes in 1964, a better understanding has been attained of the transient reduction of strength in soils.

Seismic-Resistant Design.—Advancements in design earthquake determination, finite element analysis, and dynamic testing of soils have enabled prediction of the behavior of embankments under vibratory loading. The San Fernando earthquake in 1971 provided a full-scale opportunity to verify the validity of these techniques. Correlation with observed performance of the Van Norman (San Fernando) Dams was encouraging.

In the design of embankment dams, zoning is an important key to built-in protection against failure. Selection of the right materials for each zone, and ensuring their proper placement, will allow control of concentrated leakage arising from distortion of the fill or from foundation displacement. One of the most effective lines of defense is a comparatively wide transition or filter zone composed of a well-graded mixture of sand and gravel. If a dam is sheared by an earthquake, the intermediate section between the core and the downstream zone can adjust to control leakage to tolerable amounts and to prevent detrimental piping of materials. With proper gradation, the sand and gravel will tend to seal the cracks which might open in the dam or its foundation.

The upper part of an embankment is especially vulnerable to seismic forces. It is susceptible to cracking and to separation at the contact with the abutment. Since seepage paths are shorter near the top of the dam, and because the internal embankment pressures are generally too low to close cracks, the potential for dangerous leaks is considerable. Therefore, in determining the zoning for this part of the embankment, a well-graded sand and gravel mixture placed on the upstream side of the impervious core should be favored as a stopper for cracks. Assurance against failure during earthquake is also provided by a substantial freeboard between the normal

Figure 5-2.—Upstream face of Lower Van Norman (San Fernando) Dam after failure (Courtesy, Calif. Dept. of Water Resources). P-801-D-79383.

water surface and the crest. This may be of decisive benefit in case of slumping or cracking of the crest. It will also provide a measure of protection against overtopping by a water wave generated by a seiche or a landslide into the reservoir.

An embankment with an ability to adjust safely to differential movements would, therefore, be one which has an impervious zone composed of a well-graded mixture of clay, silt, sand, and gravel; ample transitions and drains; thoroughly compacted gravel or quarried-rock zones; and liberal freeboard. One of the least resistant would be a dam with a thin, sloping core of silt or other easily eroded soil, thin filter or transition zones, and dumped rockfill. Dumped rockfill may have questionable merit in a high dam because it is susceptible to considerable settlement under severe shaking.

Seepage Control

A dam will alter the natural balance of conditions at its site. As water is brought into storage, an adjustment will begin which develops a new flow net through the barriers that confine the reservoir.

Unless seepage is intercepted and safely conveyed away, it may exert detrimental pressures or remove erodible materials. The integrity of a dam therefore depends upon the functioning of a properly designed and well maintained drainage system.

When excessive seepage conditions threaten the safety of a dam or reservoir, various procedures may be considered; the first being to lower the storage level. This will reduce the hydraulic gradient and is a prudent interim step until permanent corrective work is completed.

The measures for controlling seepage through pervious foundations depend upon several factors. In some cases, a combination of several kinds of seepage control measures may be adopted. A positive cutoff, achieved by excavation to an impervious foundation, is regarded as most desirable under an embankment. Such a cutoff should be sufficiently broad to ensure a seepage gradient low enough to avoid damaging the embankment material, and should have excavation slopes flat enough to avoid stress concentrations. If such positive protection cannot be attained by excavation and backfilling with impervious material, consideration should be given to other seepage control measures singly or in combination such as:

- Impervious earth blankets extending upstream from the embankment.
- Slurry trenches.
- Grout curtains.
- Concrete cutoff walls.
- Vertical drains.
- Relief wells.

Seepage Analysis.—Seepage records are valuable indicators of the condition of a dam. When examined on a long-term comparative basis, the records may provide insight into changes occurring within the structure or its foundation. A uniformly stable condition would be unusual. Normally, the seepage channels will tend to become either constricted or enlarged as time passes. This will be reflected in the changes indicated by the flow rates.

Monitoring of water pressures in the embankment and in its foundation should be continued for several cycles of reservoir operation to document cyclic changes. The ground-water system may adjust slowly to the reservoir's presence. Significant seepage may not be apparent until several years after commencement of reservoir impoundment. Increased flow of natural springs in the vicinity can be read as a possible indicator of reservoir seepage. Measurement of changes in water temperature and chemical quality in wells and springs may also provide valuable data. Specific seepage paths must be traced. This can be accomplished by additional drilling, dye or isotope tests, or water temperature surveys. Flow nets drawn from observed water levels are useful in analyzing seepage problems. Ground-water levels can be measured at single and multistage wells and pore pressure cells. Multistage wells may contain two or more perforated pipes extending to selected depths and each isolated in a single aquifer by plugs of bentonite or other expansive materials, so that several aquifers can be monitored in a single drill hole. Due to the difficulties sometimes encountered in such construction, single wells at different depths are preferable to multistage wells in some cases.

Incipient piping failures of earth embankments sometimes can be detected and remedied before they become serious. Appearance of soil or rock particles in seepage must be recognized as a sign of a dangerous condition. During initial reservoir impoundment, for example, water will find its way into foundation channels that may contain erodible materials. A conspicuous manifestation of the flushing of such conduits would be a sand boil at the exit. This would usually call for quick remedial action. Sometimes, however, the signs of trouble are not so obvious. Internal erosion may progress so gradually that removal of material is not discernible by visual examination. Other monitoring devices must be used to assure that deterioration does not go unnoticed.

For the detection of weaknesses developing in an embankment, observation of discharge and turbidity of seepage is quite important and is comparatively easy. However, it has been neglected at many projects. To identify sources of seepage, drainage systems preferably should be divided into separate zones. Systems should be designed to distinguish rainwater, foundation water, and flow through the embankment.

Chemical analysis is essential where the foundation contains soluble solids. Comparison of the salt constituents in the reservoir water with the composition of the drainage water will show if a significant amount of material is being dissolved and carried away. At sites where solution is a potential problem, monitoring should include a correlation of seepage patterns with reservoir levels, piezometric pressures, and settlement. Fortunately, deterioration caused by this process is characteristically gradual, usually allowing time for remedial action.

Filters and Drains.—At a site with soils of adequate strength and impermeability, the embankment may be constructed as a single homogeneous mass. Many old dams were built in this way. However, in present practice, the embankment is more likely to consist of an impervious core enclosed by pervious shells. Internal drains are often placed in the downstream shell to intercept and carry away seepage. These may be relatively narrow, vertical or inclined zones immediately downstream of the core, a blanket on or near the foundation (including abutments) under the shell, a zone at the toe of the embankment, trenches filled with pervious material, perforated pipes, or combinations of these measures. Horizontal drain blankets preferably should be used as companions and extensions of inclined drain zones or chimneys placed just downstream of the core.

Transition zones properly designed and constructed should be able to control leakage through a crack in the impervious core. An effective defense will be provided by a zone of coarser material such as cobbles or rockfill just downstream of the transition. With its greater permeability and its filter compatibility with the transition (so that transition materials cannot enter the rockfill voids), this zone should convey leakage safely.

If a crack occurs in a fine-grained core, the filter must prevent transport of material through the opening. An ideal filter on the downstream side of the core will adjust rather than sustain the cracking. This capability is also important in the filter upstream from the core so that it can function as a "crack filler" if the core crack tends to remain open.

Careful consideration must be given to the selection of permeable material for drainage of an embankment or of a natural reservoir barrier. Aggregate drains must function as filters to retain soil or rock particles and as conduits to convey water safely to discharge points. To meet the first requirement, a graded filter — a coarse aggregate layer protected by one or more layers of finer aggregate — is effective. In controlling large seepage flows, filter aggregates fine enough to resist piping are not usually sufficiently coarse to meet the full discharge requirement. The necessary capacity can be provided either by pipes or by the coarse element of a graded filter. If the seepage outfall must extend over a relatively long distance, water can be collected by open-jointed, slotted, or perforated drainpipe and conveyed to a closed pipe discharge system. All pipe openings must be sized to prevent the entrance of the surrounding aggregates.

Trench or finger drains may be used as alternatives to continuous blankets, especially where drain materials are very expensive. If such drains are used, the material must be thoroughly compacted to ensure that it does not consolidate on saturation. Otherwise an open seepage conduit may develop in the top of the trench, bridged by the overlying embankment. Enough testing should be performed on the compacted drain material to ensure that it will not be subject to detrimental consolidation. In view of this potential weakness, finger drains probably should be avoided unless other alternatives are unavailable or prohibitively expensive. Care should also be taken to avoid contamination of the drain materials.

The possibility of gradual adulteration of originally cohesionless embankment zones cannot be disregarded. It could happen through migration of clayey fines or depositing of chemicals from seepage. Whether such conditions commonly develop enough to impair filtering and draining capability is not easily verified. A generally accepted view is that clean, hard crushed rock and sand and gravel in embankments will not undergo significant changes during the life of a reservoir. In contrast, however, some weathered alluvial materials and soft rocks susceptible to deterioration and/or recementation should not be used where cohesionless zones are specified.

A drain must be stable enough to withstand the surging which may be necessary to remove clogging by chemical deposits or bacterial growth. Drain stability tests should be conducted for vibration and surging effects.

In some cases, sinkholes have appeared on the crests and slopes of embankments composed of coarse, broadly graded soils of glacial or alluvial origin. The fines in the subject soils tended to be nonplastic or to have low-to-medium plasticity. The fines apparently were not compatible with the coarser particles from the standpoint of filter requirements. The sinkholes were believed to be caused by erosion at a concentrated leak, causing the finer soil particles to migrate out of the compacted soil mass, exiting through cracks in the foundation rock or through filters which were too coarse to retain the fines. Fine-to-medium sand filters should be considered for dams with thin cores of such materials. Emphasis must also be placed on the sealing of cracks in rock foundations under dam shells consisting of these materials.

The above experiences would caution designers to consider that such coarse, broadly graded soils may not necessarily possess the self-healing properties sometimes supposed. These soils may not be as well-graded as a concrete aggregate or many deposits of river sand and gravel. Theoretically, such well-graded materials may have just the correct quantity of each particle size to fill the voids of the progressively larger particles. The particle-size distribution of the typical coarse soils which have been associated with the sinkhole phenomenon shows that the volume of fine particles is greater than the volume of the voids of the coarse sand and gravel fraction, and the coarser particles, therefore, may be floating in a matrix of fines.

Another important observation is that construction of a coarse filter without some particle segregation is difficult. Segregation assures that there will be locations at the boundary between the protected material and the filter where the filter consists only of gravel-sized particles. At these segregated locations, fines can enter the filter.

Consideration should therefore be given to placing a sand filter downstream from the core, especially for major dams with thin impervious cores and thin filter zones. The difficulties described have occurred at only a small number of dams composed of the suspect soils. Internal erosion of coarse, broadly graded soil cores apparently develops only when an unfavorable combination of the following conditions exists: (1) thin core, usually vertical; (2) downstream filter of coarse sand and gravels, with little or no fine sand sizes (usually not a processed material, except for screening to remove cobbles); (3) steep or jagged rock foundation not adequately sealed; and (4) a very rapid reservoir filling. For such conditions, the usually accepted filter criteria may be inadequate.

Some engineers experienced with embankment dams would say that sometimes inadequate recognition is given to the as-constructed grain size distribution in the core. They would point out that segregation of the coarser material to the bottom of each lift may tend to create pervious seepage horizons through which water and fine soil particles may pass. Unless these strata are prevented by thorough mixing of the fill material as placed, the theoretical grain size relationships suggested by laboratory testing may be misleading. The laboratory gradation is determined from a nearly perfectly mixed specimen, as opposed to field blending which is seldom uniform.

In this regard, in a discussion of a paper by James L. Sherard at the 13th International Congress on Large Dams in New Delhi, India, in 1979, Consulting Engineer Thomas M. Leps communicated the following view:

"Provided that the material is not badly gap-graded, an adverse condition which by itself would normally eliminate use as adequate core material, broadly graded soils will generally, *when well blended*, consist of gravel and cobble particles suspended in a matrix of soil. Hence, it should be perfectly clear that any filter zone, to be effective, must be designed to protect against migration of the core's soil matrix without benefit from the 'suspended' coarser particles. Additionally, the essential need to filter against the soil matrix is doubly important when one recognizes that a broadly graded core material cannot practicably be placed without an important degree of segregation; i.e., such material is most unlikely to be *well blended* as placed. Unfortunately, this dominant fact of practical construction experience leads inevitably to the creation of exactly the three conditions *** postulated as leading to piping through the core:

  i. An avenue of heavy leakage (which is provided by segregated streaks of gravel and cobbles);
  ii. Adjacent streaks of segregated soil fines (which can be rapidly and progressively attacked by the relatively heavy leakage noted in i); and,
  iii. Streaks of coarse gravel in a segregated coarse filter, of sufficient coarseness to convey away the fines eroded as noted in ii.

"In summary, in the writer's judgment, the essential point for the designer to bear in mind is that it is difficult, if not impracticable, to so blend broadly graded materials under construction conditions that they will behave in the same manner that their perfectly blended counterparts exhibit in laboratory tests. The practical solution is to require enough processing of core soils to eliminate the coarser fraction, say the plus 5 cm sizes. This can be simply achieved for non-plastic soils by use of either a vibrating grizzly or jaw crusher. It should be supplemented on the fill by required disking and harrowing to provide reasonable blending.

"Similarly, specification and use of broadly graded filter materials, for which field blending on the fill is usually impractical, should be avoided, particularly when it is appreciated that it is the soil matrix of the core which must be protected, not the coarser fractions. In this, the writer fully agrees *** that, in the cases cited, the downstream filter zones (the main line of defense against piping) were too broadly graded and hence locally too coarse. The simple solution is a two-zone filter drain, the zone next to the core being a fine to medium sand.

"Finally, it seems apparent to the writer that it could be unfortunate if the concept prevailed that broadly graded glacial fills are 'internally unstable.' Instead, it should be realized that these materials are basically excellent, except when man disarranges their gradation balance by the inevitable mechanism of segregation."

Relief Wells (Foundation Drains).—Pressure relief wells (foundation drains) are often placed under concrete dams, usually immediately downstream from the grout curtain. They are also used in the foundations of other dams. When they are drilled in erodible materials, their design must incorporate features to prevent piping. A combination of upstream impervious blanket and downstream drain wells can provide effective seepage control.

Relief wells are used not only in combination with upstream impervious blankets but also with various other schemes to control hydrostatic pressures in the downstream zones of the embankment which could lead to piping or slope instability. Relief wells are sometimes connected with a drainage gallery under the dam. There is need for regular surveillance and maintenance of relief wells.

Grouting.—Grouting is done to seal subterranean channels as well as cracks in structures. It is not always effective by itself but often is a dependable safeguard when combined with adequate drainage systems. The grout must be mixed to proper proportions for the site conditions and has to be injected under controlled pressures to prevent damage to the dam or the foundation. In establishing a grout curtain under an embankment, several rows of grout holes are generally preferable to a single row. The curtain should be supplemented with a drainage system and a series of piezometers to check the efficiency of the grout barrier.

Cutoffs.—Underseepage is controlled most effectively by extending a cutoff into impervious foundation. This should be combined with a drainage system to intercept any seepage that may still find its way through the rock or the dam. Where a complete cutoff is not feasible, satisfactory control may be assured in some cases by adding relief wells or toe drains.

When a cutoff is to be constructed at an existing dam, its influence on stability must be carefully analyzed. If used in remedial seepage control for an embankment, the preferable location for the cutoff is at or near the upstream toe. This normally requires the draining of the reservoir.

Properly constructed slurry trench cutoffs are essentially impervious and plastic and have engineering capabilities similar to stiff clay. Their effectiveness has been demonstrated on many projects where they have adjusted to embankment or foundation deformations without cracking or differential settlement.

Advances in slurry trench construction methods have broadened the applicability of cutoffs by enabling greater depths with limited volume of excavation. The slurry trench has been used successfully at major dams in several countries. Although it has been incorporated into the original design of the dam, it can also be an effective corrective measure on an existing structure. The common procedure involves excavation of a narrow trench, keeping it filled with bentonite slurry to support its vertical walls. After the trenching has been extended to final depth, it is backfilled by dumping earth materials into the slurry pool.

To obtain and retain a uniform slurry mix, backfill components of clay, well-graded sand and gravel, and bentonite preferably should be weight-batched into mixers for blending with a predetermined quantity of water. Although on some projects the trench backfill has been mixed by windrowing, dozing, or blading; mechanical mixing with aggregate or concrete mixers is generally superior.

Careful control must be maintained of the fine-grain content in the backfill of the slurry trench. A generally acceptable range for materials passing a 75- μm (200-mesh) sieve is from 10 to 30 percent. This is intended to ensure impermeability without excessive settlement.

Techniques have also been developed for the installation of concrete walls or diaphragms through the use of slurry trenches. Tremie concrete has been placed successfully in such construction to achieve a positive cutoff. An early example was the upstream impervious wall installed in 1964 at the Allegheny Reservoir Dam on the Allegheny River in Pennsylvania. This was composed of concrete 0.76 meter (2.5 feet) thick and extended to a maximum depth of 56.4 meters (185 feet) into the alluvial foundation. The wall is about 335 meters (1100 feet) long. It was constructed using the ICOS method, involving excavation in slurry-filled holes and displacement of the slurry with tremied concrete. In more recent years concrete diaphragms have been built at several other dams, including Manicouagan 3 in Quebec in 1972. This 107-meter (350-foot) high earthfill structure is founded on alluvium over 107 meters deep, consisting of sand, gravel, cobbles, and boulders. The cutoff through this material was established by two parallel concrete walls 0.6 meter (2 feet) thick, 3 meters (10 feet) apart from centerline to centerline, composed of interlocking piles and panels.

One of the least effective alternatives for seepage reduction is sheet piling. Although this was installed at many early dams, it has not proven to be dependable. Steel sheet piling cannot be regarded as a positive way of controlling seepage. Vibrating pile hammers and other measures may improve pile alinement, and bentonitic slurry may be helpful in sealing the piling interlocks. However, sheet piling in most applications cannot be expected to provide a watertight barrier.

Blankets.—Blanketing is another useful alternative for seepage control. A complete blanket would extend into an impervious contact along its full boundary. Partial blanketing is sometimes done to lengthen the path of percolation in the foundation. Blankets are most often constructed of earth. Other materials, such as plastic sheets, have been used with varying results. Some of these liners have tended to be susceptible to damage and accelerated deterioration. Techniques of placement require careful attention. To be most effective, an earth blanket normally should have at least a 0.9-meter (3-foot) thickness and should be thoroughly bonded to the adjoining impervious elements. The mix of the earth should be designed to assure watertightness. In locations where the blanket may be subjected to erosion, it should be covered by protective material.

Engineering of Conduits and Structures

Many of the defects disclosed at dams have been in appurtenant conduits or structures such as outlet works and spillways. One hazard which has been well demonstrated is where a rigid structure is placed upon a yielding foundation or across a shear or a fault zone. Voids or fractures at such places may result in uncontrolled release of water.

Conduits and structures under high fills may be subjected to seriously damaging movement and cracking, particularly where the foundation is relatively soft. All such facilities should be placed on sound rock if possible. Even with firm support, structural elements may be broken by embankment adjustments, inducing lateral pressures. Accommodation of such movement must be enabled by proper location and shaping of the buried structure. Parts projecting appreciably into the fill should have gradual changes in geometry to safeguard against rupture of either the soil mass or the structure.

A primary requirement for a conduit under a high embankment is that it be strong enough to carry the most severe loads that may be imposed by the fill. Its strength can be enhanced by placement in a notch or trench in the foundation. The bottom and sides of this excavation must be thoroughly cleaned. Any earth backfill around the conduit must be compatible with the surrounding material. Both within and outside of the core limits the backfill should be selected and compacted to minimize settlement. Often the best practice is to backfill completely around the pipe with concrete.

Conduits which pass under an embankment must be constructed with an effective watertight barrier around them. Means of achieving this include tight bedding and installation of seepage collars. A common cause of failure is piping of material along the outside of the conduit. Not infrequently, this can be attributed to poor compaction of backfill around the conduit.

Outlets.—Every reservoir should have an outlet with a capacity proportionate to the reservoir storage. Emergency draining time will often be the controlling factor in sizing these works. The capability of rapid lowering of a reservoir during a crisis can be extremely important. On such occasions properly functioning outlet works are essential. Just being able to lower the water level a few feet quickly could make the difference between saving and losing a dam. An ideal objective, although not always obtainable at high dams and large reservoirs, would be to have an outlet capacity which would permit reducing the water depth at the dam by one-half within 1 or 2 weeks.

Where the conduit feeds directly into the distribution system, a blowoff or short bypass preferably should be installed close to the reservoir to release the full capacity of the outlet.

Gate or valve control at or near the upstream end of an outlet is highly desirable so that conduit water pressures can be limited within or under the dam. This is especially important for an outlet under an embankment or through a reservoir rim that would erode readily if the conduit ruptured. Where such control is not provided, an upstream bulkhead should be included to enable unwatering of the conduit under full reservoir head.

There are advantages in providing more than one means of controlling the release through the outlet works. Facilities equipped with only one valve or gate require careful maintenance and periodic testing to assure that they are operable.

Location of outlet pipes within walk-in conduits is desirable. Where this is done, a ventilating system should be provided to reduce condensation and consequent corrosion of metal.

Painstaking care is required to assure that the outlet pipes themselves are constructed properly. When a metal pipe is buried, concrete encasement must be provided and adequately reinforced to withstand embankment loads. The reinforcing steel cage must be centered on the pipe. Poor workmanship can result in less encasement than provided in the design. The thickness of the encasement usually should not be less than 200 millimeters (8 inches) even for small conduits.

If conduits must cross faults or shear zones, some means of accommodating displacement should be provided. Protection against detrimental differential movement is needed also where a conduit must cross both fill and rock foundation. For buried pipe, one way is to place a closed jacket or carrier conduit around the pipe. Flexibility is an essential requirement for the pipes at the point where they leave any rigid encasement.

The need for seepage or cutoff collars around an outlet pipe warrants careful attention. In certain circumstances, they should not be used. At one damsite, for example, the outlet was placed in a hard, massive granitic rock; yet the designers called for collars. The foundation was fractured badly below grade by improper use of explosives in an attempt to excavate for the collars. Any possible benefit of cutoffs in this setting was outweighed by the damage to the foundation. Collars around conduits may also do more harm than good where they cause difficulty in compaction. This objection can be met partly by spacing the cutoffs so that heavy compactors can be run easily between them. Strict inspection is important, too. Generally, where the conduit is placed in a trench excavated in rock, collars may serve little purpose. Better protection against seepage is provided by placing the concrete encasement of the conduit to fill the whole space between the pipe and the trench.

Where seepage collars are subjected to embankment loading, one practice has been to place the collar integrally with the conduit, using reinforcing steel to tie the two elements together. Some designers have suggested that the collars should be separated from the conduit by means of asphaltic joint filler, so that the cutoff is free to move and still maintain a water seal. Collars are usually best protected against embankment distortions if they can be located in the center third of the dam base, where movement will be primarily vertical.

Various kinds of collars or cutoffs may be used to safeguard a dam and its appurtenances from the adverse effects of seepage. While their performance records include both successes and failures, they are of recognized value under certain conditions. For example, cutoffs can serve a useful purpose where there is a possibility of separation of a structure from its backfill, such as at a high wall adjoining an embankment.

Spillways.—Increasingly conservative criteria for sizing flood control works are tending to raise some project costs appreciably. At some sites these facilities have been almost as expensive as the dam itself. But false economizing on spillways has led to many problems. Spillway chutes have been terminated too high and too close to the dam. Erosion from a perched spillway can be very hazardous. Flipbuckets are poor substitutes for stilling basins where rock is not resistant to erosion. More than one dam has been endangered by lack of waterstops in spillway chutes. Even a small spillway discharge has been known to overtax underdrain systems where waterstops were omitted.

Spillways preferably should not be placed across active faults nor below potential slide areas. Landslides are especially likely to be activated during storms in the wet season and have been known to clog spillways when they were most needed.

Outlets and spillways may be built as tunnels in the dam abutments or through other barriers forming the reservoir rim. From the standpoint of safety, this is one of the preferred methods of construction. However, the condition of the geologic formations may have an important influence on the selection of alinement. If tunneling is done in soft rock abutments, every precaution must be taken to prevent damage to the foundation through rock yielding. Where this is violated, by inadequate support of the excavated faces or by other negligence, the weakened zone may constitute a hazard of major proportion.

Figure 5-3 and 5-4 show a damaged and repaired spillway at Bartlett Dam. Figure 5-5 shows the damaged spillway concrete at Lahontan Dam caused by freezing and thawing. Figures 5-6 and 5-7 show the Island Park Spillway which was damaged by alkali-aggregate reaction.

Galleries.—In designing gallery systems within or under dams, consideration should be given to the need for working room for maintenance and repair. Economy-minded designers have provided galleries at some sites which will accommodate only small, strong-hearted inspectors. Most importantly, an access system should enable rapid movement and mobilization of men and equipment in an emergency. An alcove or two in the gallery system, for example, would be valuable for equipment storage and work space. Motorized conveyance devices can also pay dividends during emergency. Since future drilling for grouting and drainage can be expected from some dam galleries, the passageway should be sized to permit setting up drill rigs so that holes can be alined at desirable angles.

Drains.—Drainpipes under embankments and reservoirs deserve special attention. The type of pipe is important. Any conduit or structure buried under an embankment must not be susceptible to rapid deterioration. Since most drainage systems are buried permanently, any failure may be difficult to detect and to remedy.

In designing a pipe system for drainage of the abutments under an embankment, two preferred guidelines are: (1) provide two outfalls connected together so that, if one outfall fails due to crushing under construction or movement or plugging during operation, there will still be a reserve discharge line, and (2) extend the upper end of each abutment drain to serve as a cleanout access opening. This will permit the introduction of water into the drain system for cleaning and testing.

Underdrains in zones subject to movement should be divided into sections with separate outfall systems so that areas of leakage can be identified. Drains on one side of a foundation shear can be isolated from those on the other side to avoid possible fracturing of lines at the shear.

Rigid pipes commonly used for drains, such as clay tile and asbestos cement, require extra care in handling and bedding since they are relatively brittle and easily damaged. Some metal pipes are very susceptible to corrosive attack, particularly when located in moist embankment.

Figure 5-3.—Spillway damage at Bartlett Dam (Courtesy, Salt River Project). P-801-D-79384.

Figure 5-4.—Repaired spillway at Bartlett Dam. P-801-D-79385.

Figure 5-5.—Lahontan Dam, spillway damage. P-801-D-79386.

Figure 5-6.—Island Park Dam, spillway damage. P-801-D-79387.

Figure 5-7.—Island Park Dam, closeup of spillway damage. P-801-D-79388.

Aluminum pipe, for example, usually has a very short service life in such an environment. Certain classes of stainless steel pipe, though usually rustproof in most applications, may be subject to crevice-type corrosion if used in a continuously flowing drainage system; however, stainless steel containing nickel would be very resistant to corrosion in such an adverse setting.

Common grades of steel pipe, even though galvanized, asbestos-bonded, and asphalt-coated, may not have a long service life under these conditions. Dipping and coating do not significantly extend the life of the pipe. Steel pipe drainage systems therefore normally should not be installed at a dam. Steel pipe with mortar lining and coating, however, may be used for drain outfalls where acids or sulfates are not present in high concentrations.

Strict control must be exercised in the backfilling of drainpipes and other conduits under embankments. Backfill should meet the specifications for the surrounding zone of the dam. Usually, adequate compaction can be achieved by pavement breakers or other heavy compactors.

Shear Keys.—For many years shear keys have been commonly accepted features of joints in concrete structures. They look effective on a drawing, but many construction engineers can attest to the difficulties of building them properly. A joint in a relatively thin structural member such as a spillway wall can become quite cluttered by the time waterstop and reinforcing steel are in place. To complicate the joint further with a shear key may be inviting honeycombed concrete which defeats the effectiveness of the waterstop. Proper vibration of the concrete in a key is difficult. The resulting poor grade of concrete at the joint may offer less shear resistance than an unkeyed joint.

Foundation Projections.—The use of concrete cutoff walls and other structures buried under embankments must be carefully considered. At one high embankment, on steep granitic abutments, low concrete walls were constructed up the abutments to provide a cutoff in case the nonplastic decomposed granite fill pulled away from the foundation. Slope indicators showed that the fill did indeed settle away from the abutments. There is a remaining question, however, whether such walls, projecting into the fill like knives, would not have a detrimental effect through shearing of the embankment and the creation of cavities as the embankment adjusts under load.

Core Walls.—Early American dams sometimes incorporated a puddled clay core. However, because of low stability and difficulty of placement, the puddled clay core was superseded by the rigid core wall, first composed of masonry and later of concrete. These walls had a tendency to crack and were not always dependably watertight. American engineers therefore gradually endorsed the concept of the rolled-earth core. Today, there are very few designers who would advocate the use of a concrete core wall in an earthfill or rockfill dam.

Superimposed Conduits and Structures.—Wherever possible, construction of conduits through or over embankments should be avoided. Difficulties have been experienced with spillways, outlets, and fish ladders in such locations. Conduits can be expected to leak at least a little. Leakage may be obscured by the structure until it has caused damage. Figure 5-8 shows a destroyed spillway on an embankment. An inadequate drainage system under the Middle Fork Dam spillway floor was one of the prime causes of the failure.

Not infrequently after an urban reservoir has been in operation for a few years, a proposal is made to construct a sewerline or water pipe or other conduit across the embankment. Trenching into the dam, coupled with improper backfilling, may create a plane of weakness in the structure. Also, discharge from a ruptured pipe may quickly wash out a dam. Where no feasible alternatives exist, the utility conduit should be placed within a flexible carrier pipe, preferably located along the upstream face of the dam rather than buried in the fill. If the conduit operates under pressure, it should be provided with valve control so that the section on the dam can be shut off from the system in case of a break.

Emergency Access.—No design of a dam, nor a plan for operation, is complete without provisions for protection during an emergency. Outlet works and spillways should be designed and maintained so that they are always accessible. Auxiliary power should be provided. Equipment and supplies for handling adverse conditions are essential. Floodlighting of critical facilities should also be considered.

The safety of a dam during an emergency requires dependable means of access. Roads to the site must enable entry of equipment necessary for servicing the dam during any adverse conditions. The road grades and bridge spans should be above the projected high waterline. As additional safeguard, alternative means of access should be provided where possible.

Surveillance

In the engineering of dams, provisions should be made to cope with changing conditions over the years of operation. The potentialities of deterioration must be closely watched. Attention must be given to these possibilities during design and should be reflected in access systems, maintenance facilities, and instrumentation. Surveillance should cover the entire reservoir area.

Engineers responsible for a new dam have opportunities to know its foundation and its materials and to determine and execute their treatment, processing, and placement. They know where the site and the structure are strong and where they may be weak. For the sake of later analysis, their knowledge and its limits must be thoroughly documented. Otherwise, those engineers responsible for the dam as it advances in age will be seriously handicapped. An old dam that has outlived its creators may be a puzzle to those others who see it for the first time. In addition to the possibility that complete design and construction records may be lacking, it is not likely to have much instrumentation to indicate its condition. There may be some survey data that offer clues to its history of movement, but the reliability of such records may be doubtful. New instrument installations may be advisable.

Figure 5-8.—Damaged spillway on embankment—Middle Fork Dam (Courtesy, Calif. Dept. of Water Resources). P-801-D-79389.

Evaluations of the safety of existing dams must pursue all aspects of design, construction, and operation. The task may be more comprehensive, more demanding, more tedious, and sometimes more puzzling than the original design effort. The engineer analyzing an old dam may have only limited knowledge of interior materials, details, or foundation of the structure and its appurtenances. Information may have to be secured from meager records of site exploration, construction, and operational surveillance. Documentation of the original designer's intentions and efforts may be incomplete. Therefore, even the most exhaustive search and analysis may not uncover everything that needs to be known to make the structure safe. Areas of uncertainty can be reduced by new drilling, testing, and measurement.

Visual examination is a fundamental and reliable way to detect malfunctioning or deterioration. Uneven settlement, discoloration or increase in seepage, and embankment sloughing are manifestations of potential failures and should be investigated by an experienced engineer. Emergency conditions, however, usually can be averted through continuous surveillance procedures. Many cases can be cited of observation programs diligently started but subsequently neglected with the passage of time. If the responsibility for examination passes to an untrained or indifferent employee, nobody should be surprised if adverse signs pass unheeded until too late.

Observers.—The selection of personnel for monitoring of dams must be done carefully. The inspector and the analyst must be practical and dedicated diagnosticians who thoroughly examine every clue in their scrutiny of the behavior of the dam. A person who becomes uninterested, complacent, or overwhelmed when surrounded by voluminous collected data should not be assigned to this demanding duty. On the other hand, an analyst concerned with quantity rather than quality of data or fascinated with overly sophisticated techniques may overlook obviously adverse trends that may be apparent by scanning data or by simple charting. The key to striking a proper balance is selection of someone who knows what to look for and is perseverant in his search, discerning in his interpretations, and communicative of his findings.

Responsibility for surveillance should be clearly designated. The need for results determined from an unvarying basis requires that, whenever possible, the same people should be assigned each time to specific tasks, although the findings should be checked independently. Fragmented or dispersed responsibility is not conducive to obtaining reliable measurements and accurate analyses. A competent observer guided by established operating and instrumentation criteria must be assigned to each major dam to detect abnormal behavior and to analyze promptly the significance of deviations.

To assure that a dam remains in good health, surveillance must be professional and continuous. The designer cannot walk away as operation is begun; he must share his valuable knowledge of how it was intended to perform. With the help of instrumentation, the designer, the operators, and the professional inspectors can judge its actual performance against the design expectations and the other contingencies that have been considered. If further protective measures are needed after the first stage of operation, the designer will have the necessary background and insight to tailor them to design requirements.

Construction and operations engineers are usually best positioned to recognize the unexpected. They must understand how the design is intended to work. Otherwise, the significance of conditions that vary from design assumptions may not be noticed. While they must enforce compliance with plans and specifications, they should advise the designer when revisions are necessary.

The importance of well-informed operations and maintenance personnel must be stressed. The value of their presence and of their routine inspections will depend upon how well they understand the design and the vulnerability of the structures. Unless they are trained to distinguish the important indicators from the unimportant, they may tend to be unperceptive even as conditions worsen. If a dangerous defect develops slowly, a resident inspector who is not attentive may not notice subtle changes.

A regular examination by experienced personnel is an indispensible element. The inspector must be able to recognize signs of possible distress such as: (1) structural joint movement; (2) piezometric fluctuations; (3) seepage variation; (4) settlement and horizontal misalinements; (5) slope movement; (6) cracking of concrete; (7) erosion; and (8) corrosion of equipment and conduits.

Monitoring.—Before filling a major reservoir, records of piezometric levels, ground elevations, and background seismic activity at the site should be compiled so that comparison can be made with the effects of water loading. For high dams and large reservoirs, installation of a sensitive seismograph network may be justified. As soon as water impoundment begins, which may happen before construction is complete, an inspection and maintenance program for structures and operating equipment must be instituted. During the first filling, this will include daily patrol of the dam and its abutments and daily observation and graphing of seepage flows and piezometric levels. Instrumentation to detect structural or foundation movement should be read monthly. These readings should be plotted and correlated with concurrent reservoir water surface levels.

Dams are especially susceptible to failure during the first 2 or 3 years after the initial filling of the reservoir. Surveillance should be aimed at detection of any tendency toward change in behavior of the dam. The search must focus on the anomalies as opposed to the norm. This requires establishment of an observational data base as early as possible in the life of the structure. Emphasis must be placed on quick processing of data.

During the initial impoundment, reservoir water may penetrate and flush out foundation openings that were not discovered during construction. This may be signaled by increases in seepage flow and turbidity which should alert surveillance forces.

Although the most critical time in the life of a reservoir may be during its first filling, several years may pass before foundation and structures have fully adjusted to loading. Thereafter, deformation will continue in response to cyclical load variations.

Attention should be focused on examination and data collection during relatively rapid changes in reservoir water surface elevations. Conditions year-to-year at high and low seasonal levels should be compared. Special monitoring should be conducted when the pool exceeds the historic high level. Abnormalities indicative of deteriorating conditions must be met with quick corrective action.

Failures may develop very slowly and the adverse conditions may not be apparent for a long time. This may be misleading and conducive to careless surveillance.

The failure of a dam is likely to be preceded by observable or measurable deformations. If its materials are brittle, however, the final rupture may be sudden, with minimal advance warning. Foundations also may fail abruptly and thus deprive the dam of vital support. These possibilities demand that surveillance systems be developed painstakingly and be strictly enforced.

Anomalies in observational data must be subjected to relentless scrutiny. Nothing should be taken for granted or explained away by casual assumptions. In the surveillance of dams, any observation that appears unusual should be reported to someone who can analyze it properly. The judgment of its significance should not be left to those who may misunderstand it and dismiss it.

Performance data should be examined not only for deviations from reading to reading, but also — and especially — for slow trends which may have subtle meanings. The implications of such long-term changes are sometimes overlooked.

The value of timely and painstaking data analysis in search of changing trends cannot be overemphasized. On larger projects where numerous instruments must be read, data retrieval and processing systems should be designed so that rapid evaluations can be made. The importance of timely actions in all parts of the surveillance process needs to be emphasized again and again. The reading of instruments must not be allowed to lag. The channels for communicating the data should be kept short, and the information provided must be examined when it arrives. Anything unusual must be called to the immediate attention of those in responsible charge. These requirements are imperative and should be obvious. Yet, there are three forces (at least) that interfere: (1) complacency due to a long-term satisfactory performance record; (2) insistence that information creep single-file and be ponderously processed through lengthy organizational conduits; and (3) desire to double-check and beautify each report so that it will withstand the scrutiny of the ages. Responding to each of these deterrents, the professional surveillant must recognize that (1) the absence of adverse signs should challenge observers to greater vigilance; (2) communications of vital surveillance information should flow in parallel channels to anyone who needs to know without delay; and (3) the value of surveillance reports must be judged more on timeliness than on appearance. They can be polished later for the archives.

To assure timely and perceptive analysis, those who read surveillance data must be selective. They must be able to sort out what may be important and study it quickly. Otherwise, there may be a tendency to bog down in the voluminous detail that can be generated by a comprehensive system of observation. After initial review, those data that may indicate questionable trends should be examined in depth.

Engineers who analyze data from dam instrumentation must strive for a clear perspective that enables prompt recognition of adverse conditions as they sift through the sometimes voluminous records. Quality of data can be more important than quantity. A single fragment of information may not be meaningful in itself, but when grouped with other data may establish a norm. Departures from the norm can then serve as indicators of the condition of the dam. The graphical summarization of data often facilitates understanding the significance of factors that might adversely affect the dam.

Once established, a monitoring regimen must still be given periodic professional review. The surveillance system should be adaptable to changing circumstances, and revisions should be made promptly to cover unforeseen variations. Measuring devices should be recalibrated on a regular basis. Instrumentation that gives faulty readings should be modified or replaced.

Deficiencies should be corrected without delay. Unhealthy dams do not improve by themselves. Their conditions must be diagnosed quickly and then decisive action must be taken. Sometimes the cost of remedial treatment will appear prohibitive. Another alternative, which is also costly, may be to remove the dam or to breach or otherwise modify it so that it is no longer a threat. There may be disagreement about who is responsible and, once that is resolved, the party charged with financing the corrective action may continue to argue that the cost is too high. Haggling over the price of safety should not be prolonged. The cost of failure may be even higher.

General Guidelines for the Observer.—There are so many conditions which might endanger a dam that great care must be taken lest some be overlooked. For this reason, a checklist of questions such as the following should be used.

1. Have changes occurred in the environs of the reservoir that may necessitate reexamination of the design or of the surveillance program (e.g., industrial activities such as deep excavation, trenching, tunneling, building construction, or storage of explosives or flammable materials)?

2. Are there utilities such as oil, water, or sewerlines near or crossing the dam or its appurtenances that would jeopardize safety if they were broken?

3. Are access roads and communication lines to the damsite located and constructed so that they will not be disrupted during extreme emergency?

4. Are the structural analyses of the dam satisfactory, or should new analyses be made using the latest design technology?

5. Is the outlet capacity adequate to lower the reservoir rapidly during an emergency?

6. Is the spillway capable of discharging floodflows projected on the basis of up-to-date hydrological records?

7. Is there danger of spillway discharge undercutting the structure?

8. Are adequate auxiliary power and other redundant systems provided for hoist operation or other requirements during an emergency?

9. Is the spillway channel constructed and maintained so that there will be no dangerous erosion, or debris deposited, in the river channel?

10. Is adequate ventilation provided in shafts, tunnels, and galleries to prevent corrosion and to protect personnel from noxious gases?

11. Is essential machinery operable, especially such items as gates, valves, and hoists?

12. Are drainage sump pumps, if any, operable?

13. Are automatic alarms and telemetering devices functioning?

14. Is riprap, soil-cement, or other revetment intact as constructed?

15. Is all instrumentation in satisfactory working order?

16. Is there vegetation on embankments or abutments that might obscure adverse conditions from the inspector's view?

17. In the case of concrete dams, is there any reason to doubt the strength of the concrete? Has this been confirmed by nondestructive tests or tests of cores?

18. Are intake works for outlets and spillways free from silt and debris?

19. Are adequate emergency supplies and equipment available for handling adverse situations at the dam?

20. Have operating mechanisms that operate infrequently been checked or exercised to verify that they function properly?

21. Are vulnerable facilities protected against vandalism or sabotage by installation of fencing, locks, and intrusion-detection devices?

22. Are competent, trained personnel assigned to surveillance?

23. Do operations personnel have proper instructions and authority for actions to be taken during an emergency?

24. Are piezometer readings and water levels in wells reasonable, steady, and consistent with reservoir height?

25. Are additional piezometers, wells, or weirs necessary for proof of safety?

26. Are reservoir linings, if any, performing as designed?

27. Are surveillance data receiving timely analyses?

28. Has the dam crest settled and thereby reduced the freeboard for flood discharge?

29. Is leakage of water excessive? Is it increasing or decreasing? Is it clear or turbid? Are there large variations in individual drain discharges?

30. Are wet spots visible on the downstream face of the embankment or at abutment groins or immediately downstream?

31. Is there evidence of dissolution of foundation rock by seepage?

32. Is potentially dangerous seepage apparent in the vicinity from sources other than the reservoir, such as in the abutments at high level?

33. Are signs visible of any sloughing or slumping of embankments, abutments, or the reservoir environs?

34. Is piping evident, especially where fills have been placed against or covered by structures?

35. At dams with concrete face slabs, is there visible warping or other distress?

36. Has cracking developed in structures, embankments, or foundations?

37. Are there any signs of erosion of the embankment or its foundation?

38. Has any change occurred in alinement of parapet walls or retaining walls?

39. Has any recent seismic activity been recorded in the area? If so, are there any signs of detrimental effects on the reservoir or its environs?
40. Is subsidence evident at the site or in its margins? Are any petroleum or water extraction or mining activities in progress that could cause subsidence?
41. Is there any progressive joint opening in the concrete?
42. Is water seeping through horizontal lift joints indicating possible dangerous uplift?
43. Is there excessive erosion of concrete, such as in the spillway or stilling basin?
44. Is chemical deterioration of the concrete manifested, such as by leaching, crumbling, cracking, or spalling?
45. Do drain outlets show any adverse signs such as leaching of cement?
46. In frigid climates, has any damage occurred from ice thrust or freezing?
47. Have any structures been undermined?
48. Is there evidence of chemical alteration of foundation materials?
49. Are uplift pressures within the design assumptions, or is it necessary to drill more relief holes into the foundation?
50. Are deflection records adequate, and are deflections consistent with changes in reservoir level and temperature?
51. Have all adverse or questionable conditions been promptly reported?
52. Do operations and maintenance personnel examine the dam often enough?
53. Have deficiencies been remedied without delay?

Instrumentation.—Widespread attention is now being given to the installation of more extensive instrumentation for study of the behavior of dams and reservoirs and forecasting of any adverse trends. Instruments strategically implanted in the vital zones can provide meaningful clues. Once the symptoms are identified and the cause determined, the necessary treatment can be prescribed.

The extent of internal distress in a dam cannot always be directly measured. However, diagnostic procedures are available which can help in identifying most ailments. A key to these procedures is effective instrumentation. Piezometers, strain gages, slope indicators, accelerographs, sensitive seismographs, tiltmeters, and survey networks preferably should be installed early so that data are recorded during construction. After the project is operational, the monitoring should continue. A comparison of operating conditions with design assumptions will help to determine whether the structure is performing satisfactorily.

Instrumental readings may give warning of an impending disaster. However, instrumentation alone is not a complete safeguard. The number of devices installed in a dam is less important than the selection of the proper types of instruments, their location at critical points, and the intelligent interpretation of the data they provide.

In the selection of equipment, service requirements must be carefully weighed. An instrument of rugged construction that gives reasonably accurate results may be preferable to a more precise but delicate instrument. Some of the characteristic problems stem from having to place certain measuring devices deep within the structure. The ideal solution would be to install a durable, wireless, remotely controlled and interrogated instrument to indicate conditions within the dam.

Everything reasonably possible should be done to protect instrumentation from damage. For example, electrical leads and tubing embedded in concrete have been severed when the enclosing structures were fractured by movement under load. Where feasible, wires and tubing should be placed in conduits for protection.

Appurtenant structures may provide ready means for installation of instrumentation. For example, walk-in outlet conduits can be used to establish strain gages to record any displacement of the outlet conduit that might be indicative of foundation or dam movement. Extensometers in hillside drain and grout tunnels can measure abutment deformation.

No general rules can be given for the kinds of measurements to be made at dams. Not only are there many kinds of dams and many different site conditions, but each dam may have a different ailment. Some kinds of instrumentation are recommended for nearly all dams, while others are recommended only in special cases.

Professionals do not all agree on the number of instruments to be used and, as operational experience is accumulated, by how much the frequency of observations should be reduced. No doubt there are some dams and reservoirs in which too many instruments were installed and where intensive measurements have been continued longer than necessary. Unless a reasonable balance is struck between safety and economy, the project will be burdened by unnecessary accumulations of data that may even interfere with sensible pinpointing of problems. Measurements of construction pore pressures, for instance, will have served their primary purpose once initial reservoir impoundment is completed.

Provisions should be made for periodic deflection measurements. Where topography permits, this can be done by theodolite from fixed bases, using either line-of-sight over the top of the dam or by turning angles to targets on the downstream face and at the crest. At concrete dams, the deflections should be consistent with changes in reservoir water surface level and in temperature and should not change appreciably from year to year. At earth and rock dams, vertical deflections are important; and these are best measured by level readings on bench marks, mainly along the crest.

To determine whether a foundation or abutment is being compressed, joint meters may be installed in series by drilling holes into the foundation to various depths, inserting pipes, and cementing them at the deep end. Cables from the joint meters ("foundation meters") can extend to the crest of the dam or to any place convenient for measurement. The meter should be mounted at the dam end of the pipe, and provision should be made for resetting if the range is exceeded. The gage should be accessible for examination and servicing if there is any doubt about its proper functioning.

A primary indicator of the performance of any dam is the water pressure distribution in the body of the structure and in the foundation. The value of determining pore pressures in embankments is widely acknowledged. Fluid pressures within a soil may represent a significant part of the total stress. Pore pressures in a soil mass will decline naturally over a period of time. The rise in these pressures as the earthfill is built higher, superimposed on their natural tendency toward dissipation, can be related to changing patterns of stress and settlement. Variation in pressures is therefore a measure of the behavior of the embankment. When construction is completed and reservoir impoundment begins, the piezometer system functions also to show the seepage pattern in the dam. The piezometer is therefore one of the most dependable indicators of the condition of an embankment or its foundation.

One of the simplest piezometers is a perforated pipe installed vertically. It should be kept capped when not in use. A tape can be used to measure the vertical distance to the standing water level. In many cases such a simple piezometer will serve adequately, especially in relatively pervious materials. Where embankment material might move through the perforations into the pipe, a graded filter should be provided by placing the pipe in a cased hole and backfilling the annular space with drain material, such as sand.

A typical hydraulic piezometer of comparatively more sophisticated design consists of a thick plastic disk fitted with one or two porous filters. Plastic inlet and outlet tubes connect this embedded sensing device with an instrument panel on which Bourdon gages indicate the internal pressure. The tubes are purged by pumping. By maintaining gages on both lines for each piezometer, along with a gage on the manifold, measurements can be easily verified. Since the instruments are exposed directly to pore pressures, readings can be taken at any time.

With the pneumatic piezometer, the gages do not monitor the water directly. Within the embedded instrument is a diaphragm which responds to pore pressure. Introduction of air or nitrogen into an inlet tube applies pressure on the rear face of the diaphragm. When this balances the pore pressure on the front face, the diaphragm flexes and thus enables entry of the actuating gas into the outlet tube. This leads to stabilization of the gas pressure in the system. The gage indication of this level is accepted as the pore pressure.

For remote measurement, several other types of piezometers have been used. One is an electric-resistance cell that separates internal water pressure from intergranular pressure by means of a porous disk and senses the pore pressure by the flexure of a diaphragm.

Concrete Dams.—Supplemental instrumentation may be advisable to assess questionable behavior detected during the examination of a concrete structure. Measurement of the overall movement of the structure may be needed or of movement between monoliths along joints or displacement at cracks, as well as hydrostatic pressures in joints and cracks and under the dam.

Deformation measurements of a concrete dam must provide a coherent understanding of movements of the dam, its foundation, its appurtenances, and its environs. Measurements are made by such devices as pendulums, inclinometers, rock meters, and dilatometers. These determine relative movements between parts of the dam or foundation blocks. Absolute deformations can be measured by geodetic methods, which measure the dam and its site in all their interrelations. Geodetic measurement has been advanced significantly in recent years by the introduction of new instruments and computer technology. Precise apparatus for electro-optical distance measurement and invar wire measurement, as well as compensation programs, have made important contributions to more efficient measurement and evaluation, to increasing accuracy and, hence, to more reliably useful results.

The common geodetic methods include triangulation, expanded now to triangulateration (triangulation plus trilateration) by the new precise instruments; traverses to measure absolute movements of points; trigonometric leveling for ascertaining absolute changes in elevation; and optical plumbing by vertical angles to measure movements of points in horizontal profiles.

Instrumentation useful in evaluating the performance of concrete dams includes:

- *Joint meters* are embedded across joints or cracks to measure changes in openings. Measurements can be read remotely via cable.
- *Inclinometers* are used for measuring the inclination from vertical at selected elevations with a high degree of sensitivity.
- *A tilt measuring instrument* consists of a portable sensor mounted on a metal plate placed upon reference plugs or a plate embedded in the structure to sense changes in rotation.
- *Extensometer or micrometer points* may be installed in a triangular pattern of three embedded plugs, two on one side of a crack or joint and the third on the other, to enable measurement of relative movement. Borehole extensometers can be used to detect internal movement at cracks. Precise electronic distance measuring instruments are also useful for monitoring structural displacements.
- *Observation wells* are open holes drilled into the structure and/or its foundation for measurement of water levels.

In usual Bureau practice, a system of foundation drains is installed during the construction of a concrete dam. The system usually consists of 75- or 100-millimeter (3- or 4-inch) diameter pipes placed at approximately 3050-millimeter (10-foot) intervals, in the floors of the foundation gallery and foundation tunnels. Periodic measurements of flow from the individual drains are made and recorded. When drain flows are minimal, measurements may be made using any suitable container of known volume. When flows are greater, weirs may be installed in the drainage gutters of the galleries and adits of a dam.

When drainage flows are large enough to be measured by weirs, the measurements are usually made on a monthly schedule. Any sudden increase or decrease in drainage is noted and correlated with the reservoir water surface elevation and any change in conditions at the individual drains. Drains should be kept free of obstructions.

In some concrete dams, internal strains give valuable information regarding structural action. Although internal strain meters are best embedded during construction of the dam, they can be grouted into drilled holes for a dam in service. The alinement of the holes in the exact orientation of maximum expected stress may be impractical, but a near-optimum direction can be provided. Long strain meters are recommended for this purpose for two reasons: (1) the long meter provides a measurement over a larger amount of concrete, and (2) the longer strain meter has greater sensitivity and accuracy. In the past, most strain meters were 250 millimeters (10 inches) in length. More recently, many strain meters of 500-millimeter (20-inch) length have been used.

The opening and closing of joints provide sensitive measurements of the action of a concrete dam because movements are magnified at a joint. Where the joint is accessible, inserts can be installed on both sides of a joint, and mechanical gages can be used with dial indicators reading to 0.01 millimeter or less. When the joint is not easily accessible, or when remote reading is preferred, electrical joint meters can be used. One type of joint meter (electrical resistance) measures both joint opening and temperature. It can be obtained in almost any desired range and sensitivity; but the greater the range, the less the sensitivity.

Two types of strain meters have been used widely, one depending upon the change of frequency of a vibrating wire within the meter and the other depending upon the change of electrical resistance of very fine music wire. The vibrating-wire meter has been used mainly in Europe, while the resistance-type meter has had widest application in the United States.

In many concrete dams safety depends upon maintaining a low uplift pressure, not only at the base of the dam but also on horizontal joints. A simple method of measurement is to attach a Bourdon gage to a pipe cemented to the top of a hole drilled into the foundation from a gallery. There may be many such holes drilled for the purpose of reducing uplift pressure. If, when a hole is capped, the pressure is found to be excessive, additional drilling may be indicated.

Embankments.—In designing or evaluating instrumentation at an embankment, emphasis should be placed on foundation settlement, seepage, and pore pressures as indicators of incipient problems.

The crossarm settlement device designed by the Bureau of Reclamation and used extensively for monitoring performance of embankments usually gives reliable readings in both earthfills and rockfills.

Hydraulic leveling devices are also used to measure settlements inside embankments. They are regarded as accurate and reliable.

Inclinometers (slope indicators) have been used for the measurement of settlements and horizontal displacements. An inclinometer must be properly installed and carefully maintained to ensure accuracy and reliability. It is a relatively sensitive measuring device and can give misleading results if such attention is lacking.

Linear extensometers are simple and reliable devices that measure movement between reference plates by electrical resistance. They have been used successfully in large embankments.

Offstream Reservoirs.—Offstream reservoirs in urban areas require especially careful surveillance. A typical reservoir of this type may be constructed on a foundation so weak and permeable that the excavated natural ridges forming part of the rim of the reservoir cannot be considered adequate water barriers alone. The foundation must be kept dry to ensure stability. One of the best safeguards is to use an impervious lining underlain by a drain system to monitor any leakage through the membrane. This will generally require a continuous impervious liner under the drain system to prevent any loss of leakage into the foundation. Also, piezometers are usually installed around the reservoir rim as a second warning line.

Remote Monitoring.—Any of the commonly used monitoring devices can be connected to onsite or remote annunciators. Piezometers can be incorporated into an alarm system that can be monitored 24 hours a day. Automatic, continuous water level recorders may be used in observation wells. Sensors can be set in the wells at preselected danger levels.

Similarly, drainage weirs with float-actuated switches can signal unusual variations in flow. The subdivision of drain systems into sections with separate outfalls facilitates analysis by identifying and quantifying the sources of seepage. To be an effective indicator the drain must be capable of collecting and conveying substantially all seepage to the measuring point.

Some large dams are equipped with seismological observatories to monitor earthquakes. At Oroville Dam, for example, earthquakes are measured by accelerometers and pore pressure cells in the embankment at various levels and by strong-motion accelerographs at top and toe, in the concrete core block, and in the grout gallery portals. These can be adjusted to actuate and record a broad spectrum of embankment response. Oroville Dam was also instrumented with large stress cells in the downstream shell to measure transient stresses during seismic events. There are 15 of these cells, 127 millimeters (5 inches) thick and 762 millimeters (30 inches) in diameter. Each cell was equipped with two transducers, one to measure both static and dynamic stresses, and one to measure static stress only. A recorder in the instrument terminal at the dam could monitor all operable cells at the same time, being activated automatically through a signal from strong-motion instruments at the damsite. This complex of stress cells provided useful data (with 9 cells operating) in an earthquake at Oroville of 5.7 Richter magnitude in 1975. However, in the following year their dynamic sensing capability was eliminated by a lightning strike. Most of the cells are still operable for measuring static stresses, which have become essentially constant. Oroville Dam instrumentation was updated in 1977 and 1978 by provision of new accelerographs, accelerometers, earthquake force monitors in the area control center, power-supply and calibration-signal conditioning equipment for the pore-pressure cells, and an advanced digital recorder in the control center. All instruments are set to be actuated by a common trigger, and all records will have a common time base.

Bureau of Reclamation Practices.—In the Bureau of Reclamation, planning, installation and control of a measurement program are supervised by the design offices. The program starts with the installation of the measurement systems during construction.

Cooperation between the design offices and the project construction office, and later the operations and maintenance organization is vital in obtaining reliable installations of instruments and accurate and timely information from the measurement program.

A schedule for installation of instrumentation and for obtaining readings at a dam begins almost at the outset of construction, and extends into the operating stage.

An instrumentation report is sent to the Bureau's design office each month. This includes instrument readings taken during the previous month and other pertinent information such as daily records of air and water temperature, reservoir and tailwater elevations, and comments on the operation of instruments. Photographs and sketches are used freely to convey information.

Schedules for measuring structural behavior are varied. Embedded instrument readings are usually required more frequently immediately after embedment than in later periods. The reading frequency may be weekly or at 10-day intervals during construction and semimonthly after construction. In some cases, a monthly reading frequency is allowed. Although this is not desirable for strain meters, it is acceptable for stress meters, reinforcement meters, joint meters, pressure cells, and thermometers. During periods of reservoir filling or rapid drawdown, more frequent readings are preferred.

Weekly data from deflection measuring devices at concrete dams, such as plumblines and collimation, are preferred. During events of special interest, such as a rapidly rising or falling reservoir, readings at closer intervals may be desired.

Data from uplift pressure measurement systems may be obtained monthly except during the initial filling of a reservoir, when readings are taken at weekly or 10-day intervals. Pore pressure gages may be read monthly. At concrete dams where drain flow is of a sufficient quantity to be measured, these data are usually obtained at monthly intervals.

Target deflection and pier net triangulation measurements at concrete dams should be taken at least semiannually during the periods of minimum and maximum air temperature to determine the extreme deformed positions of the dam. During early stages of reservoir filling, additional measurements are desirable and are made approximately midway between those of minimum and maximum air temperature conditions. These are useful in noting deformation trends and to correlate collimation and plumbline information.

Periodic leveling is conducted in the vicinity of and across the top of a dam to detect any vertical displacement of the structure.

The measurement program for a concrete dam initially should cover a time period which will include a full reservoir plus two cycles of reservoir operation, after which monitoring is considerably reduced. However, some measurements, such as those from plumblines, collimation, foundation deformation meters and gages, and from clusters of embedded meters, which are considered essential for long-term monitoring, are continued indefinitely. For these measurements, the frequency of readings may be extended.

Measurement systems require a well-defined program for processing data. Great volumes of data can accumulate and would be of little value without dutifully followed schedules and guidelines for analysis, reporting, and remedial action where necessary.

Processing of large volumes of data is done efficiently by the computer at the Bureau's E&R Center in Denver. Certain output is drawn by an electronic X-Y plotter. Reports are prepared from these results.

Testing—Any program for monitoring the performance and condition of a dam should provide for materials testing when needed. Soniscope inspection, for example, is an effective nondestructive method of determining the integrity of concrete or rock. By measuring sonic wave velocities, the soniscope delimits defective portions of a concrete dam or of concrete slabs on an embankment. Deterioration is indicated by abnormally low velocities or abrupt variations in readings as the instrument is moved over the dam.

Specimens of concrete cored from the structure can be examined to determine strength and petrographic characteristics. Testing of cores is of special value where concrete deterioration is attributed to alkali-aggregate reaction. Weakening of a dam under such chemical attack is usually a slow process requiring continuous surveillance, including periodic testing, to determine when strength has declined below allowable limits.

Scope of Surveillance—Engineering and surveillance of a dam should not be confined to the narrow limits of the damsite. Conditions in the environs may have a significant bearing on the safety of the reservoir. To illustrate, during an emergency at an important dam, malfunctioning of the outlet works necessitated spillway discharges into a virgin unlined channel downstream from a gate structure in a topographic saddle. Roads to the dam across this channel were washed out at relatively low flow, cutting off the dam from quick access. Moreover, hundreds of thousands of cubic yards of earth and rock were deposited in the river resulting from erosion of the natural spillway channel. The consequent blockage of the river caused objectionable backwater against the dam, its outlet works, and the powerplant. This was compounded by heavy leakage through the dam. Remedial work, although eventually successful, was hindered by these events.

Bureau of Reclamation Program for the Safety Evaluation of Existing Dams (SEED)

The Bureau conducts periodic safety evaluations of its existing dams and reservoirs to identify problems and to determine what repairs, operational restrictions, or modifications may be required. The evaluation comprises comprehensive review of the design, the construction methods and materials, and the operational history; examining the performance and condition of the dam and reservoir; and analyzing any apparent inadequacies.

The Bureau's examination team first completely reviews all the essential data on the dam, then makes an examination, analyzes and compares all the data, prepares or updates a data book. A written examination report containing conclusions and recommendations is prepared.

While certain uniform practices can be applied in safety evaluations, each dam must receive special consideration based on unique characteristics of the site or of the structure itself.

To identify the weaknesses in a dam, examination team members must thoroughly understand the modes and causes of failure. Case histories of previous failures should be mandatory reading. Team members should use all sources available to them for reports of failures. The Bureau's Engineering and Research Center Library, Correspondence and Records Section, and the Archives are excellent repositories of such information.

Clues to the defects in a dam or in its foundation may be found in the design and construction records, in the instrumental data and personal observations from operational surveillance, and in inspection by experienced professional engineers and geologists.

Records of most newer dams can be expected to be adequate for a safety evaluation. However, when data on a dam are deficient, this should be documented and steps should be taken to correct the inadequacy.

Examination Team.—The examination team for a dam should have expertise in civil and mechanical engineering and in engineering geology. Team members should be experienced in design, construction, and operational surveillance of dams, and knowledgeable of the causes of dam failure.

Bureau of Reclamation practice calls for a team composed of a civil engineer, preferably with dam design experience, an engineering geologist, and a mechanical engineer. Less experienced personnel sometimes accompany the team on a training basis.

The Bureau has developed a training program which supplements its SEED Manual. The program is designed to enable a trainee to learn effective examination and analysis techniques.

Data Sources.—

Geologic.—Geologic information pertinent to evaluation of an existing dam and its reservoir is obtained from:

- Mapping and details of shear zones, faults, fractures, joints, seams, fissures, caverns, landslides, compressible or liquefiable materials, and weak bedding planes.
- Exploration logs.
- Drill cores.
- Geophysical data.
- Ground-water levels in the vicinity before and after reservoir filling.
- Materials test data.
- Aerial photographs of site.
- Regional geologic data.
- Foundation treatment details.

Seismic.—Considerations relevant to seismic potential at the dam and reservoir include:

- Seismic and tectonic history of site and region.
- Location of active and potentially active faults.
- Earthquake effects which could influence structures, such as shaking, foundation displacement, slides, settlement, liquefaction, and seiches.

- Maximum credible earthquake location and magnitude.
- Potential for reservoir-induced seismicity.

Hydrologic and Hydraulic Data.—Hydrologic and hydraulic data pertinent to evaluation of dam safety are obtained from:

- Project hydrologic records.
- Spillway and outlet drawings and design records.
- Data on upstream reservoirs and diversions.
- Probable maximum flood (PMF) calculations.
- Flood routing studies.
- Operation and maintenance records, including spillway and outlet operation procedures and schedules.
- Reevaluation of spillway capacity based on present technology and hydrologic records.

Concrete Dams.—In assessing the safety of existing concrete dams, valuable information may be obtained from:

- Plans and specifications and other design records.
- Records of foundation treatment.
- Geologic and seismological reports on the site and the region.
- Exploratory data, including logs, photographs, and specimens.
- Laboratory test records.
- Construction records, including photographs.
- Comparisons of foundation and concrete materials design properties with data from exploration, construction, and laboratory tests.
- Results of stress and stability analyses.
- Seepage and drainage records.
- Instrumentation records.
- Operations and maintenance observations, including examination reports.
- Comparisons of original design criteria and analytical methods with present practices.
- Records of any mineral or oil or water extraction which could adversely affect the dam or reservoir.

Foundation—Useful information on the foundation of a concrete dam can be obtained by studying:

- Characteristics of the rock mass at the site.
- Regional geology.
- Exploratory logs and specimens.
- Foundation treatment records.
- Seepage and drainage records.
- Mapping and details of rock joints, fractures, faults, shears, and seams.

Materials Data—Materials data pertinent to the evaluation of existing concrete dams include:

- Strength and durability of concrete.
- Modulus of rupture and elasticity of concrete.
- Type of cement, cement factor, admixtures, aggregate mix, and water-cement ratio.
- Lift height and method of placement.
- Treatment of contraction joints and lift surfaces.
- Actual time history of concrete placement and joint grouting.
- Heat generation characteristics of the concrete mixes.
- Physical, chemical, and mineralogical characteristics and sources of aggregates used.

Analytical Data—Analytical data useful in the evaluation of existing concrete dams can be obtained from:

- Records of stability and stress analyses.
- Materials testing records.
- Foundation studies.
- Assumed loading conditions.
- Assumed temperature variations.
- Timing of grouting of construction joints in the construction sequence.
- Extent of cooling that occurred prior to grouting.
- Results of analysis of pressure distribution within the foundation.
- Details of shear keys, if any, in contraction joints.
- Results of abutment analyses.
- Comparison of computed and measured stresses and deformations in dam and foundation.

Embankment Dams.—Information valuable to evaluation of the safety of existing embankment dams can be obtained from:

- Plans and specifications and other design records, including stability analyses.
- Geologic and seismological data.
- Records of foundation treatment.
- Laboratory test records.
- Construction records, including photographs.
- Seepage and drainage records.
- Instrumentation records.
- Operations and maintenance observations, including examination reports.
- Comparison of original design criteria and analytical methods with present practices.
- Comparisons of foundation and materials design properties with data from exploration, construction, and laboratory tests.
- Records of any mineral or oil or water extraction which could adversely affect the dam or reservoir.

Foundation—Useful information on the foundation of an embankment dam can be obtained by studying:

- Characteristics of the formations at the site.
- Regional geology.
- Exploratory logs and specimens.
- Foundation treatment records.
- Seepage and drainage records.
- Mapping and details of rock joints, fractures, faults, shears, and seams.

Materials Data—Materials data pertinent to the evaluation of existing embankments include:

- Classification, gradation, and Atterberg limits.
- Densities.
- Moistures.
- Freeze-thaw tests of riprap.
- Consolidation and settlement.
- Solubility.
- Dispersive clay tests.
- Permeability.
- Petrographic and mineralogical analyses.
- Fill placement procedures, including lift thickness, compactive effort, and method of compaction.
- Embankment construction control tests.
- Static and dynamic strength properties.

Analytical Data—Analytical data useful in the evaluation of an existing embankment dam can be obtained from:

- Records of stability analyses.
- Materials testing records.
- Foundation studies.
- Assumed loading conditions.
- Pore pressure distribution analyses.
- Analyses of seepage distribution within the dam and foundation.
- Abutment studies.
- Comparisons of computed and measured deformations in dams and foundation.
- Records of any foundation uplift or fracturing caused by grouting.

Geologic Review.—Geologic and seismic hazards warrant very careful study. Considering the data available, the reviewer must assess the seismic risk and decide whether a detailed site evaluation and dynamic analysis should be given high priority.

Reservoir boundaries must be studied to identify any slopes that may become unstable. Analyzing aerial photographs may be a useful initial step in defining such hazards. Then, ground surveys and even subsurface exploration may be advisable to determine the extent and characteristics of the unstable masses.

Essential parts of the evaluation of a dam are the comprehensive studies necessary to assess the existence of, or potential for, faulting at the site and its environs. Recent improvements in fault detection procedures have enabled identification of faults previously unknown. Some of these faults undoubtedly exist in the foundations of old dams.

The evaluator of an existing dam may find sound rock exposures at the site but insufficient drilling records to verify these visual indications that the foundation is adequate. If the dam has been performing satisfactorily for many years, and the foundation has never shown signs of distress, new exploration of the site may be unwarranted. Such judgments must be made by professionals who have all available information in hand.

Hydrology Review.—Many ruins of dams attest to the uncertainties of predicting floods. Even the most thorough analyses of hydrological potential have sometimes been shown inaccurate by later events. Projected runoff must be based on probabilities extrapolated from historical record, but floods once regarded as improbable have destroyed dams.

Many old spillways have been found to be undersized, often attributed to the use of primitive methods based largely upon judgment. In the absence of adequate hydrological data or suitable analytical techniques, design capacities were sometimes calculated as arbitrary multiples of the maximum flood peak that had been measured or estimated up to that time. The flood records of those early days were typically short term, incomplete, or inaccurate. An embankment with a spillway of insufficient capacity must be analyzed to predict the probability and the consequences of its overtopping. Where the hazards are significant, expeditious measures must be taken to improve the capability to pass floods safely.

If in evaluating hydrologic hazards, the flood calculations are not current, then precipitation and runoff data should be updated. A new flood routing should be done if changes in data are significant.

Design Review.—The design of the dam should be reviewed to compare actual performance with intended performance. Records of construction should be studied to determine whether structures were constructed as designed or that appropriate alterations were made to accommodate unforeseen conditions.

The design should be examined to ascertain if the facilities are capable of meeting structural and hydraulic requirements. Procedures used in the design should be compared with currently accepted criteria. If previous analytical methods are questionable, the design should be subjected to analysis under latest technology.

The data on which the original design was based must be checked for adequacy, including verification that all possible adverse circumstances were considered. Design criteria should be weighed against changed conditions. New data such as recent seismic or hydrologic records should be studied to determine their possible effects.

Unsafe designs may result from overlooking a possible loading condition, insufficient materials sampling and testing, erroneous assumption of materials properties, or inadequate foundation exploration.

Construction Review.—Unanticipated conditions discovered during construction such as foundation weaknesses or high groundwater are indicators that should receive attention in the review. Laboratory test records also may point to suspected deficiencies in materials or construction methods.

Records of the original condition of the foundation and the dam, and any subsequent changes, should be studied during the preexamination review. This should help to identify areas and conditions that warrant special attention during the examination.

Defects in a dam may be caused by faulty construction, such as inadequate foundation treatment, poor control of material placement, or failure to reject or remove frozen or otherwise deleterious materials. In some cases, inferior construction has resulted from failure of field forces to recognize conditions unforeseen by the designers. The construction records may provide clues to such deficiencies.

Excavations at the damsite during construction may have disturbed the natural slope equilibrium. Reservoir water changes the ground-water regimen and thus may affect slope stability. Reviewers must be alert to these possibilities.

Operations Review.—Records of operations and maintenance such as instrument readings and analysis, application of Designers' Operating Criteria, and deviations from prescribed operating practices should be examined to identify any problems that may threaten the dam or reservoir. The effectiveness of operational surveillance and the adequacy of instrumentation should be evaluated.

In reviewing operations and maintenance records for spillways and outlets, special note should be made of functional abnormalities in gates, valves, control systems, intake structures, conduits, and energy dissipators.

Consideration must be given to any potentially adverse effects of dams upstream or downstream of the dam being evaluated.

Review of Adverse Conditions.—Some of the more common adverse conditions considered in the Bureau of Reclamation's Safety Evaluation of Existing Dams are:

CONDITION	CAUSE
Foundation deterioration	Removal of solid and soluble materials Rock plucking Undercutting
Foundation instability	Liquefaction Slides Subsidence Fault movement
Defective spillways	Obstructions Broken linings Overtaxing the capacity Faulty gates and hoists

Defective outlets	Obstructions
	Silt accumulations
	Faulty gates and hoists
	Unsatisfactory gate position and location
	Inadequate capacity
Concrete deterioration	Alkali-aggregate reaction
	Freezing-thawing
	Leaching
Concrete dam defects	High uplift
	Unanticipated uplift distribution
	Differential displacements and deflections
	Overstressing
Embankment dam defects	Liquefaction
	Slope instability
	Excessive leakage
	Removal of solid and soluble materials
	Slope erosion
	Settlement
Reservoir margin defects	Erosion
	Perviousness
	Instability

Field Examination.—Following review of records, a site examination by experienced engineers and geologists facilitates analysis of any suspected anomalies. Structures and auxiliary equipment, as well as the reservoir environs, should be examined for areas of distress, including adverse movement, seepage, or equipment malfunctions. The examination may confirm or eliminate concerns arising from review of operations and maintenance records.

The Bureau of Reclamation uses site-specific guidelines in its examination of existing dams. The application of uniform criteria provides continuity from examination to examination throughout the dam's history. These criteria are useful as long as the guidelines remain flexible, and are not to be substituted for sound judgment.

A checklist of examination items should be prepared in advance and followed carefully, but the scope of the examination should not be limited to the list.

Operations personnel familiar with the dam should accompany the examination team and should be prepared to explain current conditions and trends. When questionable conditions are known in advance, professionals or consultants with special expertise may be included on the team.

Deficiencies often are indicated by changes in the condition of structures, foundation, or reservoir periphery. Before the examination, the most recent performance measurements should be studied. These should be in hand during the examination for ready comparison with field observations. Any instrumental inaccuracies or inadequacies should be noted and corrected expeditiously by repair, replacement, or supplemental measuring devices.

Examination of the dam should be made periodically when the reservoir is full, so that loading is at the static maximum condition. Intermediate and formal examinations are accomplished at 3-year and 6-year maximum intervals. They usually consist of visual examination of the accessible structural features. Submerged structures may be examined by divers. As the dam is placed in service, a program of frequent examinations should be started so that initial performance can be evaluated and corrective measures can be taken if necessary. Examination reports should describe any questionable conditions, and should recommend remedial action.

Photographs taken during the examination may be of considerable value for future comparisons of conditions. They should be regarded as an essential part of the permanent record of the dam and reservoir.

Examination team members should make complete on-the-spot records of their observations in the field as well as notes on significant points drawn from the review of records. Entries must be as precise as possible regarding the location and nature of any questioned conditions. Any apparent departure from the norm, even if it seems to be minor, should be identified as a possible indicator of incipient trouble.

The Environs.—The margins of the reservoir must be given continuing attention to detect landslides, active faulting, erosion, or leakage paths. Field surveillance may be augmented by aerial inspection, photographic interpretation, ground surveys, or instrumentation.

The Bureau conducts a comprehensive program for monitoring and analysis of potential landslides in the peripheries of its existing reservoirs. Such areas are inspected following unusual events such as heavy rainfall and large volumes of runoff and extreme reservoir fluctuations. During the first impoundment of a new reservoir, close surveillance is maintained until there is assurance that the dam and its reservoir are safe throughout the full range of operation. The results are recorded in the Bureau's Landslide Register. To assure that this information is kept current, the examination team should examine recent geologic reports, aerial photographs, and operating records for any clues to instability in the reservoir rim. Developments in the reservoir environs such as roads, sewers, and water drainage may induce slope instability by steepening or loading of slopes, changing the drainage pattern, or raising ground-water levels.

Pertinent characteristics of a landslide include its slope, areal extent and depth, age, kind of material, water content, rate and pattern of movement, and proximity to structures.

The hazard posed by a landslide may depend on both its size and location. Even a small slide could damage a spillway or outlet and thereby threaten the dam itself.

Examination team members must understand the causes and modes of slope failures. Landslides can often be detected by escarpments, bulges, leaning trees, or misalinement of facilities.

Precise vertical and horizontal control measurements of the suspected slopes should be made periodically. These measurements usually involve triangulation from fixed points in the vicinity.

The study of seepage patterns and variations must be a basic part of any reservoir surveillance program. Leakage potential is of particular concern where the reservoir rim or dam foundation are alluvial or composed of unconsolidated material or certain kinds of sedimentary or volcanic rocks. Seepage paths may be found in gravel strata, fault zones, rock joints and cracks, or solution channels.

Leakage is not unusual in dam foundations or in reservoir rims of volcanics or limestone or aeolian sediments or alluvium. The latter two kinds of formations can be especially susceptible to erosion. Natural barriers that confine the reservoir must be examined thoroughly to assess such potential weaknesses. Lightly consolidated sediments and lightweight materials such as volcanic cinders and diatomaceous shales are also to be suspected.

In arid and semiarid regions, formations underlying or surrounding reservoirs may suffer appreciable loss in strength upon saturation. During first reservoir impoundment, close surveillance should be maintained to detect any movements due to such weakness. In some cases, the adverse conditions may develop slowly, necessitating long-term monitoring.

Concrete Dams.—Concrete dams should be examined for signs of excessive stress or instability. Survey points and plumblines should be measured regularly and the results plotted to determine trends in structural movement. Examiners should check the alinement of parapets and handrails. Contraction joints should be checked for displacement between blocks.

Concrete surfaces should be examined for overstressing, weathering, chemical reaction, cavitation, and erosion. Cracking and spalling should be studied, including comparison with conditions reported from earlier inspections. New cracks and spalls should be analyzed to determine their cause and characteristics. An attempt should be made to correlate interior distress in galleries with that apparent on exterior surfaces.

Seepage should be traced to its sources such as waterstop failure, unbonded lift surfaces, and cracks. Rate of seepage should be compared with previous measurements for corresponding reservoir elevation.

Dam and foundation drains should be examined for any indication of possible obstructions and the records should be examined for significant changes in discharge.

Cracking is likely to be the first symptom of concrete distress. Associated conditions may include seepage through the cracks, deposits from leaching, or spalling of edges, or offsets. Abnormal conditions may indicate questionable physical or chemical properties of the concrete. These symptoms may include leaching, pattern cracking, scaling, or freeze-thaw weathering. To assess the damage, core drilling of the concrete may be advisable. Cores from drill holes should be logged to record the depth and nature of deterioration, unusual deposits, staining, voids, construction joints, and foundation contact.

Embankments.—An embankment should be examined for signs of cracking, sliding, sinkholes, erosion, seepage, animal burrows or undesirable vegetative growth. The embankment must be examined regularly for signs of settlement or horizontal movement. Cracking may develop first at the crest or in the face near the crest. The embankment and its foundation must be examined frequently enough to discover and to monitor wet areas, springs, and boils. These may appear in the fill slope, in the abutments, or in the channel even at appreciable distances from the dam. Flows from any of these sources, as well as from embankment and foundation drains, should be measured regularly and analyzed for constituents. Variations in seepage, either increasing or decreasing, must be studied carefully to understand their causes and any potential detriment to the dam or reservoir.

Displacement of an embankment may be detected by misalinement of parapet walls, guardrails, or appurtenant structures. The dam crest should be checked for settlement that may reduce freeboard. The embankment slopes should be examined for evidence of cracking or bulging. Cracking must be studied carefully to understand its origin. Surficial drying and shrinkage may be harmless. On the other hand, a longitudinal crack or escarpment may delineate the upper extreme of a slide mass. Transverse cracking, especially in the fill near the abutments, may indicate tension or shearing due to settlement. Cracking parallel to the crest may also be evidence of differential movement between embankment zones.

The downstream face of the embankment and the terrain and channel immediately downstream should be examined for damp areas, boils, springs, evaporites, and abnormal vegetative growth. If such signs of excessive seepage are found, they should be studied to determine whether the water comes from the reservoir or from other sources. Examiners must make accurate records of the location and characteristics of these evidences, using maps, photographs, and narrative description. If the suspected areas are to be placed under surveillance rather than undertake immediate remedial work, these data will be valuable in future comparison of conditions to determine trends.

Survey monuments, piezometers, inclinometers, and internal movement installations in the dam should be examined for disturbance by erosion, vehicles, vandals, or frost heave. Open-well piezometers should be checked to assure that protective caps or enclosures are secure. Seepage measuring devices should be examined for disrepair and obstructions.

The upstream face of the embankment and the reservoir margins should be examined periodically when the reservoir is drawn down. If this is not feasible, examination may be made underwater.

During the examination of an embankment, special attention must be given to areas where structures are within or abutting against the fill. These interfaces may be conducive to internal erosion.

Grass on the embankment and in its immediate vicinity should be mowed to permit observation of any cracking, sliding, or seepage. Trees and bushes must not be allowed on embankments, not only because they limit access and visibility, but also because they pose potential hazards due to toppling in windstorms, fill cracking by root invasion, or opening of seepage paths by root decay. The embankment must also be kept free of burrowing animals.

Spillways and Outlets.—Inlet and discharge channels of spillways and outlet works must have stable slopes and be free of debris and vegetative growth. Channels should be examined for erosion, sinkholes, boils, and potentially damaging eddy currents. The discharge channel should be examined for degradation which might create adverse tailwater conditions.

Spillways and outlets should be examined to assure that capacity is not impaired by unauthorized devices such as screens or flashboards. The structures should be examined for damage by weathering, cracking, chemical reaction, erosion, cavitation, or vandalism. They should be checked for evidence of displacement. Surfaces of channel walls and floors at transverse joints should be free of offsets to prevent dislodging of panels during high flows. Joints, weepholes, and aeration slots should be maintained clean of silt, debris, and vegetation. Conduits should be examined for cracks, bulges, displacement, and excessive leakage.

Backfill and cut and fill slopes adjacent to structures should be inspected for signs of instability, including subsidence. Interfaces between structure and fill should be examined for piping.

Guides for trashracks and gates must be well maintained. Drains should be kept free of obstructions that impair their function. Seepage should be directed away from corrodible metal such as electric conduits, pipes, and fixtures. Stilling basin drain air vents must be checked to assure that screens are in place and vents are open. Stains on walls of structures should be studied for indication of flow patterns under the range of recorded discharges. Channel protection at energy dissipators should be examined regularly, with special attention given to any indication that material may be washed in or out of the structure during operation.

An examination of water conveyance facilities should cover all parts of the system. In some cases, this may require placing temporary dikes, bulkheads, or stoplogs to permit access by the examining team.

The walls or linings of open channels should be checked for alinement and for drainage. Blocked water in the backfill or foundation may freeze and cause damaging expansion.

Embankment or channel slope protection should be examined regularly to assure that it has not deteriorated so that its function is impaired.

Some detrimental conditions that may impair water passages, such as structural or foundation displacement, or faulty underdrains can be identified by examination. Also apparent from field examination are obstructions such as gravel deposits and slide masses in the channel downstream. Examiners should be alert, also, for actual or incipient landslides into the spillway or outlet approach channels. Silt deposits, driftwood, or vegetative growth in the channel can impede flow. Similar obstructions may hamper operation of energy dissipators and discharge channels. Movement of abrasive materials during operation may cause erosion of concrete in stilling basins.

Electrical and Mechanical Equipment.—As part of the regular examination, electrical and mechanical hoisting equipment should be tested by operation through the full range under actual conditions. Examiners should look for signs of poor lubrication, binding, vibration, overheating, and inadequacy of remote control systems and of the main and auxiliary power supply. The equipment should be examined for deteriorated, corroded, loose, worn, or broken parts. Gate seals should be inspected for deterioration, cracking, wear, and water leakage. Hydraulic hoists and controls should be checked for oil leakage. Fluidways, leaves, metal seats, and valve seals should be examined for corrosion, cavitation, wear, misalinement, and leakage. Sump pumps should be operated to assure dependable performance. Air vents for gates and valves should be checked to verify that they are open and protected.

During the examination, attention should be given to whether clear, complete, and convenient operating instructions for the equipment have been posted in every location where they might be needed. Equipment controls should be checked for safeguards against unauthorized operation. Ice prevention systems should be tested for proper operation. Electrical and mechanical equipment must be examined for damage caused by inadequate weather protection. Heating and ventilating systems should be tested for their capability to maintain adequately dry environments for equipment. Stoplogs, bulkhead gates, and lifting frames or beams must be readily available.

Examiners should question operations and maintenance personnel about any unusual conditions or problems with the equipment. Equipment maintenance and exercising procedures should also be discussed to verify that they are in conformance with requirements.

Accessibility of controls for emergency operation of vital facilities must be ascertained. Need for remote controls should be considered. If conditions during an inspection preclude examination of an outlet or operation of gates or valves, these actions should be taken later and should be thoroughly reported.

Auxiliary power supply should be provided for emergency operation of equipment during outages of the main source of power. The reserve fuel supply should be adequate to operate the standby unit for the duration of a severe outage. The auxiliary power station should be operated periodically to exercise gates and valves. Operating instructions for the standby unit should be posted for convenient access.

Regarding mechanical equipment problems on old dams, the following advice is offered by Fred Engstrom, Chief of the Mechanical Branch at the Engineering and Research Center of the Bureau of Reclamation:

"These vital pieces of equipment, including gates, valves, hoists, controls, trashracks, and conduits, must be operable to insure the safety of the dam. Equipment may be unable to perform its intended function for many different reasons: design deficiencies, deterioration, broken parts, faulty operating procedures, vandalism, icing, silting, settlement or shifting of the supporting structure, or a failure in the source of power.

"Periodic examinations and regular exercising of mechanical equipment are essential to ensure that the equipment will operate when needed.

"OUTLET GATES

"Too often outlet gates do not get examined because downstream flow requirements cannot be interrupted, the stream must be kept alive, or it takes too much time and equipment to pump out the stilling basin. Examiners are not eager to enter a recently dewatered outlet either, where the re-steel may hang in space or cling to the concrete like a snake and where the beam of a flashlight is swallowed by the pitch blackness of the conduit. It is also difficult to forget that a relatively thin sheet of metal is all that stands between you and perhaps several hundred feet of water.

"HIGH-PRESSURE SLIDE GATES

"Cast iron, high-pressure gates are the Bureau of Reclamation's most used outlet gates for heads up to 200 feet and they have proven to be excellent gates."

"The most common problem with high-pressure gates in old dams is the destruction of the bottom surface of the gate leaf by cavitation. The early gate leaves had a rather broad, flat bottom, with the result that the flow, particularly at high heads or small gate openings, tends to spring free from the upstream side of the gate, causing cavitation and damage farther downstream on the leaf.

"Regulating gates extensively damaged from cavitation are sometimes used for years in the damaged condition since they still perform their intended function of regulating flow and are not required to completely shut off the flow, as guard gates are provided for that purpose. Guard gates are used either wide open or completely closed and do not ordinarily suffer from cavitation damage. Guard gates should never be used for regulation because they may not be designed to operate at partial openings and, most important, the emergency protection, for which they have been designed and installed, is lost.

"It is usual that the rate of cavitation damage increases because the damaged area creates another flow disturbance and additional cavitation. Therefore, repairs should be made as soon as possible.

"Eroded areas of cast iron gate leaves are usually filled with epoxy because it is not practical to repair cast iron by welding. The leaves on some cast outlet gates are cast steel and may be satisfactorily repaired by welding.

"After the eroded areas are filled, a stainless steel bar is attached to the bottom of the gate with screws and bolts. The shape of the leaf is not effectively improved by this repair, but the stainless steel is about one hundred times more resistant to cavitation attack than cast iron is.

"A newer design of the high-pressure gate has practically eliminated cavitation by making the bottom of the gate leaf very narrow. At small gate openings there is still cavitation potential and the critical area of the leaf is protected by an overlay of stainless steel."

"SPILLWAY GATES

"Spillway gates are equivalent to the safety valve on a boiler, serving the same important function for the dam. As compared to outlet gates, they usually have the advantage of being under relatively low head and of being more easily accessible for examination. However, they are usually large and design loads can be high. The most critical problem with spillway gates has been settlement of the gate structure or expansion of the piers. This structural movement often binds the gates in the gate slots.

"Corroded wire ropes may break when a long-unused gate is hoisted against a full reservoir. Regular examinations, lubricating the ropes, and the use of stainless steel rope can eliminate this hazard.

"Improperly designed or maintained float-operated spillway gates have opened unexpectedly, endangering fisherman and waders, and have also failed to open when they were needed. This type of erratic operation has been caused by unusually high waves overtopping the piers, filling the floatwells; by floatwell intakes or outlets becoming partially plugged with debris; or by control valves improperly set.

"CORROSION IN GATE SHAFTS

"The Bureau has not had too much of a problem maintaining adequate protective coatings for gate leaves and water passages. However, this is not true of equipment housed in gate shafts. Seepage, and an occasional spew, through gate shaft and chamber walls and ceilings, produce severe corrosion on hoists, pipes and access ladders and metalwork. Corroded ladders often look stronger than they are and examiners must always be alert to this hazard.

"Shields of corrugated fiberglass supported above the equipment or sheets of polyethylene draped over outlet pipes, ventilation pipes, and other equipment will protect them from leakage. Often a ventilation system to reduce the humidity in the shaft is required to maintain an adequate coat of paint."

"EXERCISING EQUIPMENT

"To be sure that a piece of equipment will operate when it is needed and that it will perform its intended function, it must be periodically exercised. Insofar as possible, the equipment should be exercised under actual operating conditions. A guard gate, for example, should be exercised under unbalanced, not balanced, head.

"The Bureau has had many experiences where a gate or valve would not open or close fully when tested. Most frequently the cause is poor adjustment of the wheels, guide shoes, seals or seal actuators, or poor alinement of the gate seats.

"Many of the Bureau's older gates and valves are water operated. These valves are very susceptible to mineral deposit and silt buildup which can make the valves inoperable if they are not exercised periodically."

"OPERATING INSTRUCTIONS

"The importance of adequate operating instructions posted in a conspicuous location cannot be overemphasized. Operating personnel tend to minimize the significance of them since they are familiar with normal, everyday procedures and do not need to refer to printed instructions. However, if the operator is incapacitated or absent from the site during an emergency situation, the lack of readily accessible operating instructions could create a serious situation. Even at an installation where the operating procedure for a gate may be obvious — for instance, with open, close, and stop pushbuttons — it is still necessary to have operating instructions. It may be important to limit gate travel because of low reservoir level or to operate side-by-side gates to equal openings to prevent undesirable flow patterns in the stilling basin. These conditions should be shown in the instructions.

"POWER FAILURES

"Equipment which is critical to the operation and safety of the dam should have an alternate source of power.

"When power is needed the most, as during a severe rainstorm, transmission lines are subject to severe damage and some alternate source of power should be available. The alternate source of power can be a set of batteries, an internal combustion engine, or an engine generator set. Needless to say, these emergency power sources should be exercised frequently to assure reliability.

"CONCLUSION

"The most effective defenses against problems with mechanical equipment on old dams are maintenance, periodic exercising of equipment, regular examinations, written operating instructions, and an alternate power source."

Guidelines for the Examiner of Existing Dams and Reservoirs.—Examinations of existing dams and reservoirs must be aimed at detection of any weakness that might threaten the integrity of vital facilities. Weaknesses may be attributable to inadequacies of construction materials, foundation defects, adverse conditions in the environs, design deficiencies, or improper operation and maintenance. To assure that these weaknesses are disclosed as early as possible in their development, the following list, drawn from Bureau of Reclamation practice, is recommended for searching out deficiencies.

Construction Materials.—

- *Concrete*—Alkali-aggregate reaction, leaching, frost action, abrasion, spalling, general deterioration, or strength loss.
- *Rock*—Disintegration, softening, decomposition, or solution.
- *Soils*—Degradation, solution, loss of plasticity, strength loss, or mineralogical change.
- *Soil-cement*—Loss of cementation, or crumbling.
- *Metals*—Electrolysis, corrosion, stress-corrosion, fatigue, tearing and rupture, or galling.
- *Timber*—Rotting, shrinkage, combustion, or attack by organisms.
- *Lining fabrics*—Punctures, deterioration, disintegration or separation of seals, or loss of plasticity.
- *Rubber*—Hardening, loss of elasticity, heat deterioration, or chemical degradation.
- *Joint sealers*—Loss of plasticity, shrinkage, or melting.

General Conditions Evidencing Distress.—

- *Seepage and leakage*—Increasing or decreasing discharge rates, turbidity and piping, color, changing temperature, taste, or appearance of boils.
- *Drainage*—Obstructions, chemical precipitates and deposits, impeded outfall, or bacterial growth.
- *Cavitation*—Surface pitting, sonic evidence, implosions, or vapor pockets.
- *Ice action*—Decreasing stability of structures, lifting of gate hoists, or obstructing of gate leaves or other mechanical equipment.
- *Stress and strain*—Cracks, crushing, displacements, shears, or creep in concrete; cracks, extensions, contractions, bending or buckling in steel; crushing, buckling, bending, shears, extensions, or compressions in timber; or cracks, displacements, settlement, consolidation, subsidence, or zones of extension and compression in rock or soil.
- *Instability*—Tilting, tipping, or sliding of structures; or bulging, sloughing, slumping, sliding, cracks, or escarpments in slopes.

Operation and Maintenance Deficiencies.—

- *Electrical and mechanical equipment*—Broken or disconnected lift chains and cables, unreliable primary or auxiliary power sources; undependable access to control stations; poorly functioning lubrication systems; or inadequate ventilation and temperature control of damp, corrosive environment of equipment in gate chambers, galleries, and conduits.
- *Accessibility and visibility*—Vegetation obscuring examination; inadequate access ladders and lighting in galleries; inadequate access roads and bridges; or poor remote control lines and communication systems.
- *Vegetal growth and burrowing animals*—Brush and trees on embankments; vegetation in structural joints; vegetal growth in channels; or holes and tunnels dug in embankments by animals.

Evidence of Deficiencies at Concrete Dams.—

- *Stress and strain*—Cracks, crushing, or offsets in concrete monoliths, buttresses, face slabs, arch barrels, galleries, operating chambers, and conduits; stress and temperature cracking patterns in buttresses, pilasters, diaphragms, and arch barrels; or stress decline in posttensioned anchorages and tendons.
- *Instability*—Excessive or maldistributed uplift pressures; differential movement of adjacent monoliths, buttresses, arch barrels, or face slabs; movement along construction joints; or uplift on horizontal lift surfaces revealed by seepage on downstream face and in galleries.
- *Seepage at discontinuities and junctures*—Embankment wraparound sections; waterstops in monoliths and face slabs; or reservoir impounding backfill at spillway control sections and retaining walls.
- *Foundation*—Piping of material from solution channels or rock joints; clogged drains; movement at faults or shear zones; sliding along bedding planes; or consolidation of weak interbeds.

Evidence of Deficiencies at Embankment Dams.—

- *Stress and strain*—Settlement; consolidation; subsidence; cracks, displacement, joint openings in concrete facings on rockfills; extension or compression along dam crest; crushing of rock points of contact; or differential settlement of embankment zones.
- *Instability*—Cracks, displacements, openings, sloughs, slides, bulges, escarpments on embankment crest and slopes, or on abutments; sags or misalinements in parapet walls, guardrails, or conduits; irregularities in embankment slopes; or bulges in ground beyond toes of slopes.
- *Seepage*—Wet spots; new vegetal growth; seepage or leakage; boils; saturation on slopes, hillsides, and in streambeds; depressions and sinkholes; or evidence of high escape gradients.
- *Erosion*—Loss, displacement, or deterioration of upstream face riprap, or its bedding, or downstream slope protection; or beaching.
- *Foundation*—Piping of material from solution channels or rock joints; clogged drains; movement at faults or shear zones; sliding along bedding planes; consolidation of weak interbeds; subsidence; or materials susceptible to quick conditions.
- *Utility hazards*—Pressure conduits on embankments; or channels along abutment slopes.

Evidence of Deficiencies at Spillways.—

- *Approach channel*—Obstructions; or slides, slumps, or cracks in slopes.
- *Log booms*—Submergence; uncleared accumulated drift; breaking or loss of anchorage; or inadequate slack for low reservoir stages.
- *Hydraulic control structure*—Instability; reduction in capacity rating; erosion at toe; unauthorized installations on crest; defective gate piers; ineffective trash control systems; or inadequate aeration.

- *Gates*—Unauthorized position; wedging; gate trunnion displacements; loss of gate anchorage posttensioning; undesirable eccentric loads from variable positions of adjacent gates; gate-seal binding; erosive seal leakage; failure of lubrication system; or lack or inadequacy of bulkhead facilities for unwatering, or of cranes and lifting beams.
- *Operating deck and hoists*—Broken or disconnected lift chains and cables; exposure of electrical or mechanical equipment to weather, sabotage, or vandalism; or defective structural members or connections.
- *Shafts, conduits, and tunnels*—Vulnerability to obstruction; pressure jets, distorted cross sections, cracks, or displacements; materials deterioration, cavitation, or erosion; rockfalls; severe leakage about tunnel plugs; defective support systems for pressure conduits in walk-in tunnels.
- *Bridges*—Possibility of collapse with consequent flow obstruction; or inadequacy for operational and emergency equipment transport.
- *Discharge conduit*—Vulnerability to obstruction; large wall deflections, cracks, or differential deflections at vertical joints; drummy surfaces, buckled lining, or excessive uplift; cross waves, inadequate freeboard, wall climb, unwetted surfaces, uneven distribution, ride-up on horizontal curves, negative pressures at vertical curves, pressure flow, or deposition; inadequate drain systems; air ingestion and expulsion; tendency for jump formation in conduits; buckling or slipping of slope lining; or erosion of channels.
- *Terminal structures*—Inadequate energy dissipation; hydraulic jump sweepout; undercutting; retrogressive erosion; loss of foundation support for flip bucket; unsafe jet trajectory and impingement; or erosive endangerment of adjacent dam or other critical structures.
- *Return channels*—Impaired outfall; obstructions; slides, slumps, or cracks in slopes; erosion or deposition creating dangerous tailwater conditions; or destructive eddy currents.

Evidence of Deficiencies at Outlets.—

- *Approach channels*—Siltation; or underwater slides and slumps.
- *Intake structures*—Lack of dead storage; siltation; potential for burial by slides and slumps; damage or destruction of emergency or service bulkhead installations; or lack or inadequacy of bulkhead, cranes, or lifting beams; or inadequacy of access bridges.
- *Trashracks and raking equipment*—Clogging of bar spacing; lodged debris; or collapse.
- *Gate chambers, gates, valves, hoists, controls, electrical equipment, and air demand ducts*—Inadequate access to control station; poor ventilation; unauthorized gate or valve positions; binding of gate seals; seizing; erosive seal leakage; failure of lubrication system; inadequate drainage and sump pump serviceability; or vulnerability to flooding under reservoir pressure through conduits, bypasses, and gate bonnets surfacing in chamber.
- *Conduits and tunnels*—Seepage or leakage; extension strains in conduits extending through embankments; inadequate capacity and ser-

viceability of air relief and vacuum valves on conduits; vulnerability to obstruction; pressure jets, contorted cross sections, cracks, or displacements; materials deterioration, cavitation, or erosion; rockfalls; severe leakage about tunnel plugs, defective support systems for pressure conduits in walk-in tunnels.

- *Terminal structures*—Inadequate energy dissipation; hydraulic jump sweepout; undercutting; retrogressive erosion; loss of foundation support; unsafe jet trajectory and impingement; or erosive endangerment of adjacent dam or other critical structures.
- *Return channels*—Impaired outfall; obstructions; slides, slumps, or cracks in slopes, erosion or deposition creating dangerous tailwater conditions; or destructive eddy currents.

Adverse Conditions in the Reservoir Environs.—

- *Reservoir*—Depressions or sinkholes in exposed reservoir surfaces; slope instability indicated by leaning trees, escarpments or hillside distortions; flood pool encroachments; siltation adversely affecting loading on dam or obstructing approach channels or other waterways.
- *Reservoir linings*—Depressions or sinkholes; erosion; or disruption by animals.
- *Downstream proximity*—Reservoir-connected springs; endangering seepage or leakage; or river obstructions creating unanticipated tailwater elevations or interference with outfall channel capacities of the spillway and outlets.
- *Watershed*—Surface changes that might significantly affect runoff characteristics.
- *Regional environs*—Sinkholes, trenches, settlements of buildings, highway or other structures; or mineral, hydrocarbon, or ground-water extractions.

The Data Book.—The data book summarizes all records pertinent to safety of the dam. It must be sufficiently comprehensive to serve as the basic reference for all future evaluations without need for additional searching of old records. Information must be kept current by adding essential observational and analytical data as they are acquired.

During initial preparation of the data book, all planning, design, construction, and operations records should be located. Engineering and geologic data are likely to be plentiful for a newer dam but scarce for an old one. Research of technical literature and augmentation of field data may be necessary.

In preparing the data book, the evaluators must draw information from all useful sources and add their own interpretations and analyses. Photographs and reduced drawings should be included. Data sources should be clearly referenced. The data book should contain data from records of design, construction, and operation, and from reports of examination teams.

The Bureau of Reclamation's data book for each dam includes a statistical summary and a management summary. The latter is a computer printout which identifies important data, problems, required studies, activities currently in progress, and special conditions and restrictions placed on operations.

The data book is updated following each examination, when significant change in conditions occur at the dam, when new criteria on design earthquakes and floods are developed, and after any repair or modification. Each succeeding examination team has the responsibility to ensure that the data book is accurate and complete. Any questionable data must be examined rigorously, checked against original sources, and corrected if necessary.

An evaluation of the safety of a dam must be based upon a sufficient volume of data to assure that all important factors have been taken into account. Investigations to augment the data base should be initiated without undue delay.

The SEED Report.—In Bureau of Reclamation practice, upon completion of evaluation of the dam, a written report is prepared for consideration and action by the Assistant Commissioner-Engineering and Research and the Regional Director. The report presents the conclusions as well as recommendations for special studies, repair, modification, or operational restrictions considered to be necessary. The report is then incorporated into the data book on the dam to update the record.

Remedial Work

General.—With proper analysis, the problems requiring repairs or modification will have been pinpointed so that remedial work can be concentrated where it is needed. Also, the extensive cumulative knowledge of the site, the structures, and the available materials should enable quick focusing on the alternatives most likely to be effective, which may include:

The Site.—

- Drainage of unstable reservoir slopes.
- Earthwork to unload and/or buttress reservoir slopes subject to sliding.
- Foundation grouting.
- Impervious blanketing.
- Cutoffs, including slurry-trench curtains.
- Horizontally drilled drains in abutments.
- Vertically drilled drains into rock foundation of concrete dams.
- Drain tunnels.
- Relief wells.
- Toe drains.
- Rock bolting of unstable foundation blocks.
- Removal of obstructions in drains.

Embankments.—

- Placement of ballast filter to protect embankment from piping.
- Repair of eroded areas in embankment.

- Raising embankment level in areas that have been subjected to settlement.
- Buttressing of unstable embankments by additional fill.
- Sealing of cracks in embankments.
- Repair of slope protection (e.g., riprap, soil-cement revetment, slabs).
- Sealing of upstream face with membrane or other seepage barrier.
- Removal and replacement of defective embankment material.
- Addition or extension of drain zones.
- Elimination of burrowing animals.
- Removal of detrimental vegetation.

Structures.—

- Removal and replacement of defective concrete in dam or appurtenant structures.
- Posttensioning of concrete structures.
- Strengthening of concrete with special chemicals (e.g., polymers).
- Repair at cracks or erosion in concrete.
- Addition of buttressing concrete members.

Spillways and Outlets.—

- Repair or enlargement of existing spillway.
- Addition of auxiliary spillway.
- Removal of unauthorized barriers in spillway.
- Modification of energy dissipators to improve performance.
- Removal of channel obstructions (e.g., slides or debris).
- Aeration of conduits to improve hydraulic performance.
- Construction of additional outlet.
- Provision of outlet bypass or larger valves.
- Removal of silt interfering with outlet operation.
- Repair or replacement of faulty electrical or mechanical equipment.
- Installation of auxiliary power facility.
- Improvement of access to dam and appurtenant works.

Foundations.—Harmful hydrostatic pressures may develop beneath a dam resulting from seepage along foundation joints or cracks. To reduce such pressures to acceptable limits, additional foundation sealing and drainage may be needed.

Drainage is often used as part of a seepage control plan. Relief wells or drainage trenches located immediately downstream from an embankment will reduce seepage pressures and direct flows so that they are safely controlled. Abutment drainage tunnels or horizontal drain holes also have served well for these purposes. Ideally, seepage should be controlled close to its source. Pressure grouting of the dam foundation, as well as the reservoir walls in some cases, has been a widely accepted way to do this. Its effectiveness has been subject to some question in recent years. At most sites, it can be useful if combined with other defensive measures. Usually the dam foundation is sealed and consolidated near its surface by blanket grouting in shallow holes at low pressures. Then deeper lines of holes are grouted to form a curtain or curtains paralleling the dam axis.

A mixture of portland cement and water is commonly used for grouting dam foundations. Where the grout take is large, materials such as sand, clay, bentonite, sawdust, organic fibers, and nut hulls have been added. In Bureau of Reclamation practice, to fill large voids, and to limit grout travel, as much as 28 cubic meters (1000 cubic feet) of thick cement grout or cement-sand-bentonite grout with calcium chloride as an accelerator may be injected at one time, following which grouting is halted temporarily to allow the grout to set up. Then the process is resumed, with intermittent injection of like quantities of grout. Sodium silicate and soda ash are also effective accelerators for cement grouts.

Where the voids or passages to be sealed are too small to accept portland cement grout, a sometimes effective alternative is chemical grout. This includes resins, polymers, and sodium silicates. Resins attain high strengths. Sodium silicates have been used with a reagent such as aluminum sulphate, calcium chloride, or sodium bicarbonate to promote gelation. A vinyl polymer (AM-9) which has been commonly used for chemical grouting, has a very low viscosity, practically equal to that of water. The viscosity remains low until shortly before the time of gelation, which can be delayed up to several hours if desired. This polymer produces a high impermeability but does not attain the strength that is possible with resins.

Some specialists in foundation treatment prefer sodium silicates for low-viscosity grouts because when combined with organic reagents their strengths can be varied to meet a range of requirements.

Concrete.—The intensity and extent of concrete deterioration will depend upon the mineral composition of the aggregate, the content of the cement, the moisture, and the age of the concrete. The rate of deterioration is not easily predictable. Effective to a limited extent in slowing the regression of concrete strength are the filling of cracks with grout or other sealants and the waterproofing of exposed surfaces. As deterioration continues, there may be a need to provide additional structural support or to remove defective parts.

Cement grouting in an attempt to improve defective concrete may be questioned in some cases, especially if high pressures are used. An increase in tensile strength would not be assured. Other materials, such as polymers or epoxy, can be used in grouting inferior concrete to improve tensile strength as well as other properties.

Various techniques have been used in attempting to make concrete structures more impervious. Bonded membranes such as bituminous coatings, fiberglass, and synthetic felts generally have not adhered well. On the other hand, applications of thin coatings like polyacetate provide excellent bonding. Synthetic resins are often applied to upstream concrete faces for waterproofing or for protection from aggressive waters. For treatment of joints and cracks, much progress has been made since the new resins were introduced. Polyurethane mastic is now used extensively.

The Bureau of Reclamation has accomplished important advances in the use of polymers to improve the quality of concrete. Techniques developed in the Bureau's Denver laboratories have been used on a number of projects in the United States.

The first large polymerization project in the country was at the Corps of Engineers' Dworshak Dam in Idaho. Polymer-impregnated concrete now provides protection against the erosion and cavitation of the outlet tunnel and stilling basin that occurred at this 219-meter (717-foot) high gravity structure during the first few years of its life. Prior to the repair, the more extensively damaged outlet showed evidence of severe cavitation 15 to 23 meters (50 to 75 feet) downstream from the inlet gate. Reinforcing steel was exposed and holes were as much as 0.6 meter (2 feet) deep. Erosion had removed about 1530 cubic meters (2000 cubic yards) of concrete from the stilling basin. Holes were as deep as 2.7 meters (9 feet) and near the center of the basin some of the foundation rock had been eroded. Large blocks of concrete found downstream of the basin suggested that cavitation was responsible for some of the damage. Under the contract for repair during 1975, concrete in the damaged areas was cut by saw and removed to a depth of at least 375 millimeters (15 inches). Steel-fiber-reinforced concrete was used for replacement. This was then polymerized with methyl methacrylate. The result was a high-strength concrete with long fatigue life and resistance to cavitation and impact.

In the repair of distressed concrete, thin surface coatings are not effective in severe cases. Overlays several inches thick may be required. Before such placement, all concrete of questionable quality must be removed. Polymerized concrete or mortar, epoxy mortar, fiber-reinforced concrete, or concrete with very low water-cement ratio are suitable replacement materials resistant to chemical or physical actions. Symptoms of structural instability are cracks of substantial width which vary with reservoir or temperature changes, especially if the cracks are leaking significantly. In cases of mild distress, leakage can be controlled by routing the crack and injecting an elastomeric filler or an epoxy mortar. Where hydrostatic pressures are high, a drainage system may be necessary. If structural stability is judged to be inadequate, posttensioning of the dam and/or its foundation may be an effective remedial alternative. Grout or mortar must be placed around the steel strands for protection against corrosion.

Reservoirs.—Any abnormal seepage from a reservoir requires immediate attention. Rapidly increasing leakage through the dam or its foundation must be recognized as a signal for emergency action. At sites where seepage variations are unusual but more gradual, there may be time for observation and analysis to determine what remedial measures are required, if any. Studies should include examination of exploration reports and foundation treatment records.

Additional data needed to evaluate the suspected seepage problem may be provided by:

1. Supplemental measurement and plotting of leakage, correlated with the corresponding elevations of the reservoir water surface.
2. Tracing of seepage patterns by using radioactive or color dyes, or infrared photography.

3. Analysis of the escaping water for sediment and chemical composition.
4. Measurement of water temperatures in a range of elevations in observation wells to identify variations in permeability.
5. Testing of observation wells for rates of infiltration during water injection and productive capacity during pumping.
6. Installation and reading of supplemental piezometers.

Compacted earth blankets, usually 0.9 meter (3 feet) or more in thickness, have been used successfully at some sites to reduce reservoir water losses. However, such blankets may be vulnerable to erosion by waves and by reservoir fluctuations. In such cases they must be protected and carefully maintained. Effective water retention has been achieved at some smaller reservoirs by the placement of membrane linings of materials such as butyl rubber, polyvinyl chloride, or polyethylene.

Bentonite has been used to seal reservoirs in several ways, including: (1) depositing it in the reservoir water; (2) surface blanketing; (3) placement as an earth-covered blanket or membrane. The first method has been relatively ineffective in most cases. Bentonite surface blankets have sealed adequately but, unless they are placed in thicknesses of more than a few inches, their lives may be short. Covered, or sandwiched, bentonite membranes tend to seal more permanently.

Limiting Conditions.—Engineers responsible for remedial action usually will not have the full range of options that were available at the time the original project was conceived. They must cope with existing conditions, including the presence of the dam itself. Being denied direct access to the foundations under the structure and its appurtenances for inspection and corrective treatment, the problem solvers must sometimes devise imaginative ways to circumvent the handicap. Often economics will rule against, or limit the time available for, lowering the water storage to facilitate work on the upstream parts of the barrier or on the reservoir floor or on the abutments below the normal water surface. For these reasons and others, remedial work may be more difficult and more expensive than corresponding categories of work would have been at the outset of the project. This argues for the most painstaking attention to preventive and defensive engineering as early as possible.

Bibliography

"Argentine Dam Fails, at Least 25 Dead," *Engineering News-Record,* vol. 104, No. 2, p. 3, January 8, 1970.

"The August 1, 1975 Oroville Earthquake Investigations," California Department of Water Resources Bulletin 203-78, pp. 194–196, February 1979.

Babb, A. O., and Mermel, T. W., "Catalog of Dam Disasters, Failures and Accidents," U.S. Department of the Interior, Bureau of Reclamation, Washington, D.C., 211 pp., 1963.

Biswas, Asit K., "Irrigation in India: Past and Present," Proceedings, American Society of Civil Engineers, Journal, Irrigation and Drainage Division, vol. 91, No. IR1, pp. 179–189, March 1965.

———, and Chatterjee, Samar, "Dam Disasters - An Assessment," *"Engineering Journal (Canada),* vol. 54, No. 3, pp. 3–8, March 1971.

Bolt, Bruce A., and Cloud, W. K., "Recorded Strong Motion on the Hsinfengkiang Dam, China," Bulletin, Seismological Society of America, vol. 64, No. 4, pp. 1337–1342, August 1974.

Bowers, N. A., "St. Francis Dam Catastrophe - A Great Foundation Failure," *Engineering News-Record,* vol. 100, No. 12, pp. 466–472, March 22, 1928.

"The Breaching of the Orós Earth Dam in the State of Ceará, North-East Brazil," *Water and Water Engineering,* vol. 64, No. 774, pp. 351–355, August 1960.

Burr, William H., "Ancient and Modern Engineering and the Isthmian Canal," 1st edition, John Wiley & Sons, New York, pp. 4–14, 1902.

California Commission to Investigate the Causes Leading to the Failure of the St. Francis Dam, report. Calif. State Printing Office, Sacramento, 41 pp., 1928.

Cedergren, Harry R., "Seepage, Drainage, and Flow Nets," John Wiley & Sons, Inc., New York, 1967.

Chopra, Anil K., and Chakrabarti, P., "The Koyna Earthquake of December 11, 1967 and the Performance of Koyna Dam," report No. EERC 71-1, Earthquake Engineering Research Center, Univ. of Calif., Berkeley, 51 pp., April 1971.

Committee on Essential Facts Concerning the Failure of the St. Francis Dam, report. Proceedings, American Society of Civil Engineers, vol. 55, No. 8, October 1929.

Cortright, C. J., "Re-evaluation and Reconstruction of California Dams," Proceedings, American Society of Civil Engineers, Journal, Power Division, vol. 96, No. PO1, pp. 55-72, January 1970.

―――, Report to Subsecretario de Recursos Hidricos y Energeticos, Mendoza, Argentina, March 22, 1970.

Crosby, Irving B., "Geological Problems of Dams," Transactions, American Society of Civil Engineers, vol. 106, pp. 1171-1192, 1941.

"Dam Collapse Contributed to Disaster," *Engineering News-Record*, vol. 188, No. 24, p. 13, June 15, 1972.

"Dam Failure in North India Kills 100, Ruins 32 Villages," *Engineering News-Record*, vol. 179, No. 11, p. 17, September 14, 1967.

"Dam Failure Inquiry," *Engineering News-Record*, vol. 197, No. 10, p. 13, September 2, 1976.

Davies, W. E., "Buffalo Creek Dam Disaster: Why It Happened," *Civil Engineering*, vol. 43, No. 7, pp. 69-72, July 1973.

"Death from the Hills," *Newsweek*, vol. 79, pp. 28-29, June 19, 1972.

de Camp, L. Sprague, "The Ancient Engineers," Doubleday & Company, Garden City, New York, p. 50, 1963.

"Details of the Failure of an Italian Multiple-Arch Dam," *Engineering News-Record*, vol. 92, No. 5, pp. 182-184, January 31, 1924.

"Disasters: Nightmare in Rapid City," *Time*, vol. 99, p. 18, June 19, 1972.

"Earthquakes Induced by Reservoirs," *Water Power & Dam Construction*, vol. 27, No. 4, p. 162, April 1975.

Engstrom, Fred, "Mechanical Equipment Problems on Old Dams." *In* Inspection, Maintenance, and Rehabilitation of Old Dams, Proceedings, Engineering Foundation Conference, September 23-28, 1973, American Society of Civil Engineers, New York, N.Y., pp. 440-450, 1974.

"Failure of Navigable Pass, Dam 26, Ohio River," Special Board report, *Professional Memoirs,* U.S. Army Engineers and Engineer Department-at-Large, vol. V, Engineer School Press, Washington, D.C., pp. 19–24, 1913.

"Failure of Teton Dam, A Report of Findings," U.S. Department of the Interior Teton Dam Failure Review Group (IRG), 762 pp. April 1977.

"Failure of Teton Dam, Final Report," U.S. Department of the Interior Teton Dam Failure Review Group (IRG), 834 pp. January 1980.

"Failure of Teton Dam," Report to the United States Department of the Interior and State of Idaho, by Independent Panel to Review Cause of Teton Dam Failure, 580 pp. December 1976.

"Federal Guidelines for Dam Safety," Ad Hoc Interagency Committee on Dam Safety, Federal Coordinating Council for Science, Engineering and Technology, Government Printing Office, Washington, D.C., June 25, 1979.

"Flash Flood Overtops, Destroys Argentina's Pardo Dam," *Engineering News-Record,* vol. 184, No. 3, p. 20, January 15, 1970.

"Floods Overtop, Breach Indian Earthfill," *Engineering News-Record,* vol. 203, No. 8, August 23, 1979.

"Floods Peril City in India," vol. 184, No. 3, *The Sunday Star,* Washington, D.C., p. A–19, September 10, 1967.

Francis, James B., Ellis, Theodore G., and Worthen, William E., "The Failure of the Dam on Mill River," Transactions, American Society of Civil Engineers, vol. 3, pp. 118–122, 1874.

Fries, Amos A., "The Failure of the Austin Dam," *Professional Memoirs,* U.S. Army Engineers and Engineer Department-at-Large, Washington, D.C., vol. IV, pp. 108–116, 1912.

Gandolfo, J. H., "Tracing the Engineer in Early Egypt and Assyria," *Civil Engineering,* vol. 1, No. 7, pp. 632–637, April 1931.

Gillette, H. P., "The Cause of the St. Francis Dam Failure," *Engineering and Contracting,* vol. 69, pp. 173–178, April 1928.

Glaser, Eduard, "Reise nach Marib," Alfred Holder, Vienna, 214 pp., 1913.

Golzé, Alfred R. (editor), "Handbook of Dam Engineering," Van Nostrand Reinhold Company, New York, 1977.

Gruner, Edward, "Dam Disasters," Institution of Civil Engineers, Proceedings, vol. 24, London, pp. 47–60, 1963.

Hathaway, G. A., "Dams - Their Effect on Some Ancient Civilizations," *Civil Engineering,* vol. 28, No. 1, pp. 58–63, January 1958.

_____, "How Dams Serve Man's Vital Needs," Transactions, American Society of Civil Engineers, vol. CT, pp. 476–488, 1953.

Healy, J. H., Rubey, W. W., Griggs, D. T., and Raleigh, C. B., "The Denver Earthquakes," *Science,* vol. 161, No. 3848, pp. 1301–1310, 27 September 1968.

Henny, D. C., "Stability of Straight Concrete Gravity Dams," Transactions, American Society of Civil Engineers, vol. 99, pp. 1041–1042, 1934.

Hill, W. R., "A Classified Review of Dam and Reservoir Failures in the United States," *Engineering News,* vol. 47, No. 25, pp. 506–507, June 19, 1902.

Hinds, Julian, "Continuous Development of Dams Since 1850," Transactions, American Society of Civil Engineers, vol. CT, pp. 489–520, 1953.

_____, "200-Year Old Masonry Dams in Use in Mexico," *Engineering News-Record,* vol. 109, No. 9, pp. 251–253, September 1, 1932.

"Historic Accidents and Disasters," *The Engineer,* vol. 174, pp. 452–455, December 4, 1942.

Houk, Ivan E., "Discussion on Stability of Concrete Gravity Dams," Transactions, American Society of Civil Engineers, vol. 99, pp. 1081–1082, 1934.

Hunter, J. K., "Failure of the Panshet and Khadakwasla Dams, Bombay," *Water Power,* vol. 16, pp. 251–252, June 1964.

"India's Worst Dam Disaster," *Water Power & Dam Construction,* vol. 31, No. 10, p. 3, October 1979.

"International Committee on Kremasta," report, Annual Bulletin, International Commission on Large Dams, Paris, pp. 123–144, 1974.

"Investigation of Failure of Baldwin Hills Reservoir," California Department of Water Resources, Sacramento, 64 pp., April 1964.

Jacobus, William W., Jr. "Hydro-Quebec's Big, Beautiful Manicouagan Studies in the Bush," *Engineering News-Record,* vol. 171, No. 17, pp. 38–45, October 24, 1963.

Jaeger, C., "The Malpasset Report," *Water Power,* vol. 15, No. 2, pp. 55–61, February 1963.

Jansen, R. B., "Behavior and Deterioration of Dams in California," Transactions, International Congress on Large Dams, Istanbul, vol. III, Q.34, R.35, pp. 435–456, 1967.

_____, "Evaluation of Dam Safety in California," Proceedings, American Society of Civil Engineers, Journal, Soil Mechanics and Foundations Division, vol. 93, No. SM3, pp. 23–36, May 1967.

_____, Carlson, R. W., and Wilson, E. L., "Diagnosis and Treatment of Dams," Transactions, International Congress on Large Dams, Madrid, vol. IV, Communication C14, pp. 975–991, 1973.

_____, Dukleth, G. W., and Barrett, K. G., "Problems of Hydraulic Fill Dams," Transactions, International Congress on Large Dams, Mexico City, 1976.

_____, _____, Gordon, B. B., James, L. B., and Shields, C. E., "Earth Movement at Baldwin Hills Reservoir," Proceedings, American Society of Civil Engineers, Journal, Soil Mechanics and Foundations Division, vol. 93, No. SM4, pp. 551–575, July 1967.

"Javanese Dam Collapses During Monsoon Rains," *Engineering News-Record,* vol. 179, p. 16, December 7, 1967.

Kedar, Yehuda, "Water and Soil from the Desert: Some Ancient Agricultural Achievements in the Central Negev," *The Geographical Journal,* vol. 123, Pt. 2, The Royal Geographical Society, London, pp. 179–187, June 1957.

Kiersch, G. A., "Vaiont Reservoir Disaster," *Civil Engineering,* vol. 34, No. 3, pp. 32–39, March 1964.

Kirby, R. S., Withington, S., Darling, A. B., and Kilgour, F. G., "Engineering in History," McGraw-Hill, New York, pp. 16–33, 1956.

Laá, G. Gómez, M. Alonso Franco, and J. L. Romero Hernández, "Reflections on Some Incidents in Spanish Dams," Transactions, Thirteenth International Congress on Large Dams, New Delhi, India, vol. II, Q. 49, R. 47, pp. 721–740, 1979.

Leps, T. M., "Analysis of Failure of Baldwin Hills Reservoir," Proceedings, Specialty Conference on Performance of Earth and Earth-Supported Structures, Purdue University and American Society of Civil Engineers, pp. 507–550, 1972.

"Lessons from Dam Incidents," reduced edition, International Commission on Large Dams, Paris, 205 pp., 1973.

"Lessons from Dam Incidents, USA," Committee on Failures and Accidents to Large Dams of the United States Committee on Large Dams, International Commission on Large Dams, American Society of Civil Engineers, New York, 387 pp., 1975.

"Lessons Learnt from Dam Disasters," *Engineering,* vol. 197, No. 5117, pp. 681–683, May 15, 1964.

"Safety Evaluation of Existing Dams," U.S. Department of the Interior, Bureau of Reclamation, 174 pp., 1979.

"Masonry Dam Crumbles in Spain." *Engineering News-Record,* vol. 162, No. 2, p. 28, January 15, 1959.

"May Day in Zgoregrad," *Engineering News-Record,* vol. 176, No. 18, p. 27, May 12, 1966.

"Model Law for State Supervision of Safety of Dams and Reservoirs," Committee on Model Legislation for Safety of Dams of the United States Committee on Large Dams, 29 pp., 1970.

Murti, N. G. K., "Khadakwasla, the Oldest Masonry Dam in India," Transactions, International Congress on Large Dams, Istanbul, vol. III, Q.34, R.53, pp. 895–915, 1967.

Neuberger, Albert, "The Technical Arts and Sciences of the Ancients," The MacMillan Company, New York, p. 411, 1930.

"The New Encyclopedia Brittanica," Macropaedia vol. V, Encyclopedia Brittanica, Inc., Chicago, pp. 440–447, 1974.

Okeson, C. J., "Geologic Requirements of the Foundations of Large Dams," Transactions, Eighth International Congress on Large Dams, Edinburgh, Scotland, vol. I, Q.28, R.4, pp. 73–85, 1964.

Parker, O. Newell, "War-Torn Dam Gets Big Repair Job," *Engineering News-Record,* vol. 162, No. 13, pp. 38–40, April 2, 1959.

Phillips, Wendell, "Qataban and Sheba, Exploring the Ancient Kingdoms on the Biblical Spice Routes of Arabia," Harcourt, Brace and Co., New York, pp. 123–222, 1955.

"Problems in Design and Construction of Earth and Rockfill Dams," Committee on Earth and Rockfill Dams, Proceedings, American Society of Civil Engineers, Journal, Soil Mechanics and Foundations Division, vol. 93, No. SM3, pp. 119–136, May 1967.

Schnitter, N. J., "A Short History of Dam Engineering," *Water Power,* vol. 19, No. 4, 8 pp., April 1967.

Seed, H. B., Lee, K. L., Idriss, I. M., and Makdisi, F., "Analysis of the Slides in the San Fernando Dams During the Earthquake of Feb. 9,

1971," report No. EERC 73, Earthquake Engineering Research Center, Univ. of Calif., Berkeley, 150 pp., June 1973.

Sherard, James L., "Sinkholes in Dams of Coarse, Broadly Graded Soils," Transactions, Thirteenth International Congress on Large Dams, New Delhi, India, vol. II, Q.49, R.2, pp. 25–35, 1979.

Silent, R. A., "Failure of the Lower Otay Dam," *Engineering News,* vol. 75, No. 7, pp. 334–336, February 17, 1916.

Smith, Norman, "A History of Dams," Citadel Press, Secaucus, New Jersey, 279 pp., 1972.

Sutherland, Robert A., "Statistical Review of Dam Construction," Proceedings, American Society of Civil Engineers, Power Division, vol. 79, Separate 355, 57 pp., November 1953.

Takase, Kunio, "Statistic Study on Failure, Damage and Deterioration of Earth Dams in Japan," Transactions, International Congress on Large Dams, Istanbul, vol. III, Q.34, R.1, pp. 1–19, 1967.

"Three Dams on San Andreas Fault Have Resisted Earthquakes," *Engineering News-Record,* vol. 109, No. 8, pp. 218–219, August 25, 1932.

Wahlstrom, Ernest E., "The Safety of Dams and Reservoirs," *Water Power & Dam Construction,* vol. 27, No. 4, pp. 142–144, April 1975.

Wegmann, Edward, "The Design and Construction of Dams," J. Wiley & Sons, New York, 254 pp., 1903.

Willcocks, William, "From the Garden of Eden to the Crossing of the Jordan," 2nd ed., E. & F. N. Spon Ltd., London, 100 pp., 1920.

———, "The Restoration of the Ancient Irrigation Works on the Tigris or the Recreation of Chaldea," National Printing Dep., Cairo, 71 pp., 1903.

Winchester, James H., "Night of Terror in Rapid City," *The Reader's Digest,* vol. 101, pp. 81–88, November 1972.

"World Register of Dams," International Commission on Large Dams, 1010 pp., 1973.

"World Register of Dams," First Updating, December 31, 1974, International Commission on Large Dams, 294 pp., 1976.

Glossary of Dam Terminology[1]

TYPE OF DAM SYMBOL

Earth .... TE
Rockfill .... ER
Gravity .... PG
Buttress .... CB
Arch .... VA
Multiple Arch .... MV

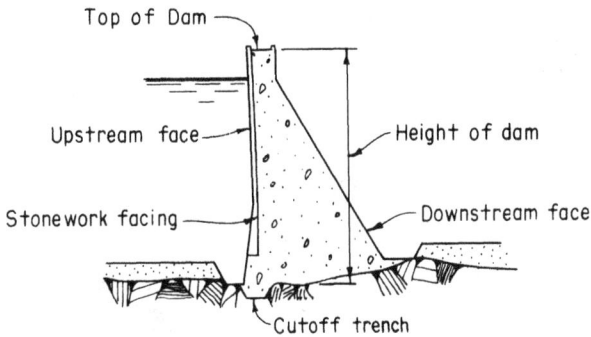

GRAVITY DAM

ABUTMENT. — That part of the valley side against which the dam is constructed. An artificial abutment is sometimes constructed, as a concrete gravity section, to take the thrust of an arch dam where there is no suitable natural abutment.

AFTERBAY DAM (REREGULATING DAM). — A dam constructed to regulate the discharges from an upstream powerplant.

AMBURSEN DAM. — See **FLAT SLAB DAM.**

ANCHOR BLOCK. — See **THRUST BLOCK.**

---

[1]Glossary is based primarily on the ICOLD (International Commission on Large Dams) "Technical Dictionary on Dams," 1978. Terms marked with an asterisk * are Bureau of Reclamation definitions.

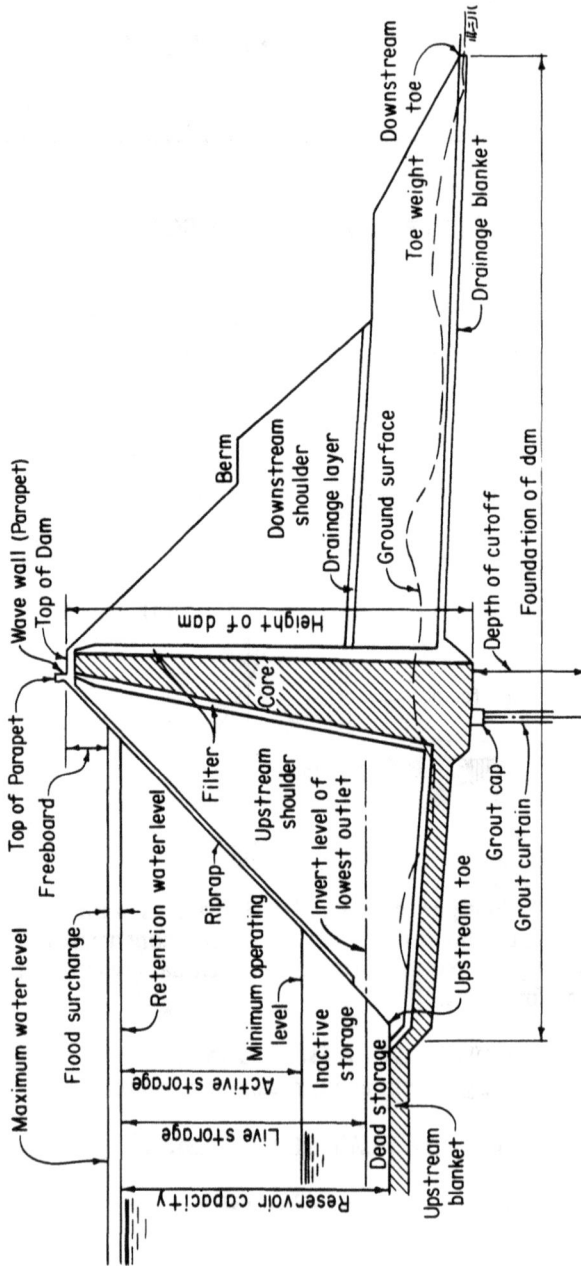

EMBANKMENT DAM

ARCH BUTTRESS DAM. — See BUTTRESS DAM.

ARCH DAM (VA). — A concrete or masonry dam which is curved in plan so as to transmit the major part of the water load to the abutments.

   CONSTANT ANGLE ARCH DAM (VA). — An arch dam in which the angle subtended by any horizontal section is constant throughout the whole height of the dam.

   CONSTANT RADIUS ARCH DAM (VA). — An arch dam in which every horizontal segment or slice of the dam has approximately the same radius of curvature.

   DOUBLE CURVATURE ARCH DAM (VA). — An arch dam which is curved vertically as well as horizontally.

ARCH GRAVITY DAM. — See GRAVITY DAM.

AXIS OF DAM. — The plane or curved surface, arbitrarily chosen by a designer, appearing as a line, in plan or in cross section, to which the horizontal dimensions of the dam can be referred.

BARRAGE (GATE-STRUCTURE DAM). — A barrier built across a river, comprising a series of gates which when fully open allow the flood to pass without appreciably increasing the flood level upstream of the barrage. The term *"movable dam"* should be avoided.

BASE THICKNESS (BASE WIDTH). — The maximum thickness or width of the dam measured horizontally between upstream and downstream faces and normal to the axis of the dam, but excluding projection for outlets, etc. In general, the term thickness is used for gravity or arch dams, and width is used for other dams.

BATTER. — Inclination from the vertical. Not in common use in English; the concept "slope" is preferred.

BERM. — A horizontal step in the sloping profile of an embankment dam.

BUTTRESS DAM (CB). — A dam consisting of a watertight part supported at intervals on the downstream side by a series of buttresses. Buttress dam can take many forms.

   ARCH BUTTRESS DAM (CB) OR CURVED BUTTRESS DAM. — A buttress dam which is curved in plan.

   FLAT SLAB DAM, AMBURSEN DAM, OR DECK DAM (CB). — A buttress dam in which the upstream part is a relatively thin flat slab usually made of reinforced concrete.

MULTIPLE ARCH DAM (MV). — A buttress dam the upstream part of which comprises a series of arches.

SOLID HEAD BUTTRESS DAM (CB). — A buttress dam in which the upstream end of each buttress is enlarged to span the gap between buttresses. The terms *"round head," "diamond head," "tee head"* refer to the shape of the upstream enlargement.

CELLULAR GRAVITY DAM. — See **HOLLOW GRAVITY DAM.**

COFFERDAM. — A temporary structure enclosing all or part of the construction area so that construction can proceed in the dry. A diversion cofferdam diverts a river into a pipe, channel or tunnel.

CONCRETE LIFT. — In concrete work, the vertical distance between successive horizontal construction joints.

CONSTRUCTION JOINT. — The interface between two successive placings or pours of concrete where bond, and not permanent separation, is intended.

CONSOLIDATION GROUTING. — Strengthening an area of ground by injecting grout.

CONTACT GROUTING. — Filling, with cement grout, any voids existing at the·contact of two zones of different materials, e.g., between a concrete tunnel lining and the surrounding rock. The grout operation is usually carried out at low pressure.

CORE, IMPERVIOUS CORE, OR IMPERVIOUS ZONE. — A zone of material of low permeability in an embankment dam. Hence the expressions *"central core," "inclined core," "puddle clay core,"* and *"rolled clay core."*

CORE WALL. — A wall of substantial thickness built of impervious material, usually of concrete or asphaltic concrete in the body of an embankment dam to prevent leakage. See also **MEMBRANE OR DIAPHRAGM.**

CREST OF DAM. — The upper part of an uncontrolled spillway. The term "Crest of Dam" should not be used when "Top of Dam" is intended.

CREST LENGTH. — The developed length of the top of the dam. This includes the length of spillway, powerhouse, navigation lock, fish pass, etc., where these form part of the length of the dam. If detached from the dam these structures should not be included. See also **LENGTH OF DAM.**

CRIB DAM (PG). — A gravity dam built up of boxes, cribs, crossed timbers or gabions, filled with earth or rock.

CROSS SECTION AT CROWN. — Cross section at crown of an arch dam which generally corresponds with the point where the height of dam is a maximum.

CURVED BUTTRESS DAM. — See **BUTTRESS DAM.**

CURVED GRAVITY DAM. — See **GRAVITY DAM.**

CUTOFF. — An impervious construction by means of which water is prevented from passing through foundation material.

CUTOFF TRENCH. — The excavation later to be filled with impervious material so as to form the cutoff. Sometimes, used incorrectly to describe the cutoff itself.

CUTOFF WALL. — A wall of impervious material (e.g., concrete, asphaltic concrete, steel sheet piling) built into the foundation to reduce seepage under the dam.

CYCLOPEAN DAM. — A gravity dam in which the mass masonry consists primarily of large one-man or derrick stone embedded in concrete.

DEPTH OF CUTOFF. — The vertical distance that the cutoff penetrates into the foundation of dam.

DIAPHRAGM. — See **MEMBRANE.**

DIKE, DYKE, OR LEVEE. — A long, low embankment. The height is usually less than four to five meters and the length more than ten or fifteen times the maximum height. Usually applied to dams built to protect land from flooding and in this case sometimes referred to as *"flood bank."* In India and the Far East the term *"bund"* is used. If built of concrete or masonry, the structure is usually referred to as a *"flood wall."* In the Mississippi Basin, where the old French word *"levee"* has survived, this now applied to flood embankments whose height can average up to 10 to 15 meters.

DIVERSION CHANNEL, CANAL, OR TUNNEL. — A waterway used to divert water from its natural course. The term is generally applied to a temporary arrangement, e.g., to bypass water around a damsite during construction. "Channel" is normally used instead of "canal" when the waterway is short. Occasionally the term is applied to a permanent arrangement (diversion canal, diversion tunnel, diversion aqueducts).

DRAINAGE BLANKET. — A drainage layer placed directly over the foundation material.

DRAINAGE LAYER. — A layer of pervious material in an earthfill dam to relieve pore pressures or to facilitate drainage of the fill.

DRAINAGE WELL OR RELIEF WELL. — Vertical wells or boreholes downstream of, or in downstream shoulder of, an embankment dam to collect and control seepage through or under the dam and so reduce water pressure. A line of such wells forms a *drainage curtain.*

EARTH DAM OR EARTHFILL DAM. — See **EMBANKMENT DAM.**

EMBANKMENT. — Fill material, usually earth or rock, placed with sloping sides and with a length greater than its height. An ''embankment'' is generally higher than a ''dike.''

EMBANKMENT DAM OR FILL DAM. — Any dam constructed of excavated natural materials or of industrial waste materials.

EARTH DAM OR EARTHFILL DAM (TE). — An embankment dam in which more than 50 percent of the total volume is formed of compacted fine-grained material obtained from a borrow area.

*HOMOGENEOUS EARTHFILL DAM (TE). — An embankment type dam construction throughout of more or less uniform earth materials, except for possible inclusion of internal drains or blanket drains. Used to differentiate it from a zoned earthfill dam.

HYDRAULIC FILL DAM (TE). — An embankment dam constructed of materials, often dredged which are conveyed and placed by suspension in flowing water.

ROCKFILL DAM (TE). — An embankment dam in which more than 50 percent of the total volume comprises compacted or dumped pervious natural or crushed stone.

*ROLLED FILL DAM (TE). — An embankment type dam of earth or rock in which the material is placed in layers and compacted by the use of rollers or rolling equipment.

*ZONED EARTHFILL (TE). —An earthfill type dam the thickness of which is composed of zones of selected materials having different degrees of porosity, permeability, and density.

FACE. — With reference to a structure, the external surface which limits the structure, e.g., face of a wall or face of a dam.

FACING. — With reference to a wall or concrete dam, a coating of a different material, masonry or brick, for architectural or protection purposes, e.g., stonework facing, brickwork facing. With reference to an embankment dam, an impervious coating or face on the upstream slope of the dam.

FILTER OR FILTER ZONE. — A band of granular material which is incorporated in an embankment dam and is graded (either naturally or by selection) so as to allow seepage to flow across or down the filter zone without causing the migration of the material from zones adjacent to the filter.

FINGER DRAINS. — A series of parallel drains of narrow width (instead of a continuous drainage blanket) draining to the downstream toe of the embankment dam.

FLASHBOARDS. — Length of timber, concrete or steel, placed on the crest of a spillway to raise the retention water level but which may be quickly removed at time of flood either by a tripping device or by deliberate failure of the flashboards or their supports.

FLAT SLAB DAM. — See **BUTTRESS DAM.**

FOUNDATION OF DAM. — The undisturbed material on which the dam structure is placed.

FREEBOARD. — The vertical distance between a stated water level and the top of a dam. Thus "net freeboard," "dry freeboard," or "flood freeboard" is the vertical distance between the maximum water level and the top of the dam. "Gross freeboard" or "total freeboard" is the vertical distance between the retention water level and the top of the dam. That part of the "gross freeboard" attributable to the depth of flood surcharge is sometimes referred to as the "wet freeboard" but this term is not recommended as it is preferable that freeboard be stated with reference to the top of dam.

GABION DAM (PG). — Special name given to a crib dam when built up of gabions.

GRAVITY DAM (PG). — A dam constructed of concrete and/or masonry which relies on its weight for stability.

   ARCH-GRAVITY DAM (PG). — An arch dam which is only slightly thinner than a gravity dam.

   CURVED GRAVITY DAM (PG). — A gravity dam which is curved in plan.

   HOLLOW GRAVITY DAM (CELLULAR GRAVITY DAM) (PG). — A dam which has the outward appearance of a gravity dam but is of hollow construction.

GROUT BLANKET. — An area of the foundation systematically grouted to a uniform depth.

GROUT CAP. — A concrete pad or wall constructed to facilitate subsequent pressure grouting of the grout curtain beneath the grout cap.

GROUT CUTOFF (GROUT CURTAIN). — A vertical zone, usually thin, in the foundation into which grout is injected to reduce seepage under a dam.

HEEL OF DAM. — The junction of the upstream face of a gravity dam with the ground surface. In the case of an embankment dam, the junction is referred to as the *"upstream toe of dam."*

HEIGHT ABOVE GROUND LEVEL. — The maximum height from natural ground surface to top of dam.

HEIGHT ABOVE LOWEST FOUNDATION OF DAM. — The maximum height from the lowest point of the general foundation to the top of the dam. See also **STRUCTURAL HEIGHT.**

HEIGHT OF DAM. — See **STRUCTURAL HEIGHT** or **HYDRAULIC HEIGHT.**

HOLLOW GRAVITY DAM. — See **GRAVITY DAM.**

HOMOGENEOUS EARTHFILL DAM. — See **EMBANKMENT DAM.**

HYDRAULIC FILL DAM. — See **EMBANKMENT DAM.**

HYDRAULIC HEIGHT. — Height to which the water rises behind the dam and is the difference between the lowest point in the original streambed at the axis of the dam and the maximum controllable water surface.

IMPERVIOUS CORE OR ZONE. — See **CORE.**

INTAKE. — Any structure in a reservoir or dam or river, through which water can be drawn into an aqueduct.

INTENSITY SCALE. — An arbitrary scale to describe the degree of shaking at a particular place. The scale is not based on measurement but on assessment by an experienced observer. Several scales are utilized (the Modified Mercalli scale, the MSK scale) all with grades indicated by Roman numerals from I to XII.

INTERNAL EROSION. — The formation of voids within soil or soft rock caused by the mechanical or chemical removal of material by seepage.

LARGE DAM. — For the purpose of inclusion in ICOLD'S World Register of Dams a large dam is defined as any dam above 15 meters in height (measured from the lowest point of foundation to top of dam) or any dam between 10 and 15 meters in height which meets at least one of the following conditions: (a) the crest length is not less than 500 meters; (b) the capacity of the reservoir formed by the dam is not less than one million cubic meters; (c) the maximum flood discharge dealt with by the dam is not less than 2000 cubic meters per second; (d) the dam had specially difficult foundation problems; (e) the dam is of unusual design.

LEAKAGE. — Free flow loss of water through a hole or crack.

LEFT ABUTMENT. — The abutment on the left-hand side of an observer when looking downstream.

LENGTH OF DAM. — The distance, measured along the axis of the dam at the level of the top of the main body of the dam or of the roadway surface on the crest, from abutment contact to abutment contact.

LEVEE. — See **DIKE.**

LINING. — With reference to a canal, tunnel or shaft, a coating of asphaltic concrete, concrete, reinforced concrete, or shotcrete to provide watertightness, to prevent erosion, or to reduce friction.

LOWEST POINT OF FOUNDATION. — The lowest point of the dam foundation excluding cutoff trenches less than 10 meters wide and isolated pockets of excavation.

MAGNITUDE. — A rating of a given earthquake independent of the place of observation. It is calculated from measurements on seismographs and it is properly expressed in ordinary numbers and decimals based on a logarithmic scale.

MASONRY DAM (PG). — Any dam constructed mainly of stone, brick or concrete blocks jointed with mortar. A dam having only a masonry facing should not be referred to as a masonry dam.

MAXIMUM CROSS SECTION OF DAM. — Cross section of a dam at point where the height of dam is a maximum.

MEMBRANE OR DIAPHRAGM. — A membrane or sheet or thin zone or facing, made of a flexible impervious material such as asphaltic concrete, plastic concrete, steel, wood, copper, plastic, etc. A "cutoff wall" or "core wall," if thin and flexible, is sometimes referred to as a "diaphragm wall" or "diaphragm."

MORNING GLORY SPILLWAY. — A circular or glory hole form of a drop inlet spillway. Usually free standing in the reservoir and so called because of its resemblance to the morning glory flower.

MULTIPLE ARCH DAM. — See **BUTTRESS DAM.**

ORIGINAL GROUND OR GROUND SURFACE. — The original ground surface at a damsite prior to construction.

OUTLET. — An opening through which water can be freely discharged from a reservoir to the river for a particular purpose.

PARAPET WALL. — A solid wall built along the upstream or downstream edge of the top of dam for ornament or for the safety of vehicles and pedestrians.

PENSTOCK. — A pipeline or pressure shaft leading from the headrace or low-pressure tunnel into the turbines.

PERVIOUS ZONE. — A part of the cross section of an embankment dam comprising material of high permeability.

PIPING. — The progressive development of internal erosion by seepage, *appearing downstream* as a hole discharging water.

PITCHING. — Squared masonry, precast blocks, or embedded stones laid *in regular fashion* with dry or filled joints on the upstream slope of an embankment dam or on a reservoir shore or on the sides of a channel as a protection against wave and ice action.

PORE PRESSURE. — The interstitial pressure of fluid (air or water) within a mass of soil, rock, or concrete.

PRECAST DAM. — A dam constructed mainly of large precast *concrete* blocks or sections.

PRESTRESSED DAM. — A dam, the stability of which depends in part on the tension in steel wires, cables, or rods that pass through the dam and are anchored into the foundation rock.

PRESSURE RELIEF PIPES. — Pipes used to relieve uplift or pore water pressure in a dam foundation or in the dam structure.

REGULATING DAM. — A dam impounding a reservoir from which water is released to regulate the flow in a river.

REREGULATING DAM. — See **AFTERBAY DAM.**

RELIEF WELL. — See **DRAINAGE WELL.**

RICHTER SCALE. — A scale proposed by C. F. Richter to describe the magnitude of an earthquake by measurements made in well-defined conditions and with a given type of seismograph. The zero of the scale is fixed arbitrarily to fit the smallest recorded earthquakes. The largest recorded earthquake magnitudes are near 8.7. This is the result of observations and not an arbitrary upper limit like that of the intensity scale.

RIGHT ABUTMENT. — The abutment on the right-hand side of an observer when looking downstream.

RIPRAP. — A layer of large uncoursed stones, broken rock or precast blocks placed *in random fashion* on the upstream slope of an embankment dam or on a reservoir shore or on the sides of a channel as a protection against wave and ice action. Very large riprap is sometimes referred to as *"armoring."*

ROCKFILL DAM. — See **EMBANKMENT DAM.**

ROLLED FILL DAM. — See **EMBANKMENT DAM.**

RUBBLE DAM (PG). — A masonry dam in which the stones are unshaped or uncoursed.

SADDLE DAM. — A subsidiary dam of any type constructed across a saddle or low point on the perimeter of a reservoir.

SEEPAGE. — The interstitial movement of water that may take place through a dam, its foundation, or abutments.

SEEPAGE COLLAR. — A projecting collar of concrete built around the outside of a tunnel or conduit, under an embankment dam, to reduce seepage along the outer surface of the conduit.

SEISMIC INTENSITY. — *Subjective measurement* of the degree of shaking at a specified place by an experienced observer using a descriptive scale.

SEMIPERVIOUS ZONE. — See **TRANSITION ZONE.**

SHELL. — See **SHOULDER.**

SHOULDER (SHELL). — The upstream and downstream parts of the cross section of an embankment dam on each side of the core or core wall. Hence the expression *upstream shoulder* or *downstream shoulder.*

SILL. — (a) A submerged structure across a river to control the water level upstream. (b) The crest of a spillway. (c) The horizontal gate seating, made of wood, stone, concrete, or metal at the invert of any opening or gap in a structure. Hence, the expressions: gate sill, stoplog sill.

SLOPE. — (a) Side of a hill or a mountain. Where the idea of orientation is intended the word *"versant"* has to be used (geographic term). (b) The inclined face of a cut, canal, or embankment. (c) Inclination from the horizontal. Measured as to the ratio of the number of units of the vertical distance to the number of corresponding units of the horizontal distance. Used in English for an inclination. Expressed in percent when the slope is gentle; in this case also termed "gradient."

SLOPE PROTECTION. — The protection of embankment slope against wave action or erosion.

SLURRY TRENCH. — A narrow excavation whose sides are supported by a mud slurry filling the excavation. Sometimes used incorrectly to describe the cutoff itself.

SOLID HEAD BUTTRESS DAM. — See **BUTTRESS DAM.**

SPILLWAY. — A structure over or through which floodflows are discharged.

STOPLOGS. — Large logs or timbers or steel beams placed on top of each other with their ends held in guides on each side of a channel or conduit so as to provide a cheaper or more-easily handled means of temporary closure than a bulkhead gate.

STRUCTURAL HEIGHT. — Distance between the lowest point in the excavated foundation (excluding narrow fault zones) and the top of dam.

THICKNESS OR WIDTH OF DAM. — The thickness or width of a dam as measured horizontally between the upstream and downstream faces and normal to the axis of the dam.

THRUST BLOCK (ANCHOR BLOCK). — A massive block of concrete built to withstand a thrust or pull.

TOE OF DAM. — The junction of the downstream face of a dam with the ground surface. Also referred to as *"downstream toe."*

TOE WEIGHT. — Additional material placed at the toe of an embankment dam to increase its stability.

TOP OF DAM. — The elevation of the uppermost surface of a dam, usually a road, or walkway excluding any parapet wall, railing, etc.

TOP OF DAM. — The crown of the roadway or the level of the walkway which crosses the dam. See also **CREST OF DAM.**

TOP THICKNESS (TOP WIDTH). — The thickness or width of a dam at the level of the top of dam (excluding corbels or parapets). In general, the term thickness is used for gravity and arch dams, and width is used for other dams.

TOP WIDTH. — See **TOP THICKNESS OF DAM.**

TRAINING WALL. — A wall built to confine or guide the flow of water over the downstream face of an overflow dam or in a channel.

TRANSITION ZONE OR SEMIPERVIOUS ZONE. — A substantial part of the cross section of an embankment dam comprising material whose grading is of intermediate size between that of an impervious zone and that of a permeable zone.

TRASHRACK. — A screen comprising metal or reinforced concrete bars located in the waterway at an intake so as to prevent the ingress of floating or submerged debris.

UPLIFT. — (a) The upward pressure in the pores of a material (interstitial pressure) or on the base of a structure. When the pressure acts uniformly around the outside of a body, e.g., the water pressure on the outside of a tunnel lining, the term *"external pressure"* is used. The degree of external pressure causing complete structural failure is termed *"collapsing pressure."* (b) An upward force on a structure caused by frost heave or by windforce.

UPSTREAM BLANKET. — An impervious blanket placed on the reservoir floor upstream of a dam. In the case of an embankment dam, the blanket may be connected to the impermeable element.

VOLUME OF DAM. — The total space occupied by the materials forming the dam structure computed between abutments and from top to bottom of dam. No deduction is made for small openings such as galleries, adits, tunnels, and operating chambers within the dam structure. Portions of powerhouses, locks, spillway, etc., may be included only if they are necessary for the structural stability of the dam.

WATER BAR. — See **WATERSTOP.**

WATERSTOP (WATER BAR). — A strip of metal, rubber or other material to prevent leakage, through joints between adjacent sections of concrete.

WAVE WALL. — A solid wall built along the upstream side at the top of a dam and designed to reflect waves.

WEIGHTING OF A SLOPE. — Additional material placed on the slope of an embankment.

WEIR. — A low dam or wall across a stream to raise the upstream water level. Termed *"fixed-crest weir"* when uncontrolled. A structure built across a stream or channel for the purpose of measuring flow. Sometimes described as *"measuring weir"* or *"gaging weir."* Types of weirs include *"broad-crested weir," "sharp-crested weir," "drowned weir"* or *"submerged weir."*

YEAR OF COMPLETION. — In the World Register of Dams the year in which construction of the dam was completed and ready for use.

ZONED EARTHFILL. — See **EMBANKMENT DAM.**

Index

M

www.ingramcontent.com/pod-product-compliance
Lightning Source LLC
Chambersburg PA
CBHW050522190326
41458CB00005B/1627